The Hidden Company That Trees Keep

The Hidden Company That Trees Keep

Life from Treetops to Root Tips

James B. Nardi

Princeton University Press
Princeton and Oxford

Published by Princeton University Press
41 William Street, Princeton, New Jersey 08540
99 Banbury Road, Oxford OX2 6JX

press.princeton.edu

Library of Congress Cataloging-in-Publication Data

Names: Nardi, James B., 1948– author.
Title: The hidden company that trees keep : life from treetops to root tips / James B. Nardi.
Description: Princeton : Princeton University Press, [2023] | Includes bibliographical references and index.
Identifiers: LCCN 2022019117 (print) | LCCN 2022019118 (ebook) | ISBN 9780691237978 (hardback) | ISBN 9780691238159 (ebook)
Subjects: LCSH: Trees—Ecology. | Forest ecology. | Plant chemical ecology. | Plant-soil relationships. | Plant-fungus relationships.
Classification: LCC QK477 .N37 2023 (print) | LCC QK477 (ebook) | DDC 582.16—dc23/eng/20220608
LC record available at https://lccn.loc.gov/2022019117
LC ebook record available at https://lccn.loc.gov/2022019118

British Library Cataloging-in-Publication Data is available

Editorial: Robert Kirk and Megan Mendonça
Production Editorial: Kathleen Cioffi
Text Design: D & N Publishing, Wiltshire, UK
Jacket Design: Wanda España
Production: Steven Sears
Publicity: Matthew Taylor and Caitlyn Robinson
Copyeditor: Laurel Anderton

Jacket illustration by James B. Nardi

This book has been composed in Proforma (main text and headings) and Futura PT (image labels)

Printed on acid-free paper. ∞

Printed in the United States of America

10 9 8 7 6 5 4 3 2 1

CONTENTS

PROLOGUE

In the mountains of North Carolina at Joyce Kilmer Memorial Forest, the magnificent old-growth trees are a testament to the grandeur that each tree can achieve during its lifetime. Joyce Kilmer expressed the promise of young tree seedlings and the dignity of old trees in the first stanza of his 1913 poem "Trees": "I think that I shall never see a poem as lovely as a tree." Kilmer could not have imagined at the time how this loveliness depended on unfailing partnerships that trees establish with countless hidden companions. To our ancestors, these hidden companions and guardians of trees existed as the hamadryads (*hama* = together with; *dryo* = tree), spirits whose destinies were intertwined with the destinies of specific trees.

You can tell a lot about a tree from the company it keeps, but most of us never see or have any idea who these companions are. They occupy an alien world, and the pages of this book reveal new discoveries from this world. While the more familiar, more conspicuous birds, mammals, snakes, and frogs of trees are also discussed and illustrated, the invertebrate and microbial companions of trees occur in the greatest numbers and have by far the greatest influence on the lives of trees. In the laboratory new discoveries about the important roles of microbial partnerships have emerged from our finding that trees, fungi, and animals all have microbial companions. However, we often know so little about the lives of these hidden companions, and this book intends to rectify that oversight. With their unmatched diversity, unparalleled life stories, and exceptional talents, these hidden creatures assume multiple jobs in a tree's web of interactions.

Trees stretch their branches to the sky and extend their roots through the soil to welcome and accommodate their entourage of countless creatures. Trees offer these companions shelter from their enemies and from the environment. Trees entice certain companions that offer important services and advantages; trees welcome agents of fertilization and seed dispersal with their flowers and fruits.

However, trees respond to attacks and intruders with an elaborate immune system and recruit many of their companions to help in their confrontations. Tree tissues injured by insects emit chemicals that attract parasitoids and predators as allies in their defenses against herbivores.

Predators of all sizes—birds, frogs, mammals, and many insects—assist in defending trees. Insect parasitoids, by contrast, are relatively small, but they make up about 10 percent of all animal species—well over 160,000 species—and choose an equally diverse assortment of hosts. Parasitoid larvae have adopted elaborate and ingenious tactics to overwhelm the immune defenses of their hosts, making them exceptional guardians and allies for trees.

Our understanding of relationships among creatures has been refined with the latest nucleic acid research. Based on the myriad nucleic acid sequences now available, new groups of organisms have been established; old groups have grown, shrunk, or disappeared. Our appreciation for the vast diversity of life on Earth grows as new species are continually discovered.

The scientific community has amassed volumes of information on some organisms, but for so many of Earth's creatures, many mysteries remain. The creatures discussed on these pages represent species that inhabit trees of Earth's temperate zones, but they share many characteristics with creatures found in the company of trees that inhabit all corners of the Earth. The life histories of so many of these invertebrate and microbial companions of trees are unknown. Careful observations by naturalists and scientists will reveal even more remarkable surprises and singular life stories. As Peter Wohlleben observed in *The Hidden Life of Trees* (2016), "Under the canopy of the trees, daily dramas and moving love stories are played out. Here is the last remaining piece of Nature, right on our doorstep, where adventures are to be experienced and secrets discovered."

The threads connecting a tree and its companions weave a rich tapestry of stories about how a tree and its companions influence each other's fate. The stories feature not only the mundane and the ordinary but also the improbable and the unprecedented.

Chapter 1 focuses on the gifts of trees and how each tree defends itself from those that exploit its beneficence and resources. While a tree must confront the onslaught of leaf chewers, sap suckers, and wood borers, it never faces these intruders alone. This first chapter introduces the different groups of organisms that ensure trees are never without support in their battles with creatures that feed on them—countless microbial and animal parasites and predators.

Subsequent chapters focus on specific parts of the tree, such as leaves, twigs, trunks, fruits, and roots. Not only the creatures that feed on these parts are considered, but also the predators and parasites that often specifically prey on these herbivores. TreeScape (fig. 1) and its associated LeafScape, BarkScape, and RootScape provide illustrated maps for the diverse company found in different tree regions from treetops to root tips.

Chapter 2 focuses on the lives that inhabit leaves, twigs, and buds—the herbivores that chew these plant tissues, their predators, and their parasites. Chapter 3 considers the inhabitants of leaves, limbs, and roots that feed not with jaws but with beaks, tapping the vast circulatory system of trees. These sap suckers

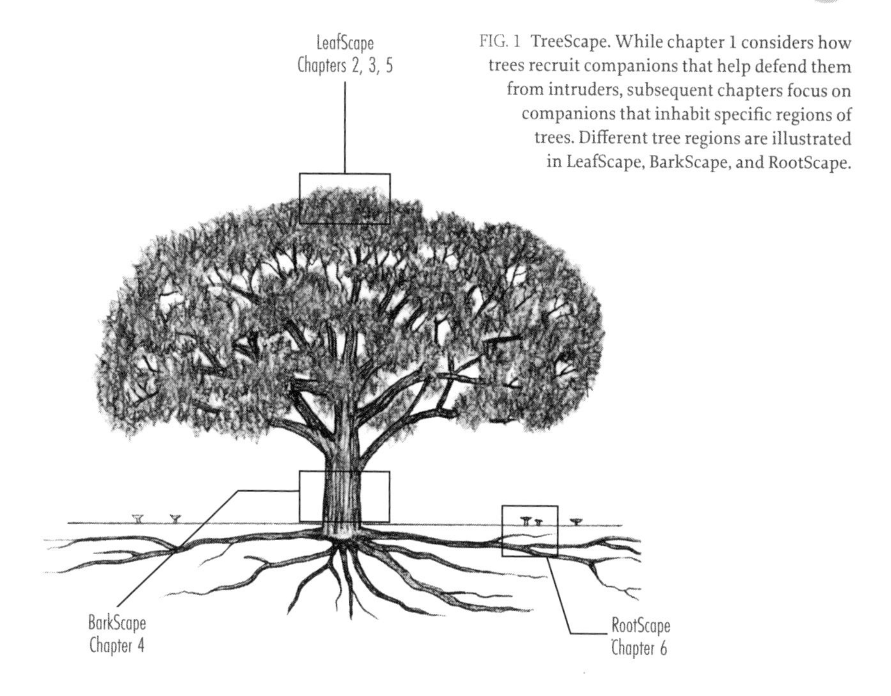

FIG. 1 TreeScape. While chapter 1 considers how trees recruit companions that help defend them from intruders, subsequent chapters focus on companions that inhabit specific regions of trees. Different tree regions are illustrated in LeafScape, BarkScape, and RootScape.

have developed special relationships with microbes, parasites, and predators. In chapter 4, life on bark, under bark, and in hollow limbs is represented by wood borers, their parasites, and their predators. Chapter 5 shows how tree flowers, with help from pollinators, ensure the pollination of flowers on many trees and their subsequent fruit and seed formation. With their enticing scents and flavors, flowers, seeds, and fruits are all subject to being eaten, but their appeal to companions of trees is also responsible for the many benefits trees derive from the pollination and dispersal that these companions provide. Flowers, fruits, and seeds of trees attract and nurture many birds, mammals, and insects. These in turn nourish their own parasites such as lice, fleas, mites, and those few flies that indirectly benefit from nourishment provided by trees. Chapter 6 describes how all life aboveground and belowground is eventually recycled. The innumerable decomposers under a tree accomplish their phenomenal job of recycling nutrients from each generation of creatures for the survival of the tree and for the survival of future generations of tree companions. As is true for other parts of the tree, all creatures beneath the tree face predators and parasites, and all these creatures eventually provide raw materials for decomposers. Chapter 7 describes how the reader can arrange firsthand encounters with the hidden company that trees keep.

Creatures associated with trees include all six kingdoms of life—animals, plants, and microbes: fungi, protists, bacteria, and archaea. Members of different groups representing different hierarchical levels of classification can be arranged in order from the least specific, most general level of classification of (1) kingdoms to the more specific levels of (2) phyla, (3) subphyla, (4) classes, (5) subclasses, (6) superorders, (7) orders, (8) suborders, (9) superfamilies, (10) families, (11) subfamilies, (12) genera, and finally (13) species. The diversity of life on trees is emphasized with profiles for these different groups.

Different groups of creatures found in the company of trees are profiled throughout the book with information on (1) their common names, (2) their scientific names, (3) the derivation of their scientific names, (4) the number of species in the group found around the world, (5) the number of species in the group found in North America (NA), and (6) the size range of species in the group.

Sometimes two or more common names have been given to a group, but each group has only one universally recognized scientific name, often derived from words of Greek or Latin origin. The origins of a scientific name are indicated in italics followed by their English equivalents. The number of species within the group that are presently known to inhabit the world are listed along with the number of species known to inhabit North America (NA). These numbers are based on the number of species presently described in scientific publications; however, new species are constantly being discovered and reported, emphasizing how much of the world's diversity is still hidden and undiscovered. Understandably, sizes of these tiny creatures are measured in millimeters and fractions of millimeters. To human eyes, creatures in this size range are most often hidden from view. Microbes, by definition, are invisible without magnification. All those animals without backbones—the invertebrates—range from microbial sizes (fractions of millimeters) to slugs and beetle larvae measuring hundreds of millimeters. Dimensions of these large invertebrates overlap with those of the more conspicuous and familiar vertebrate inhabitants of trees—birds, mammals, reptiles, and amphibians.

The more we learn about the lives of these creatures, the more we appreciate the contributions of each tree's companions. We realize how intertwined their lives are and see how each creature—no matter how small or how large—leads a life of unfathomable complexity.

ACKNOWLEDGMENTS

Each book, like each tree, has its hidden companions that contribute to its growth and wholeness.

Princeton University Press

Robert Kirk and Christie Henry offered the support that initially launched this project at Princeton University Press.

As production editor, Kathleen Cioffi patiently, skillfully, and very efficiently guided the manuscript through its gauntlets of copyediting, designing, and indexing.

As copyeditor, Laurel Anderton molded the initial manuscript to emphasize its best features and corrected oversights I most probably would have missed.

Colleagues and Friends at the University of Illinois

Dorothy Loudermilk, who directs Graphic Services at the School of Chemical Sciences, has been a steadfast partner in scanning and labeling the hundreds of illustrations, enhancing their good features, and eliminating their flaws. Preparing all these illustrations, often rescanning and relabeling, was a monumental undertaking. Dorothy's unwavering patience, understanding, and good humor made this task a joyful one.

At the Illinois Natural History Survey, Dr. Tommy McElrath oversees one of the world's largest collections of insects. Tommy granted me access to specimens with which I wanted to become more familiar. Being able to examine them at close range led to several new discoveries.

With Cate Wallace behind the controls of the scanning electron microscope at the Beckman Institute, we were able to capture an insect's view of the three-dimensional landscapes of leaf and twig surfaces.

Wen-Yen Wu's skill as a photographer and microscopist is reflected in the clarity and sharpness of the endophytes in oak leaves that he imaged with microscope and camera.

Only after my friend and Illinois alumnus Steve Wagner called my attention to the woolly aphids of beech trees did I realize just how many other creatures depend on these tiny, fluffy insects for their livelihoods. Figure 147 is based on his photograph of a woolly aphid colony.

Among the Trees of the Indiana Woods

I have shared discoveries from root tips to treetops with my wife Joy, our four-legged family, and many friends. The efforts of my friends and companions in Ouabache Land Conservancy and all those who protect and restore forests are ensuring that trees and their countless companions will continue to weave their rich, harmonious tapestries.

THE MANY FORMS AND FUNCTIONS OF A TREE'S COMPANIONS

1

THE COMPLEX WEB

You can tell a lot about a tree from the company it keeps—the birds that fly, the mammals that climb, the snakes that slither, and the frogs that hop among its branches and call trees their homes. Squirrels, monkeys, porcupines, bats, sloths, songbirds, owls, hawks, geckos, and tree frogs are all familiar residents of trees. Some, like sloths and tree frogs, settle down in one tree, almost never leaving their one arboreal home, but most travel from tree to tree in search of the food, shelter, and lookout posts that trees offer. They have forged lasting relationships with trees. These relatively large and conspicuous vertebrate companions of trees eat their leaves and fruits, disperse their seeds, chew their bark, pollinate their flowers, and control their insects. The destinies of trees are interwoven with the fates of these familiar animals that keep them company (fig. 2).

FIG. 2 A great crested flycatcher emerges from a tree hole to pursue a soldier fly that just emerged from under the bark of a tree.

All the birds, mammals, frogs, lizards, and snakes whose lives come together on a tree during its lifetime can be easily tallied, but all the smaller creatures that establish their lives on that tree are countless. The small hidden creatures among the trees' leaves, limbs, and roots are not only the ones that have the greatest impacts on the lives of trees but also the ones about which we often know the least. Among them are those that nurture each generation of nestling birds, those that satisfy the appetites of tree frogs, and those that supplement the vegetarian diets of many mammals. Trees may simply tolerate their company or may actively entice and welcome their company, but sometimes trees must aggressively defend themselves from certain uninvited companions.

Every tree, whether rooted in a great forest, city park, or backyard, has a unique story to tell about the creatures that share its company. The following pages focus on trees of Earth's temperate zones. However, the creatures that keep company with these trees have different species as their counterparts in other parts of the world. Looking closely at a tree reveals that it is a busy crossroads of activity where innumerable lives arrive, depart, or just carry on. We can think of each tree as a community where creatures are caught in a web of interactions that links them to each other and to the tree in which they live. They all share two essential needs: energy and nutrients. This web of interactions involves coexistence, cooperation, and competition among members of the community. And even though individual creatures may fall prey to predators and trees may have their leaves devoured, the members of the web manage to transact their business—giving and taking energy and nutrients—and usually balance their accounts so that few, if any, species gain or lose too much. Each species has its job to do and does it. One tree can thus support—directly or indirectly—many lives. The destiny of each tree is tied to these myriad connections with its companions (fig. 3).

Partnerships among trees and other creatures began when green plants attained the ability to capture the energy of sunlight with molecules of chlorophyll. With the energy they captured, plants began to produce nourishment and oxygen not only for themselves but also for countless other creatures. This process is known as photosynthesis, a word that means "to put together with light." Trees "put together" sugars and oxygen from the simple raw materials of carbon dioxide from the air and water from the soil, using the energy that their chlorophyll molecules capture from sunlight. This energy from the sun is transferred to the sugars that the tree produces, to the creatures that feed on the sugars, and to the creatures that feed on the creatures that have fed on the sugars of the tree. Each life form uses this energy of sunlight—directly or indirectly—to survive. When one creature eats another or eats part of the tree, it obtains energy and nutrients from the meal. This is true whether that meal is alive or dead, plant or animal or fungus. Without contributions from certain partners that supply them with mineral nutrients such as magnesium for each chlorophyll molecule and calcium for every cell wall, however, trees could not survive, even though they can perform the astounding feat of capturing the energy of sunlight.

These contributing partners include microbes and all the other creatures larger than microbes that provide recycled mineral nutrients for a tree by consuming the remains of creatures that once lived on the tree as well as dead or discarded parts of the tree itself. In other words, the tree and its many partners, including other decomposers, provide the raw materials for the recycling efforts of the decomposers. These other partners are (1) herbivores and pollinators that consume the nutrients of the tree's living tissues, as well as plants such as lichens and mosses that live on the tree; (2) fungivores that consume the tree's fungal pathogens, its fungal decomposers underneath it, and its fungal partners known as mycorrhizae and endophytes; and (3) predators, parasitoids, and parasites that survive by feeding on living animals of the tree (fig. 3).

By far the most numerous inhabitants of trees are also the tiniest. These are the microbes—fungi, protists (*protist* = very first) such as slime molds and protozoa,

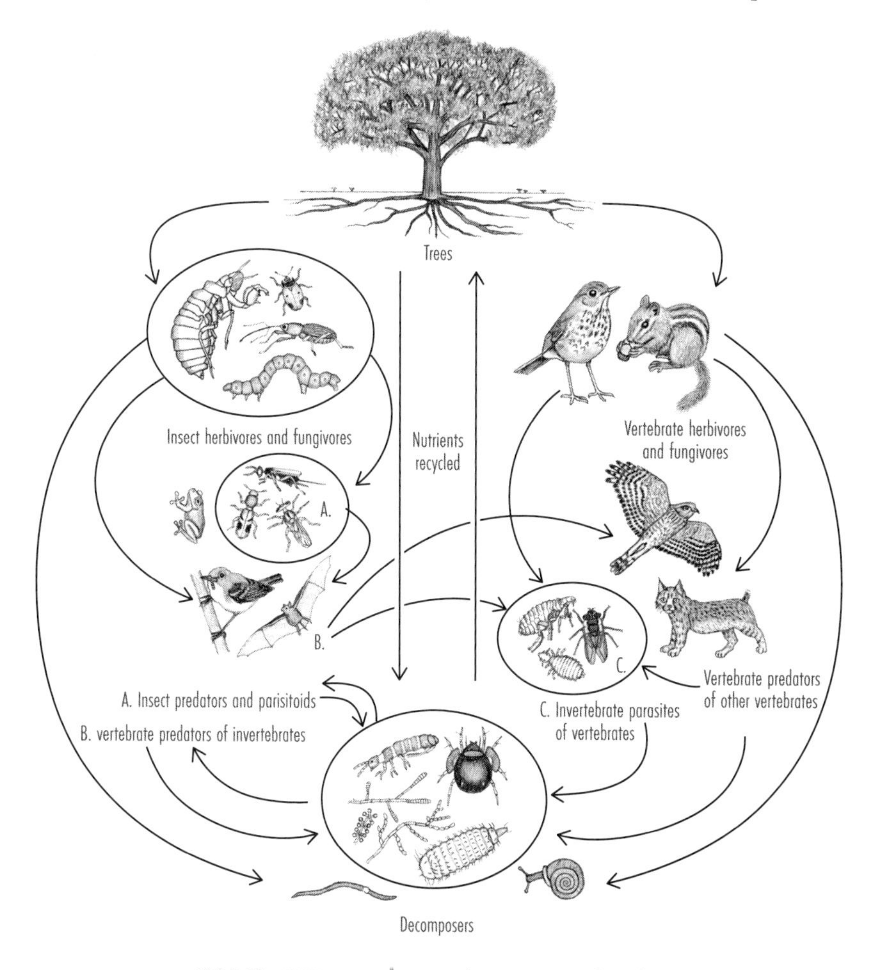

FIG. 3 The companions of a tree exchange energy and nutrients.

and prokaryotes (*pro* = before; *kary* = nucleus) such as bacteria and archaea—that intimately associate with the tissues and cells of a tree and form what is known as the tree's microbiome. Trees are not unique in having microbiomes; microbial associates also have intimate partnerships with the tree's animal companions. Bacteria and archaea are known to reside in fungal tissues, so even fungi have microbiomes (Bonfante and Anca 2009). The network of interactions in a tree extends across all kingdoms—archaea, bacteria, fungi, plants, animals, and protists. Microbes clearly touch the lives of everyone in a tree in ways that remain bewildering.

Except for microbes, arthropods are always the most numerous creatures on all parts of a tree aboveground as well as underground. Arthropods—animals with jointed legs—include the familiar insects, mites, spiders, woodlice, centipedes, and millipedes as well as the unfamiliar proturans, pauropods, diplurans, and symphylans. As a group, arthropods make up a phylum, a taxonomic group comparable to our own animal phylum—the Chordata, animals with backbones. In their forms and in number of species, arthropods are by far the most diverse animals of our planet. Eighty percent of animals are arthropods. At last count, arthropods boasted over a million species distributed among 10 classes. One of the 10 arthropod classes is the Insecta, and insects make up about 82 percent of the arthropods. Ranking second and third in number of species are the class Arachnida and the class Crustacea, representing respectively about 8 percent and 3 percent of the arthropods. All 10 arthropod classes associate with trees in some way. Many of the species that live in the litter and soil under trees live nowhere else on Earth; representatives of the remaining seven arthropod classes—Collembola, Protura, Pauropoda, Chilopoda, Diplopoda, Symphyla, and Diplura—are wed to lives underground and rarely venture far beyond its dampness and darkness.

The vertebrates to which we humans belong represent a subphylum within the phylum Chordata. Vertebrates number 57,000 of the 60,030 species in their phylum. Compared with this number of species distributed among the five classes of vertebrates—from fishes to mammals—arthropod species outnumber vertebrate species by a factor of 22. The number of new arthropod species described each year always vastly exceeds the number of new vertebrate species described.

We can appreciate the complexity of interactions among these countless creatures, but no one can know all the creatures involved and all the ways they relate to each other. This mystery is especially profound for the groups of small creatures, many of whose species are still undiscovered and unnamed. For those that we know by name, many accomplish seemingly unimaginable, Herculean feats. For others, we often know little, if anything, about how they spend their days and how their obscure lives mold the life of their tree, their forest, and beyond.

Many of these creatures, even though they are often smaller than the periods in this book, are formed from molecules, cells, and tissues like ours and are made up of organ systems like ours—hearts, nerve cords, brains, intestines. The forms of their organs may differ from those of our own organs such as brain, gut, liver, heart, and lung, but they function like our organs and share similar proteins, and all have been molded over geological time by their environments. Their body surfaces are covered with sensory bristles of all lengths and widths that touch, taste, smell, hear, and see from many angles and on many sides. Their survival depends on being acutely aware of their surroundings. At the base of each of these sensory bristles lie one or more nerve cells that convey sensations to their brains and nerve cords. Processing of sensory information in the central nervous system determines which muscles are excited and how the creatures ultimately respond. Despite their minuscule forms, insects, spiders, and mites can exquisitely monitor their environment. The beauty and exemplary functioning of their myriad forms often inspire better designs for our own instruments and machines.

No one expressed our disparate as well as shared heritage with these fellow creatures in more lyrical and moving terms than the writer-naturalist Henry Beston (1928):

> *In a world older and more complete than ours they move*
> *finished and complete, gifted with extensions of the senses we*
> *have lost or never attained, living by voices we shall never hear.*

THE GIVING TREES: GIFTS FROM TREES TO THEIR COMPANIONS

We often take for granted the many gifts of trees. With light energy captured from the sun, trees provide chemical energy and nourishment for their companions. All members of a tree's community, including the tree, at the end of their lives provide energy and nutrients as raw materials for the countless decomposers. The decomposers then liberate and recycle these mineral nutrients into forms that the tree can obtain from the soil. The cycle continues unabated; each tree takes up these nutrients from the soil for its own tissues and takes up more energy from the sun so it can share its energy and mineral nutrients with countless companions (fig. 4).

Trees are gracious and neighborly. Their forms muffle noises as well as add cheer and beauty to a landscape. They hold moisture in the soil, and they also protect it from erosion. With their roots, trees pull up mineral nutrients from deep in the soil that will be shared with other creatures in their communities. They absorb carbon dioxide and many pollutants from air. Each year when they shed their leaves, trees return many nutrients to the soil to replenish the fertility of the land. Trees offer nourishment and refuge to all visitors and companions. In the cold of winter, they block icy winds (fig. 5). In the heat of summer, they provide coolness and shade (fig. 6). Saving trees and planting trees invests in the integrity and beauty of the Earth.

Urban trees can reduce summer temperatures as much as 6.5°F (3.6°C) (McDonald 2016). When the collective transpiration of trees in a forest condenses as clouds, air containing the transpired vapor condenses and decreases in volume, resulting in reduction of air pressure. As the air pressure drops, air with less

FIG. 4 Trees create a welcoming environment for their companions and visitors.

FIG. 5 A tree sparrow huddles among the branches of a spruce tree during a winter storm.

moisture is horizontally drawn in and generates winds that cool the landscape (Pearse 2020). Thoreau, in deploring the rampant cutting of America's trees in the nineteenth century, mistakenly noted with relief, "Thank heaven, men cannot cut down the clouds." "Aye, but they can!" presciently observed the naturalist Gene Stratton-Porter in *Music of the Wild* (1910). "I never told a sadder truth, but the truth is that man can 'cut down the clouds.'" And the latest research on how trees influence temperature, cloud formation, and climate confirms her sad truth.

FIG. 6 A fox squirrel relaxes in the shade of walnut leaves.

From treetop to root tips, wherever trees grow, each tree—in life and in death—hosts its own community of creatures. Some creatures choose a particular tree for its fruits or leaves, its flowers, or its decaying hollow limbs; the tree's insect life attracts others (fig. 7). Each creature seems to find what it needs somewhere on the tree and manages to repay the tree in some way for its generosity. All these creatures share in the gifts offered by trees, and trees are accepting of all sorts of companions, some of which even devour their leaves, eat their fruits, or bore into their wood. Many companions of trees turn out to be trusted allies in defending them and ensuring that harmony and balance is maintained among trees and their myriad companions.

HOW TREES RESPOND TO COMPANY AND DEFEND THEMSELVES WHEN NECESSARY

Within the cells and tissues of trees, the synthetic reactions of photosynthesis and the breakdown reactions of respiration continually fuel the chemical reactions that occur during the life of trees. During photosynthesis, energy-rich sugar molecules are generated with energy from the sun, water, and carbon dioxide; and during respiration, the energy of those sugar molecules is released as carbon dioxide and water, which are again recycled by photosynthesis. The compounds generated by these chemical reactions combine to form the myriad chemicals that make up the metabolites of the tree. Trees are continually producing primary metabolites that are essential for their proper growth and physiology; these include the substances that make photosynthesis and respiration possible and the

FIG. 7
Trees defend themselves from insects that eat their leaves, and they are assisted in their defense by insect-eating birds such as this vireo.

variety of hormones that orchestrate the formation and functioning of each part of the tree. Secondary metabolites are compounds that are not essential for the tree's survival but that certainly influence how trees interact with other creatures in their environments. Plants are estimated to produce an astonishing 200,000 different secondary metabolites. Plant cells constantly produce these chemicals, some of which repel or sicken many insects. But there are always some insects that relish the flavors that are so repellent to others. Not only do these insects have means to detoxify these compounds, but some have even adopted these chemicals for their own defense against insect and vertebrate predators. The colors, flavors, and scents of trees can not only repel but also entice the company they keep.

In the face of insect and microbial attacks, plants superficially appear placid and passive, but under the surface of their leaves and bark, plant cells promptly recognize an attack from a foreign agent and mount a robust defense. Although Willa Cather noted that "I like trees because they seem more resigned to the way they have to live than other things do," she had no idea how assertive a tree can be when facing challenges from herbivores and pathogens. The lack of flight or any other evasive movement by trees belies their highly effective ability to circumvent attacks with their arsenal of evasive chemistry. The defensive chemicals of trees can repel, intoxicate, or disrupt digestion of their herbivore attackers.

Trees release volatile defense compounds whenever they experience mechanical damage to their tissues. Trees that are battling intruders alert their tree neighbors of dangers and threats by releasing a variety of volatile compounds that waft through the air and elicit the production of additional defenses in preparation for their own possible upcoming battles. Over a thousand different volatile compounds are known to be released by different plants as their first defensive response to invaders (Dudareva et al. 2004), and these compounds prompt a tree to release a cascade of new defense chemicals that travel to other parts of the tree, sounding the alarm from branch to branch and root to root. These volatile chemicals carried through the air recruit not only other trees but also predatory and parasitic insects as allies in battles with plant-feeding insects (fig. 8).

Just as our immune systems can distinguish self from nonself, rejecting any objects recognized as foreign—from bacteria to tissue grafts—plants also respond to foreign invasions of their tissues. After encounters with microbial pathogens, insect mandibles, insect beaks, insect ovipositors, or even the pressure exerted by

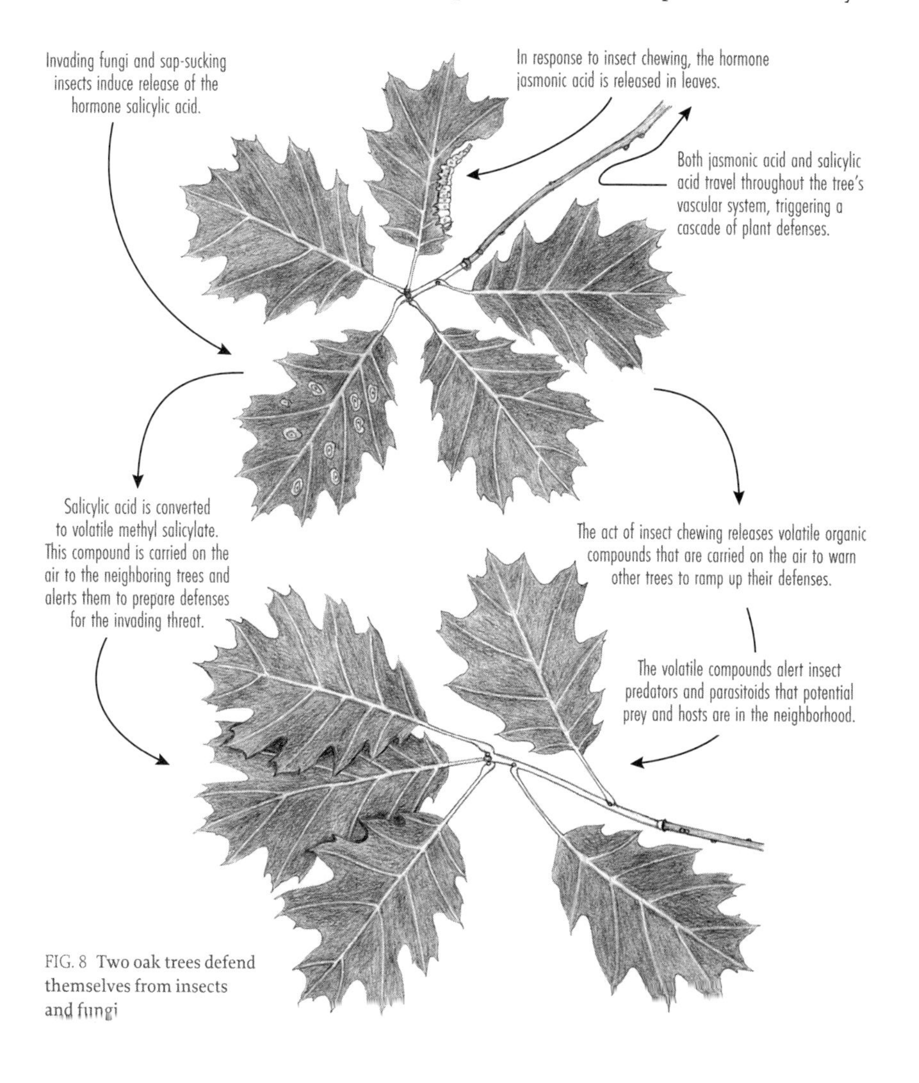

FIG. 8 Two oak trees defend themselves from insects and fungi

insect tarsi, the levels of the two master hormonal regulators of a tree's immune response—jasmonic acid and salicylic acid—almost invariably increase in the tree's tissues. Insect saliva often contains substances that boost these two master regulators of the initial wound response, but in some instances, the herbivores secrete chemicals that can tamp down the normal defense response to mechanical damage (Musser et al. 2002).

After a fungal or insect attack, the tree hormone salicylic acid is converted to methyl salicylate, which vaporizes and wafts to other leaves and other trees where it triggers the release of other defensive compounds. The release of these key compounds alerts tree cells to send out reserves from their repertoire of thousands of defensive secondary metabolites (Fürstenberg-Hägg et al. 2013).

Hormones such as jasmonic acid and salicylic acid govern all aspects of a tree's life—its birth and eventual death, its leaf sprouting and flowering in the spring, its leaf fall in the autumn, and all its growth between birth and death.

WHAT HAPPENS AFTER AN INSECT BITES?

The very first response of a tree to an insect bite is comparable to our response to a mosquito bite. Nerves immediately send a long-distance signal to the brain that a mosquito has landed and is taking a blood meal. In our bodies the signal is induced at the site of the bite by the chemical glutamate, which mobilizes the calcium that then carries the signal all the way to the brain.

To communicate that damage has been inflicted at one location on the tree to more far-flung locations on the tree, an insect bite on a leaf likewise induces the release of glutamate. Glutamate is the positively charged form of the amino acid glutamic acid, a simple compound found throughout living cells that travels from the wound throughout the plant at about 1 mm/second. As it moves through the tree, glutamate binds to receptors on surfaces of plant and animal cells. Receptors are proteins of cells whose forms perfectly complement the forms of the proteins that bind to them, in the same way that only a specific key will perfectly match a specific lock. These surface receptors represent channels in cell membranes that allow the influx of calcium into plant cells in the same manner that glutamate opens ion channels and acts as the most abundant neurotransmitter in our animal nervous systems, rapidly sending signals from cell to cell. Binding of glutamate to its receptors opens the ion channels for calcium (Muday and Brown-Harding 2018).

The sudden release of calcium activates special enzymes that cleave fragments or peptides (PEPs = plant elicitor peptides) from one or more large proteins referred to as precursors of plant elicitor peptides (PROPEPs) (Haner et al. 2019). These newly generated peptides diffuse to nearby cells, where they bind to receptors on their surfaces and by the act of binding activate and elicit the tree's master regulators of the immune system—jasmonic acid and salicylic acid or their derivatives (collectively referred to as jasmonates) in the cells of the vascular tissues. The presence of jasmonic acid exerts a ripple effect on the expression of many genes referred to as jasmonic-responsive genes. After being induced by herbivores and pathogens, these key hormones set off a cascade of defense compounds to ward off attacks from insects and fungi (Ramirez at al. 2009). In the years ahead, more details will be added to this complex story.

TREES AND THEIR ALLIES

Herbivores and microbes seem to find ways to counter or circumvent even the best defensive strategies of trees. Trees have come to rely not only on their own multiple chemical defenses but also on the ingenuity and aid of their microbe, arthropod, and vertebrate companions. Trees do not have to fight their battles alone; they can count on help from these allies in confrontations with insects and pathogens. In addition to pollinating their flowers and dispersing their seeds, allies of trees keep the numbers of invasive microbes as well as leaf-chewing, wood-boring, and sap-sucking insects in check. Trees in turn reward their allies with an all-you-can-eat buffet.

By enticing creatures to live on their leaves, bark, and flowers, trees also attract the predators that keep the numbers of these other creatures in check. Different wasps and predatory bugs stalk and eat insects that feed on leaves and wood. Many songbirds flit from branch to branch in search of their six-legged meals and the telltale signs that insects leave behind—leaf rolls, leaf mines, bored wood, and well-chewed leaves. Nuthatches and woodpeckers constantly inspect the recesses of bark for any insects that may be sheltered there (fig. 9). Hawks and owls keep

FIG. 9 By providing many hideouts for insects, the shaggy bark of shagbark hickory also provides a rich hunting ground for nuthatches.

FIG. 10 A great horned owl is scolded and mobbed by crows for attacking one of their fellow crows. The owl's very existence depends on its meals of birds and mammals that have been nourished by the countless insects, fruits, wood, and leaves of trees.

FIG. 11 Porcupines do not seem to have met many trees whose bark they do not relish, and the fisher is one of the very few predators that can evade the formidable quills of this big, furry herbivore and keep its population in check.

a sharp lookout for movements of birds and squirrels (fig. 10). Fishers search evergreens of the north woods for signs and scents of porcupines (fig. 11). The hunters and the hunted hold together the intricate web of life in a tree community.

Vertebrate Allies and Their Deceptive Insect Prey

The birds, frogs, and mammals that perch, climb, hop, flit, and ramble among a tree's leaves and branches are the least numerous of its allies but certainly the largest and most conspicuous. Many are predators of insects and purveyors of microbial allies that colonize tree tissues.

Birds and other vertebrate predators of insects, however, must deal with the deceptive tactics of the insects they stalk and eat. Insects can masquerade as any part of a tree, including lichens on bark and bird droppings on leaves, and their disguises are surprisingly deceptive (fig. 12; chapters 2 and 4). Green is always a popular color for insects that live among green leaves; green caterpillars blend well with green leaves and green twigs. Brown and gray caterpillars intermingle with brown bark and gray twigs.

As immobile pupae, moths and butterflies are even more vulnerable to attack than they are as wandering larvae. As a prelude to pupation, many caterpillars spin cocoons as shelters in which to undergo their transformations; some burrow into the soil beneath a tree, hide in leaf rolls, or crawl under loose bark or into clefts in the bark. However, birds such as nuthatches and woodpeckers patrol tree trunks and routinely dig out insects from the recesses of the bark. Migrants such as warblers and vireos have an uncanny ability to uncover pupae and caterpillars tucked away in leaf rolls. The pupa or chrysalis of the hackberry butterfly, however, has a mottled color pattern that blends not only with the green of the

FIG. 12 A dagger moth called the tufted bird-dropping moth has all-purpose camouflage patches of blue, brown, lichen green, and white and can be as easily mistaken for a bird dropping as it can for a patch of lichen.

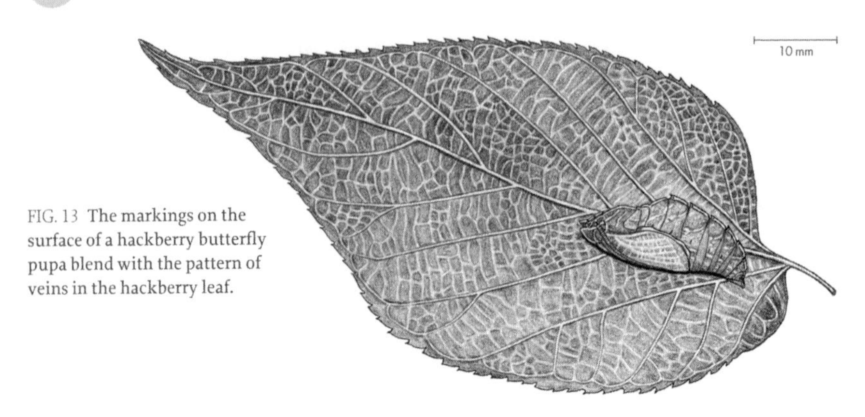

FIG. 13 The markings on the surface of a hackberry butterfly pupa blend with the pattern of veins in the hackberry leaf.

hackberry leaf but also with the cream color of the leaf veins (fig. 13). Those pupae with naturally camouflaged cuticles can rest exposed and unnoticed as they masquerade as part of the tree on the surfaces of leaves, twigs, and bark.

Birds assiduously search trees for their favorite caterpillars. They leave no leaf unturned as they scour the canopy for food. While so many caterpillars blend in with their surroundings (fig. 14), they still leave chewed leaves as telltale traces of their whereabouts. Birds use these signs to locate their prey as they scour foliage and branches. Even well-camouflaged caterpillars cannot be too careful about deceiving their predators. After their leaf-feeding sessions, caterpillars go so far as to eliminate all the partially eaten leaves by chomping through the petioles of these leaves and letting them fall to the forest floor, far from the caterpillars' dining area. These deceptive caterpillars leave few clues for any clever and hungry birds that attempt to track them down (Heinrich 2019; Heinrich and Collins 1983).

Rather than resorting to deception and camouflage to avoid predators, however, some insects use just the opposite strategy of being ostentatious and blatantly calling attention to themselves. These insects sport bright, often garish colors to announce attributes that make any close encounters with them unpleasantly memorable and preferably avoided. These attributes can include painful stings, foul-smelling secretions, irritating hairs, or toxic blood. The word soon gets out that these insects should be kept at a respectful distance (chapters 2 and 4).

Harmless mimics often imitate these insects with bad reputations for successfully intimidating predators; the mimics gain protection simply by virtue of superficial appearances without having to invest in any weaponry. Other harmless insects shock and startle anyone who threatens them with their sudden, unexpected changes in form or color (fig. 15).

For all the defenses insects have mounted against their vertebrate predators, a few predators remain unimpressed and undeterred by what these caterpillars have to offer. How can caterpillars covered in annoying, fuzzy coats possibly be appetizing to any bird? However, the cuckoos of North America seem to have a fondness for fuzzy caterpillars and can dispatch great numbers in a short time (fig. 64; chapter 2). Cuckoos are endowed with thick linings of mucus in their

FIG. 14 An inchworm poses as a branch of a wild cherry twig.

FIG. 15 When an underwing moth (chapter 4) takes flight, its flash of striking color suddenly startles a foraging possum.

stomachs that absorb the fuzz, which the birds simply regurgitate as a fuzz ball in the same way that cats cough up fur balls.

You would suspect that wasps with their painful stings would have few enemies, if any, but even these insect hunters of other insects are also hunted. Certain songbirds not only share caterpillar prey with wasps but also feed on the wasps themselves. Somehow birds such as tanagers can snatch and dispatch wasps and bees in midair without ever getting stung (fig. 16).

Arthropod Allies: Parasites and Predators

The insect predators and insect parasite companions of trees safeguard the trees by keeping in check the damage inflicted by insect herbivores. The antennae of parasitic and predatory insects pick up some of the airborne defensive chemicals released by plants attacked by herbivores. The scents of plants in distress recruit and guide parasites and predators to aid the besieged plants (Ananthakrishnan 1999). The olfactory sensors of predators and parasites are finely attuned to detecting the odors emitted by damaged plants and by the digested plant fragments of caterpillar droppings. Predators such as ground beetles and rove beetles patrol mostly under trees, but some species also search leaves and branches for insects, mites, and fungi that threaten trees' well-being. Hornets and wasps, lacewings and ladybird beetles, robber flies and long-legged flies, assassin bugs and beetle predators survey trees from the air, under their bark, and over their surfaces for insects that

FIG. 16 Tanagers are some of the few birds that have expanded their insect meals to include wasps, bees, and hornets.

become meals before they become devastating pests. Insects whose larvae are parasites on other insects are just as thorough in their surveillance. These insect parasites and predators control populations of insects that feed on trees, but they, along with the insect herbivores, also convey beneficent microbes to and from trees for the benefit of all parties—trees, insects, and microbes (Arnold et al. 2003).

Parasitic Insects

Parasitic insects exert an unheralded influence on the harmony and integrity of a tree's food web. Many carry out their early work as larvae and pupae hidden within the bodies of their hosts (endoparasites). At metamorphosis, however, parasitic insects forsake their secluded carnivorous lives for lives outdoors where they placidly sip nectar and munch on pollen grains. Parasitic insects make up around 10 percent of all animal species. Parasitic wasps are estimated to number well over 130,000 species, possibly even more species than beetles, which for so long have been claimed to be the animal order with the greatest number of species. An estimated 15 percent of all insects are parasitic on other arthropods (Forbes et al. 2018; Feener and Brown 1997; Askew 1971), and no arthropod—herbivore, fungivore, decomposer, predator, or even another parasite—seems to be spared from being a host for some parasite.

Each parasite prefers hosts of a particular developmental stage, and each developmental stage—from egg to adult—has its share of parasites. The few parasitic beetles and moths feed as ectoparasites on the outer surfaces of their hosts. Some parasitic insects are parasites of parasites—hyperparasites. The caterpillars of about 50 species of moths and the larvae of 5,000 species of beetles live as external parasites. Most of the thousands of species of wasp and fly parasites (at least 130,000 species and 16,000 species, respectively), however, feed within the bodies of their hosts surrounded by their blood cells and have developed elaborate tactics to evade the blood's immune system. The influence of so many different parasitic insects—some with broad, wide-ranging preferences for their hosts, others with very narrow, finicky preferences—ripples throughout the food web. Strands of the food web radiate from predators and parasites to practically all living animal members of a tree's food web—its decomposers, herbivores, even other predators and parasites.

The hosts of parasitic insects are often shared as the prey of predatory insects. Parasites can subdue far larger creatures for their hosts than predators can overpower as their prey. Locating a host or the general vicinity of a host is the responsibility of the parasitic insect's mother. Each time predators feast they must use energy to chase down and overpower their prey. However, once parasitic insects settle down with a host or hosts, they can conserve a great deal of energy by staying in one place and growing at the expense of the host, which continues to provide shelter and nourishment until the parasite no longer needs its still-living, but doomed, host. Because hosts of most insect parasites eventually die, the parasites are generally referred to as parasitoids to distinguish their lifestyles from those of insects or other parasites that live at the expense of their hosts but do not kill them.

The many families of parasitic wasps boast species whose hosts range from insect eggs and embryos to large beetle grubs and hefty caterpillars. Many wasps are very particular about their hosts and will parasitize only specific families or species of insects. Caterpillars are certainly favorite hosts. In the following pages, we will encounter a number of parasitic wasps that parasitize hosts found on tree leaves, under bark, underground, or inside galls. They employ their ultrasensitive antennae to track down the scent and the slightest movements of even the best-concealed hosts. As an introduction to these incredibly diverse insects, we feature an ichneumonoid wasp that is known to parasitize leaf-rolling caterpillars, and a tiny chalcidoid wasp from the leaf litter under a tree. These two wasps are members of the two most species-rich superfamilies of parasitic insects—Ichneumonoidea and Chalcidoidea (figs. 17 and 18). Other parasitic insects will be introduced as their specific habitats on trees are discussed in upcoming chapters, but the parasitic wasps in other superfamilies are neither as common nor as diverse as the wasps in these two superfamilies.

— ICHNEUMONOID WASPS, SUPERFAMILY ICHNEUMONOIDEA, include two large families of parasitic wasps: Ichneumonidae and Braconidae.
  - **Ichneumon Wasps, family Ichneumonidae** (*ichneumon* = tracker) (25,000 described species, estimated 60,000–100,000 species; 5,000 described species NA; 3–40 mm).
  - **Braconid Wasps, family Braconidae** (*brachy* = short, referring to their usually small size) (17,000 species; 1,900 species NA; 1–15 mm, many 1–3 mm). Many thousands of species, possibly as many as 30,000, remain undescribed.

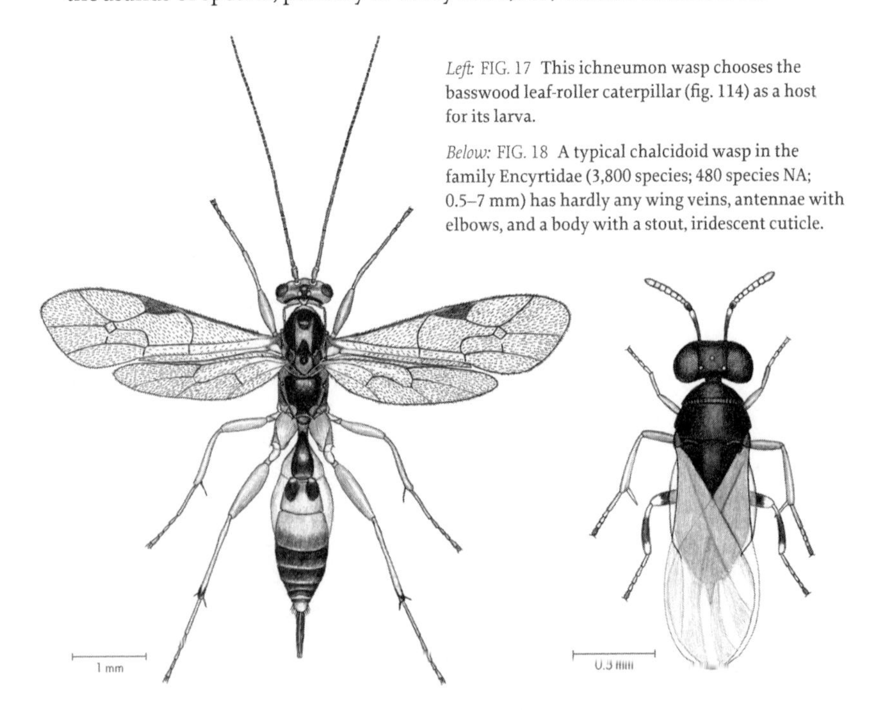

Left: FIG. 17 This ichneumon wasp chooses the basswood leaf-roller caterpillar (fig. 114) as a host for its larva.

Below: FIG. 18 A typical chalcidoid wasp in the family Encyrtidae (3,800 species; 480 species NA; 0.5–7 mm) has hardly any wing veins, antennae with elbows, and a body with a stout, iridescent cuticle.

— CHALCIDOID WASPS, SUPERFAMILY CHALCIDOIDEA, include 19 families (*chalkos* = bronze, copper, referring to their metallic colors) (22,000 described species but at least 500,000 species estimated; 2,000 species NA; 0.1–20 mm). As we explore different parts of trees, we will encounter members of several of these chalcidoid families.

Many parasitic wasps are capable of the Herculean task of subduing hosts hundreds of times their size. There are at least three secrets of success that enable these parasitic wasps to subjugate the immune responses of their larger hosts—polyembryony, teratocytes, and symbiotic viruses (fig. 19).

POLYEMBRYONY OF PARASITIC WASPS One lone wasp can accomplish a feat worthy of hundreds of wasps: polyembryony. The infrequent appearance of identical human twins, triplets, or even octuplets is accompanied by equal partitioning of a single fertilized egg to generate respectively two, three, or eight identical embryos. Insect parasitoids are capable of consistently accomplishing even more remarkable reproductive achievements. From a single fertilized egg laid in a host insect, as many as several hundred identical eggs can arise, and all can develop and eventually emerge as adults from the remains of their host. From a small maternal investment, a massive return is achieved. A tiny parasitic wasp mother makes up for her diminutive size by contributing up to 500 offspring from a single egg deposited in a single host that is far larger than she is. Thanks to polyembryony, the collective contribution of each mother wasp to the control of host populations far outweighs her minuscule size.

TERATOCYTES OF PARASITIC WASPS As every animal embryo begins its life, cells segregate into two distinct lineages—those cells destined to form all the parts of the

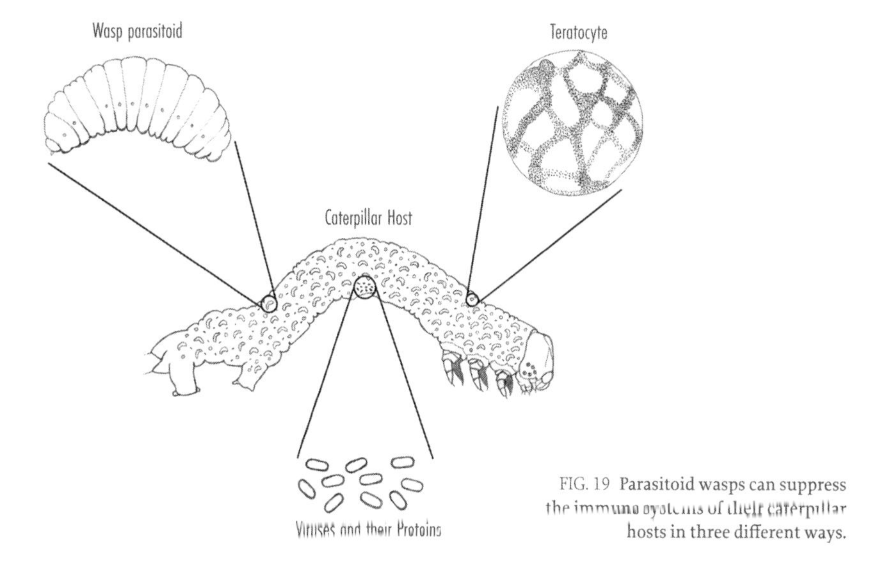

FIG. 19 Parasitoid wasps can suppress the immune systems of their caterpillar hosts in three different ways.

embryo proper, and those extraembryonic cells that form the protective membranes encompassing the embryo. The extraembryonic cells of certain parasitoids separate from newly hatched larvae and take on a new role as they float free in the body cavity of the host. They begin avidly absorbing nutrients from the blood of the host insect and rapidly expanding. In their new role they are referred to as teratocytes (*terato* = monster; *cyte* = cell). While the cells of the embryonic lineage proliferate to produce hundreds of identical embryos during polyembryony, the production of teratocytes involves the proliferation and excessive growth of the extraembryonic cells as teratocytes. All parts of the perfectly spherical teratocyte grow. Its nucleus undergoes a dramatic increase in its DNA content, adopting a complex, arabesque topology, and the diameter of the cell expands twentyfold or more to about a tenth of a millimeter, while other cells by comparison are less than a hundredth of a millimeter. These massive, nutrient-rich teratocytes supplement the diet of a parasitoid larva as it approaches metamorphosis and probably also help suppress the immune response of its larger insect host.

SYMBIOTIC VIRUSES OF PARASITIC WASPS The genetic material of symbiotic viruses is integrated in the genetic material of the parasitoid, and as the mother parasitoid injects her eggs or egg into her host, she also releases viruses or protein particles of viral origin. These either coat the egg surfaces or in some cases multiply in the host's blood cells and suppress the host's immune response. With viral particles covering the eggs, the host's blood cells are fooled into mistaking these particles for self-proteins and thus do not mount an immune response and do not encapsulate what they do not recognize as foreign. However, if these particles are first removed from the newly laid eggs before these eggs are injected into the host, the treated eggs are quickly encapsulated by the hosts' blood cells. Here is a sinister example of mimicry at the protein level tricking the host into not recognizing its foreign invader.

Members of another family of parasitic wasps lead most improbable lives either as hyperparasitoids of parasitoids or as parasitoids of predatory hornets and wasps. These colorful black and yellow parasitic wasps are peculiar in so many ways that they are considered members of their own superfamily, which contains only one family (fig. 20).

— TRIGONALID WASPS, SUPERFAMILY TRIGONALOIDEA, FAMILY TRIGONALIDAE (*trigon* = triangular; *al* = wing) (90 species; 8 species NA; 3–15 mm).

Each mother trigonalid wasp lays several thousand tough, tiny eggs on tree leaves; the eggs can lie dormant for many months. Only after an egg passes through the gauntlet of caterpillar jaws and into the caterpillar's digestive tract does it hatch. The newly hatched parasitoid larva passes through the cells of the caterpillar's gut into the caterpillar's blood space. However, once again the wasp hatchling patiently lies dormant in the caterpillar's blood space until the caterpillar ends up as a meal for a predatory wasp larva or until the caterpillar is parasitized by either a parasitic fly or an ichneumonid wasp. Only a larva of one of these parasitic insects is acceptable as a host for this hyperparasitoid. However, if the

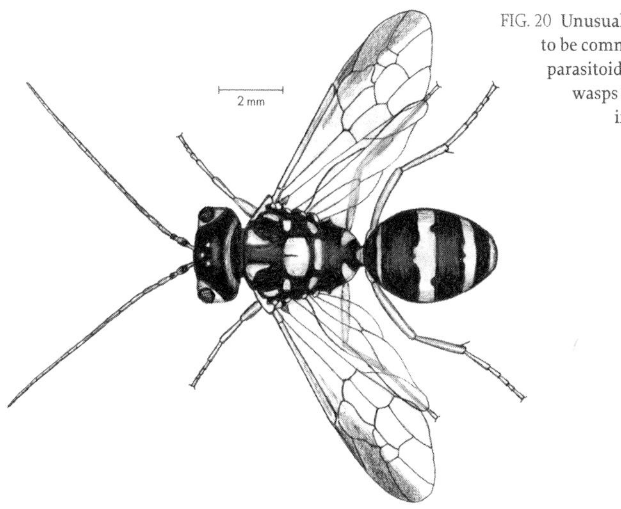

FIG. 20 Unusual life cycles seem to be commonplace among parasitoids, and trigonalid wasps lead remarkably improbable lives.

caterpillar becomes a meal for the larva of a hornet or yellowjacket, then the trigonalid larva ends its days as a parasitoid of a predatory wasp's larva.

Several families in the fly order Diptera are exclusively parasitic and usually very selective in their choices of hosts, but one family of flies, the Tachinidae, can claim the most species of parasitic fly larvae and the broadest range of host choices.

— TACHINID FLIES, FAMILY TACHINIDAE (*tachin* = swift) (> 10,000 species, with new species continually being described; 1,350 species NA; 2–18 mm).

The first impression one gets of tachinid flies is their "bristliness." Tachinids are renowned for the length and abundance of their sensory bristles (fig. 21; LeafScape, 9). Of all the flies that parasitize other insects, tachinid flies are by far the most diverse, and their choices of hosts are correspondingly diverse. While some tachinids are very choosy about their hosts, over 100 species of hosts are acceptable to other tachinid species, including such hosts as larval and adult beetles, stink bugs, moth and butterfly caterpillars, earwigs, centipedes, sawfly larvae, grasshoppers, scorpions, or larvae of other flies.

The variety of tachinid forms and colors is matched by the variety of ways in which mother tachinid flies can lay eggs or lay newly hatched larvae that are retained within an insect version of a uterus until the time of hatching. They can carefully place a few eggs on hosts, or they can scatter many eggs in an area frequented by hosts or on leaves being consumed by host larvae. These eggs have tough eggshells and can remain viable for many weeks until a suitable host finally ingests them, and the tachinid larvae hatch inside the host's gut. Other flies deposit newly hatched larvae on surfaces of leaves, logs, or soil where they are left on their own to search for a suitable host. These active larvae or planidia (*plano* = wandering) are covered with thick cuticles to prevent desiccation. Although no fly larvae have true legs, these specially adapted larvae have false legs and two long projections (cerci) at their tail ends that expedite their search for the right host (fig. 22). However, once

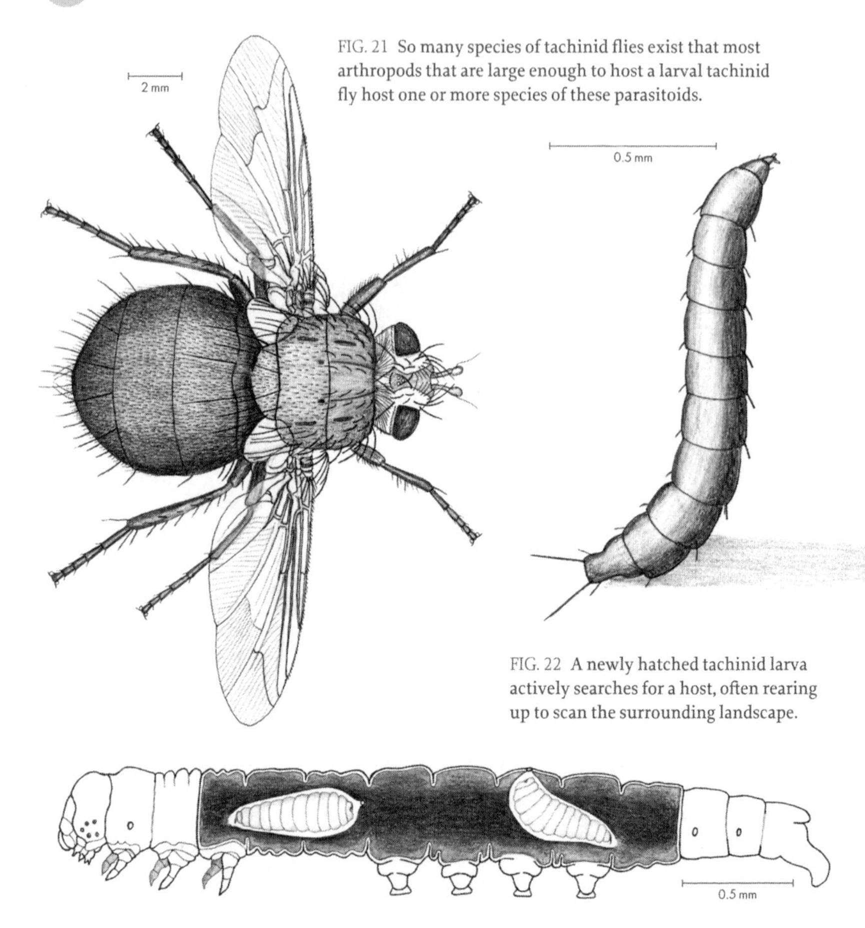

FIG. 21 So many species of tachinid flies exist that most arthropods that are large enough to host a larval tachinid fly host one or more species of these parasitoids.

FIG. 22 A newly hatched tachinid larva actively searches for a host, often rearing up to scan the surrounding landscape.

FIG. 23 A view inside a caterpillar parasitized by fly larvae: two larvae of a fly parasitoid avoid the immune response of their host by remaining encapsulated in a sheath that the host forms to seal off the intruders; however, the larvae still manage to maintain critical connections to nutrients from caterpillar blood at their front ends and to oxygen at their rear ends.

inside that host, these active larvae transform at their first molt to sedentary, legless larvae that grow inside the comfort and security of the host's body.

One group of tachinid flies that use adult beetles as hosts have elaborations of the mother flies' abdominal segments that enable them to snatch beetles in flight and pierce the firm beetle cuticles with a spine that guides the mother's short ovipositor to its exact destination. While almost all tachinid mothers lack the often long, penetrating, and impressive ovipositors of their parasitic wasp counterparts, they have adopted other strategies to ensure that their larvae begin life with certain advantages.

Once inside their insect hosts, tachinid larvae must face the onslaught of the host's immune response to their intrusion. These fly larvae do not resort to the tactics of parasitoid wasps that suppress and overwhelm the host's immune system; instead they rely on the host's healing response to isolate them from host immune cells. The blood cells that normally engulf and suffocate intruders recognized as foreign now form an encapsulating sheath that seals off the parasitoid larva from the blood space of its host. Only the larva's mouth remains free to access nutrients from the host's body. At its rear end, each larva obtains oxygen by forming a snorkel connection to the exterior air, thereby maintaining a very narrow passageway through the host's integument or through one of the host's large air tubes (tracheae) (fig. 23).

Fly Predators

— ROBBER FLIES, FAMILY ASILIDAE (*asilus* = type of fly) (7,500 species; 1,000 species NA; 3–50 mm).

— LONG-LEGGED FLIES, FAMILY DOLICHOPODIDAE (*dolicho* = long; *poda* = legs) (7,400 species; 1,300 species NA; 1–9 mm).

After a larval lifetime of stalking other insects in the litter beneath trees (fig. 326), and in some cases beneath tree bark, larvae of these flies transform into adult robber flies and long-legged flies that stalk tree leaves and branches (fig. 24; LeafScape, 25). The larvae of one genus of long-legged flies, *Medetera* or woodpecker flies, specialize in preying on the abundant larvae of bark beetles. At metamorphosis the eyeless and legless larvae transform into adult flies with large, bulging eyes and keen eyesight. With their long, spiny legs, they move swiftly to capture and firmly grasp their prey.

Robber flies choose favorite perches from which they can scan their territories for passing prey. These swift flies can snatch prey as easily from the air as they can from the surface of a leaf or branch. Some of the fuzzier members of this family look and sound like bumblebees. Only a close inspection can establish whether the fuzzy insect is a fly or a bee. Flies, however, have two rather than four wings and do not have stingers. Nor do these predatory flies have jaws for chewing, but they do have sharp, penetrating beaks with which they impale their prey, inject digestive enzymes, and then suck their victims dry. Anyone who has attempted to hold a robber fly can attest that its bite is as painful as the sting of a bee.

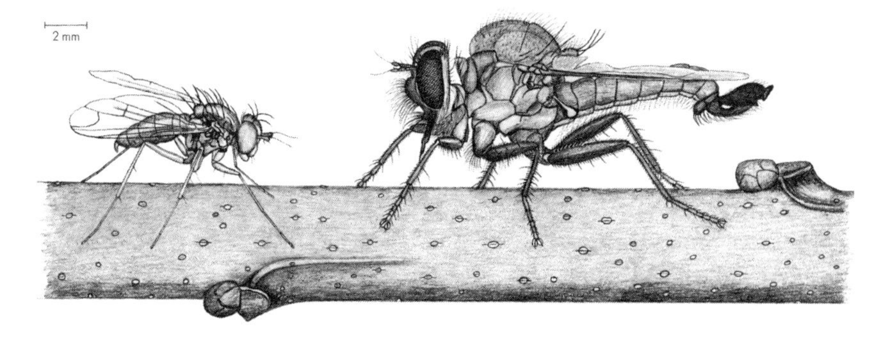

FIG. 24 Robber flies (*right*) and long-legged flies (*left*) begin life as larval predators in the soil or under bark, and as adults they patrol trees for insect prey aboveground.

Long-legged flies perch high above the bark or a leaf on their long, slender legs and survey the landscape for small arthropods. The swift, fidgety movements of these iridescent green-blue flies, which are otherwise inconspicuous, make them stand out on sunlit leaves and twigs. During their courtship, the male flies are known for ostentatious displays of finery that can embellish their legs, antennae, face, or wings.

Wasp Predators

— HORNETS, YELLOWJACKETS, FAMILY VESPIDAE (*vespa* = wasp) (5,000 species; 300 species NA; 10–40 mm), SUBFAMILY VESPINAE (70 species; 22 species NA). The success of a hornet (fig. 25) or yellowjacket (LeafScape, 11) colony and the well-being of its thousands of residents depend on the availability of a constant supply of insects to feed its growing population. Wasps catch caterpillars and other insects, chew them, and then feed them to their large broods. As many as 2,500 larvae and 500 workers can live in a single paper nest that the industrious hornets craft from chewed wood pulp (fig. 26). While hornets, such as the bald-faced hornet, establish their nests in conspicuous locations, yellowjacket nests are usually concealed underground, in hollow logs, or under bark. A yellowjacket colony can have as many as 5,000 workers and 15,000 brood chambers. With that many wasps bustling around a tree, whatever caterpillars are around will not likely strip many leaves from that tree.

The wasp queen and her daughters do all the insect collecting as well as the pulp and papermaking. Only at the end of summer do male wasps begin appearing in the colony along with newly hatched queens. After the young queens have mated with the males, each queen finds a protected spot under bark or in a hollow limb where she can pass the winter. In the spring she starts a new family and new nest alone, for none of the males or her worker daughters survive the winter. The queen uses her jaws to scrape off bits of wood from tree trunks and branches. She adds her saliva, chewing the wood to a pulpy texture and shaping it into the foundation for a new nest. After the queen has constructed several brood chambers, she lays a fertilized egg in each one and then begins to nurse daughter larvae that hatch from her eggs. Several weeks later all these larvae develop

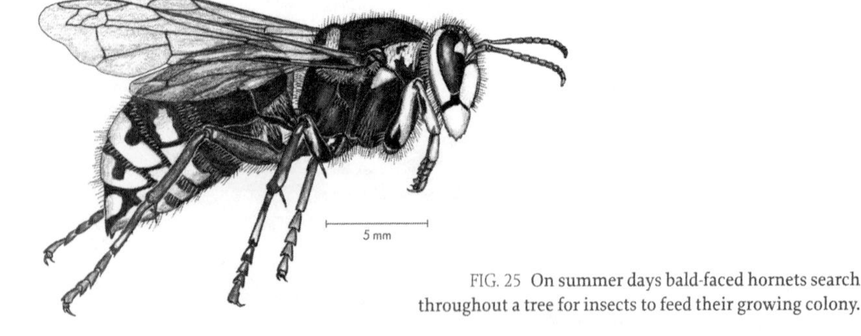

FIG. 25 On summer days bald-faced hornets search throughout a tree for insects to feed their growing colony.

FIG. 26 The construction of a hornet nest from tree pulp represents a communal effort by the hornet queen and her many daughters.

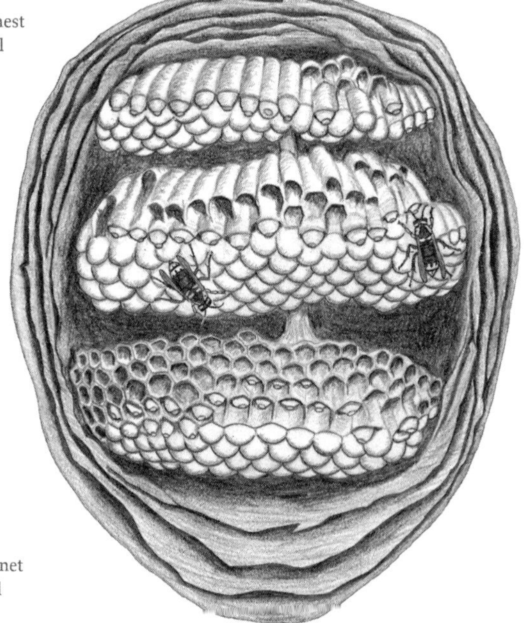

FIG. 27 The interior structure of a hornet nest is a testament to the architectural prowess of these creatures.

into workers that will take over the job of building more brood chambers and nursing more larvae. But only the queen lays the eggs. Thanks to the assistance of her many daughters, the nest grows from the few paper chambers she first built to several thousand brood chambers by the end of the summer (fig. 27).

Bug Predators

— TRUE BUGS, ORDER HEMIPTERA, SUBORDER HETEROPTERA (*hetero* = different; *ptera* = wings, referring to their mosaic wings that are membranous at the tips but hardened at the bases) (42,000 species; 4,100 species NA; 1–110 mm).

All three suborders of true bugs have piercing-sucking mouthparts that are used not only for sucking sap from plants (chapters 3 and 6) but also for sucking blood and fluids from animals. Only the suborder Heteroptera, however, includes predatory bugs. True bugs that are predators have salivary glands that can deliver an extremely painful dose of venom to anyone who carelessly handles or molests them. These venoms and their multiple components prove very effective not only in subduing but also in digesting all but the cuticular skeletons of their insect prey (Walker et al. 2016).

— ASSASSIN BUGS, FAMILY REDUVIIDAE (*reduvia* = hangnail, perhaps referring to the ominous-looking beak [proboscis] hanging from the head) (7,000 species; 200 species NA; 4–40 mm).

Assassin bugs are striking insects—some of the largest and most conspicuous of our true bugs (fig. 28). They stalk the bark, leaves, and the ground beneath trees,

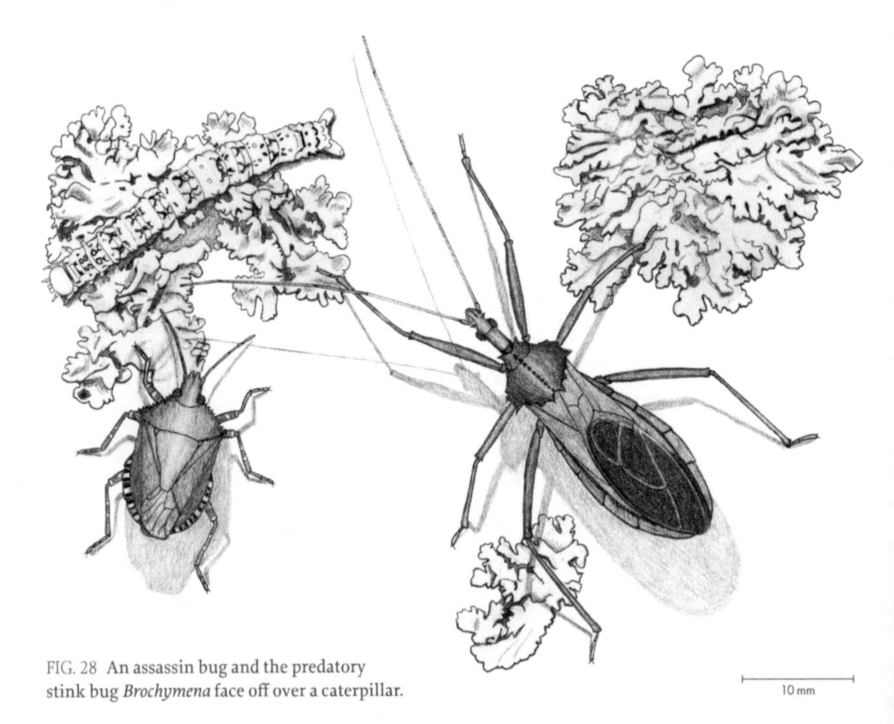

FIG. 28 An assassin bug and the predatory stink bug *Brochymena* face off over a caterpillar.

tackling some of the larger insects by impaling them with their long, extended beaks. The potent salivary enzymes that pour into the impaled victim begin digesting and liquefying the internal tissues of their prey well before the predigested contents begin passing down the bug's digestive tract.

— STINK BUGS, FAMILY PENTATOMIDAE (*penta* = five; *tomos* = sections)
(4,700 species; 230 species NA; 5–18 mm).

The few members of the stink bug family that suck insect blood rather than tree sap include the *Brochymena* stink bugs that patrol tree trunks and blend so well with their backgrounds of bark and lichens (fig. 28; BarkScape, 26).

Brochymena stink bugs roam tree trunks and branches in search of caterpillars and other insects. These stink bugs use their beaks to first stab their victims, then to inject them with digestive juices, and finally to suck out the body fluids of their prey. If these stink bugs are as inconspicuous to the eyes of other insects as they are to our eyes, they should be very successful at ambushing their unsuspecting victims.

Predatory Beetles and Mites

Many of the predatory beetles and mites are ground-based allies of trees. The predatory beetle and mite residents of the leaf litter and soil not only ward off attacks on tree roots but also eliminate many insects before they emerge from the soil to lay eggs or feed aboveground on leaves, twigs, and wood.

As larvae and as adults, ground beetles and rove beetles are known for their speed and sharp, powerful jaws. Their larvae are very similar in appearance, but at metamorphosis, the form and behavior of the two beetles visibly change. Adult ground beetles have wing covers, or elytra (*elytra* = sheath, cover), that restrict the movement of their abdomen, while adult rove beetles have elytra that leave the posterior end of their abdomen exposed. Adult rove beetles often raise their abdomen in a threatening pose even though they are incapable of stinging. The great diversity of forms in these two large families ensures that almost all invertebrate companions of trees are potential meals for these carnivorous beetles (fig. 29; chapter 6; figs. 327–329; RootScape, 5, 15, 21).

— ROVE BEETLES, FAMILY STAPHYLINIDAE (*staphylinus* = kind of beetle)
(63,000 species; 5,000 species NA; 1–35 mm).

— GROUND BEETLES, FAMILY CARABIDAE (*carabus* = kind of beetle)
(40,000 species; 2,440 species NA; 1–65 mm).

New mite species are constantly being discovered, and their number steadily grows. At least 45,000 species are known so far, but several times this number could be waiting to be named by acarologists, those who specialize in the study of mites. Mites are found in just about all habitats—soil, ponds, household dust, animal skin—so it should not be surprising that they are found on tree leaves, bark, and roots (figs. 30, 96, 109, 293, 294; LeafScape, 32; BarkScape, 30; RootScape, 42, 43). The soil and leaf litter inhabitants probably display the greatest range of forms and behaviors (chapter 6). Because many are so tiny, you need to inspect a tree closely and carefully to detect their presence. However, red velvet mites with their intense red cuticular coats stand out wherever they are found.

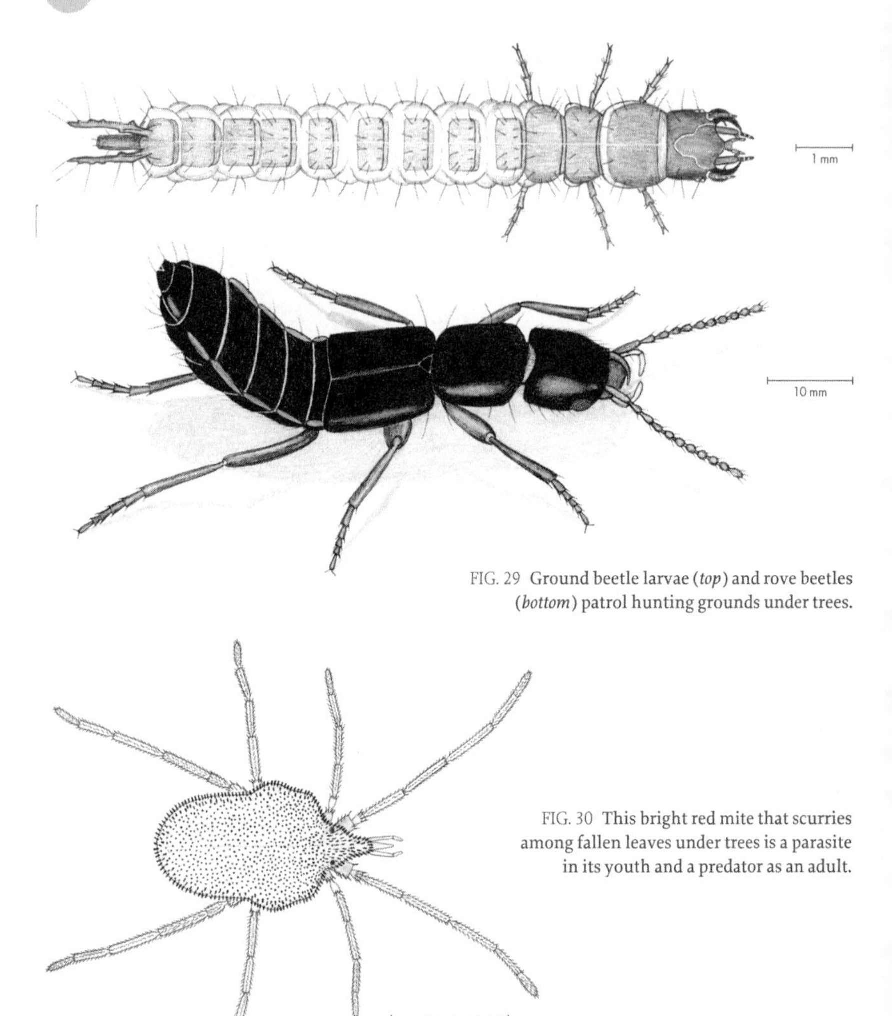

FIG. 29 Ground beetle larvae (*top*) and rove beetles (*bottom*) patrol hunting grounds under trees.

FIG. 30 This bright red mite that scurries among fallen leaves under trees is a parasite in its youth and a predator as an adult.

— RED VELVET MITES, CLASS ARACHNIDA, SUBCLASS ACARI, FAMILY TROMBIDIIDAE (*trombid* = little timid one) (~ 300 species; unknown number of species NA; 1–4 mm).

Lacewing Predators

The fragile, delicate appearance of adult lacewings (figs. 31 and 32; LeafScape, 13, 36) belies their ability to dispatch legions of aphids.

— GREEN LACEWINGS, FAMILY CHRYSOPIDAE (*chrys* = gold; *op* = face) (1,500 species; 85 species NA; forewing 4–32 mm).

Lacewing larvae have reputations as voracious and rapacious hunters. In his popular natural history book of 1902, *The Book of Bugs*, Harvey Sutherland wrote how the

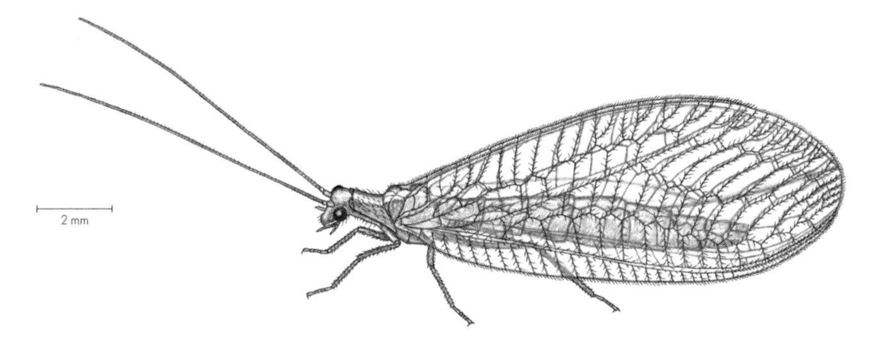

FIG. 31 Aphids seem to be the preferred food for larval and adult green lacewings.

mother green lacewing "dare not lay her eggs in one mass for the first one out would eat all the rest. She spins a lot of stalks of stiff silk and sticks one egg on the end of each, thereby giving each young one a chance for its life." Even though a female green lacewing places each of her eggs on top of a stalk stretching about 10 mm above the leaf's surface and out of reach of many small predators, her relatives the brown lacewings simply place their eggs directly on leaf surfaces, perhaps reflecting a difference in familial temperaments.

— BROWN LACEWINGS, FAMILY HEMEROBIIDAE (*hemera* = goddess of the day; *bio* = life) (600 species; 60 species NA; forewing 4–18 mm).

Larvae of a few of the approximately 1,500 species of green lacewings stake out hunting grounds on lichen-coated trunks and branches decked out in coats of lichens that they have sown together with silk threads. Lacewing silk is derived from an unusual source, their internal excretory organs—the counterparts of our kidneys known as Malpighian tubules. The tubules pour their contents into the larva's gut, and then silk is spun from the larva's anus. The spiny bristles covering its back impale and secure the lichen coats. Weighed down with its lichen attire, this camouflaged larva lumbers across the bark landscape, resorting to stealth rather than speed for its hunting success (chapter 4; fig. 195; BarkScape, 27). The

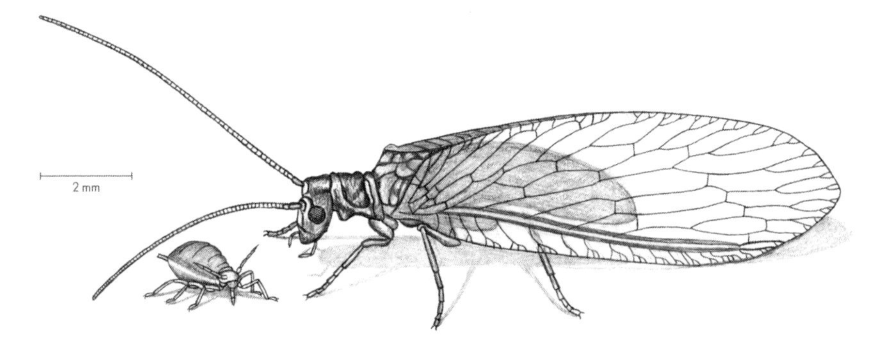

FIG. 32 The most obvious difference between brown lacewings and green lacewings is their color.

larvae of other green lacewings and all brown lacewings are unencumbered by any such bulky attire and move rapidly over leaf surfaces, dispatching slow-moving aphids and sessile scale insects (chapter 3; fig. 145; LeafScape, 13).

With the approach of autumn, all these fierce-looking larvae use their silk to fashion almost perfectly spherical cocoons in which they transform into deceivingly delicate, ethereal adults. These surprisingly hardy adults nestle in the leaf litter under trees where they survive the winds and freezing temperatures of long winters. In the spring, unblemished by months of cold, they flutter forth to lay eggs on tree leaves.

These aphid predators, the lacewings, are not without their own predators: parasitoid wasps in the family Heloridae. This entire family of wasps—albeit a family with only 12 known species worldwide and only one species from North America—feeds exclusively as parasitoids on lacewing larvae. The infected larval lacewing survives the depredations of the helorid parasitoid long enough to spin its pupal cocoon, but only an adult helorid wasp emerges from the ill-fated lacewing cocoon.

Spider Predators

— SPIDERS, CLASS ARACHNIDA, ORDER ARANEAE (*arane* = spider) (50,000 species; 3,000 species NA; body length 0.4–90 mm).

All spiders are predators, and they have adopted a variety of hunting approaches, both on and under trees. Some spiders ambush, some pounce, some weave webs that trap their prey. The silks spun from their spinnerets can have various physical properties; different silks are used to weave webs of several forms for different purposes. Members of one spider family, the Thomisidae, can produce silk but choose not to spin webs to catch their prey.

— CRAB SPIDERS, FAMILY THOMISIDAE (*thomis* = whip) (2,100 species; 135 species NA; 2–11 mm).

Wherever insects are likely to land from the air or venture forth on foot, crab spiders most often stake out their hunting territory. Most species blend in beautifully with their surroundings on tree bark, leaves, or flowers. Other species stand out by mimicking bird droppings. These spiders take on poses crabs assume and scuttle

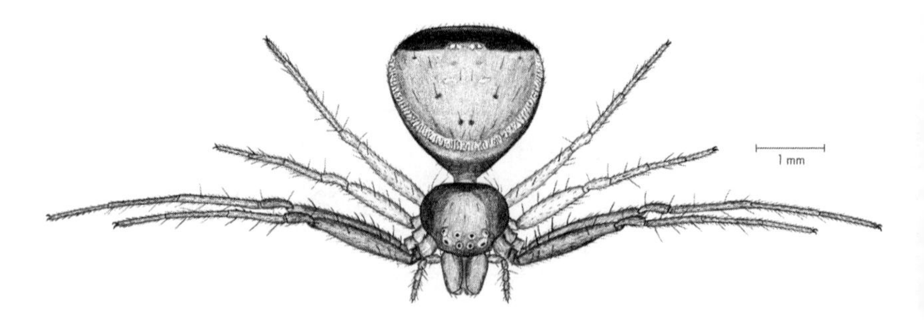

FIG. 33 A crab spider with legs outstretched prepares to ambush a fly that landed on its leaf.

sideways, backward, and forward in crablike fashion. With their outstretched front legs, they are always prepared to quickly snatch approaching prey (fig. 33).

— JUMPING SPIDERS, FAMILY SALTICIDAE (*salti* = jump) (6,000 species; 315 species NA; 1–25 mm).

This is the largest family of spiders. They stand out among spiders with their large eyes; flat, forward-facing faces; and often iridescent colors (fig. 34). Many of them owe their extravagant colors to iridescent scales like those that impart the stunning colors of butterfly wings. Those large eyes can swivel in their sockets and have remarkably good vision for appreciating the forms and colors of fellow spiders. These spiders are most active during the day, and their beauty stands out in sunlight. Certain members—the peacock spiders—have developed colorful and extravagant courtship displays; however, even the less flamboyant jumping spiders participate in elaborate courtship dances involving waving of legs and shaking of abdomens. As predators they jump and pounce on their prey. Living high in treetops, however, presents the hazard of misjudging a jump and falling to the ground. Before risking a fall, a jumping spider always secures a silk line to its leaf or twig, just in case it overshoots or falls short of its mark.

— BOLAS SPIDERS, FAMILY ARANEIDAE (*arane* = spider), SUBFAMILY CYRTARACHNINAE (*cyrto* = curved; *arachna* = spider) (66 species; 15 species NA; 2 mm males; 15 mm females).

Bolas spiders hunt by night; by day they are mistaken for bird droppings as they rest motionless on leaves with legs tucked under their bodies. The resemblance is striking: each spider's wide, oblong abdomen has a shiny cuticle that appears wet and is covered with small bumps and warts (fig. 35). The hefty females, which are

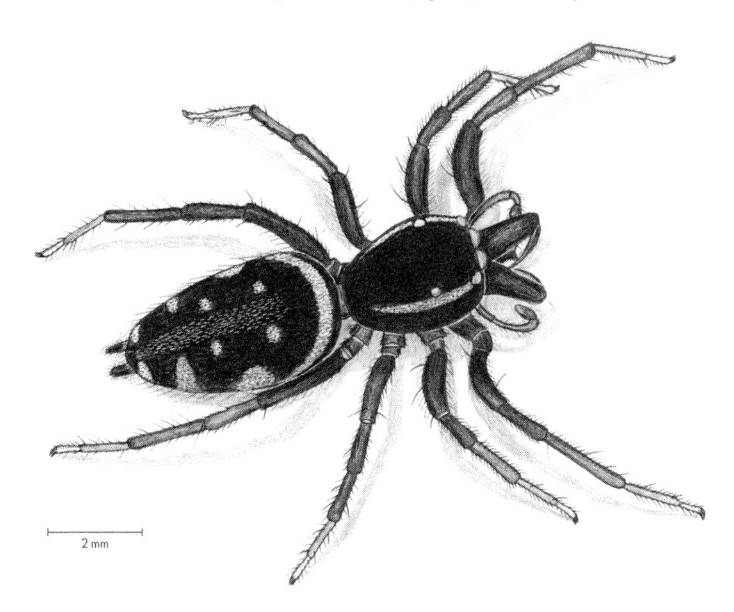

FIG. 34 Jumping spiders pounce when they hunt and dance when they court.

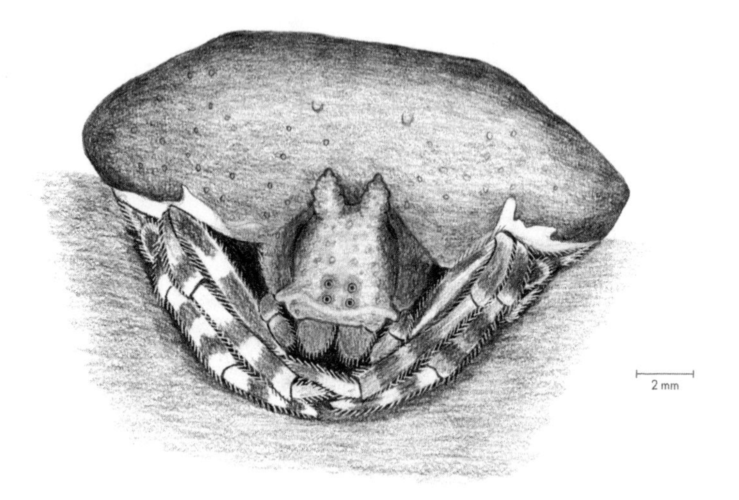

FIG. 35 This bolas spider has tucked its eight legs under its body to pose as a bird dropping.

7 to 14 mm in length, dwarf the tiny male spiders measuring only 1–3 mm. These spiders that never spin webs are exceptional members of the spider family known as orb weavers, most of whom use their silk to construct handsome circular webs. Instead, these spiders spin out only a few horizontal strands of silk from which they hang from a twig or leaf. Once positioned, the spider then drops a vertical line from her spinnerets and begins combing out a glue-like silk that forms a conspicuous droplet swinging free at the end of its vertical silk strand (fig. 36). The whole silken structure is reminiscent of the bolas of weights swung on cords that human hunters use to entangle their prey. Not all spider silks can cling to the slippery scales on moth wings, but the silk in the swinging droplet tenaciously grips scales and quickly ensnares the whole moth.

Almost all the prey items caught by this swinging bola are male moths, and analysis of the bola's droplet has revealed that it contains a chemical that mimics the scent or pheromone of many female moths and is especially enticing to male moths. The spider finds a favorite hunting spot, and from the security of its hideout draws in moths with a chemical lure that can be picked up by sensitive moth antennae, often from several miles away. During the moths' courtship season, a bolas spider can catch and eat its fill of male moth meat each day.

Spiders also catch and eat many flies, but in some cases the tables are turned. There is a whole family of fly parasitoids that eat nothing but spiders.

— SPIDER FLIES OR SMALL-HEADED FLIES, FAMILY ACROCERIDAE (*acro* = topmost; *cera* = horn) (520 species; 61 species NA; 4–12 mm).

Spider flies have oddly proportioned bodies: tiny heads that are almost all eyes, swollen abdomens, and hunched backs (fig. 37). What these fly parasitoids lack in the antennal acuity of their wasp counterparts, they make up for by having disproportionately large eyes with exceptionally sharp eyesight. Rather than dexterously placing their eggs in hosts with well-aimed ovipositors, these flies

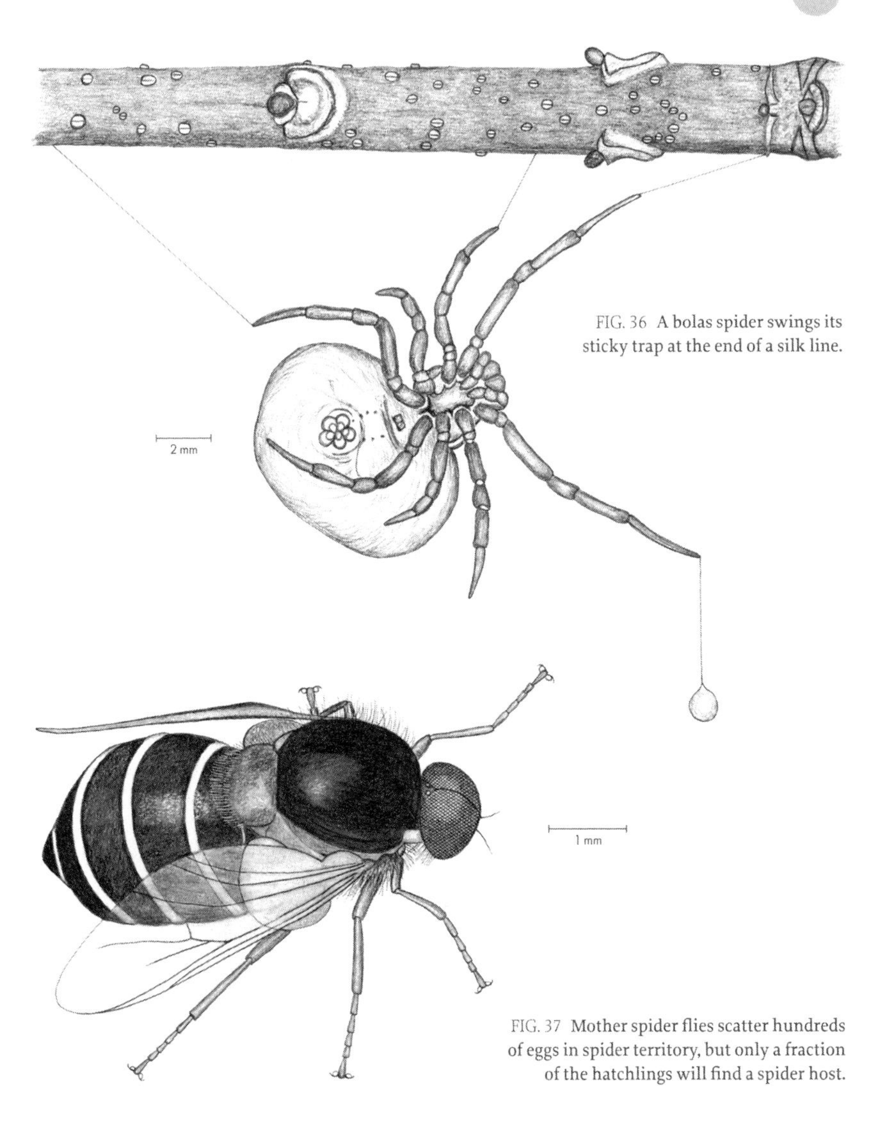

FIG. 36 A bolas spider swings its sticky trap at the end of a silk line.

FIG. 37 Mother spider flies scatter hundreds of eggs in spider territory, but only a fraction of the hatchlings will find a spider host.

place large numbers of eggs—up to 5,000—wherever their newly hatched larvae are likely to find a spider host. The larvae are on their own and must be self-reliant. Even though, like all fly larvae, these newly hatched acrocerid larvae completely lack legs, they are capable of scurrying about with their two rear appendages (cerci) that propel and support them as they wave their bodies in hope of contacting a host. Although they can survive on their own for several days without a host, each planidium larva feverishly searches for a spider leg or a silk strand of a web that could lead to a spider. At least a few of these many hatchlings will have sufficient industry and good fortune to locate a host.

Microbial Allies

Endophytes: Symbiotic Microbes Belowground and Aboveground

Nitrogen is one of the 18 essential mineral nutrients that trees and other green plants need to produce their chlorophyll, amino acids, and myriad proteins. And this essential element is ultimately derived from the nitrogen gas (N_2) that makes up three-fourths of the air we breathe. However, nitrogen gas cannot be directly used by plants but must first be converted to a form that all plants can use. Plants take up nitrogen in the form of either nitrate (NO_3^{-1}) or ammonium (NH_4^{+1}). Because bacteria and archaea are the only organisms that have accomplished the energy-demanding feat of converting N_2 (nitrogen gas) to NH_4^{+1}—the process known as nitrogen fixation—bacteria have always been essential microbial partners for green plants. The roots of most trees take up NH_4^{+1} from nearby soil where free-living bacteria carry out their nitrogen fixation. However, the roots of a few trees have established intimate symbiotic relationships with bacteria that take up residence inside root nodules, supplying even more concentrated NH_4^{+1} directly to plant cells and bypassing its transport from soil to root cells (fig. 38).

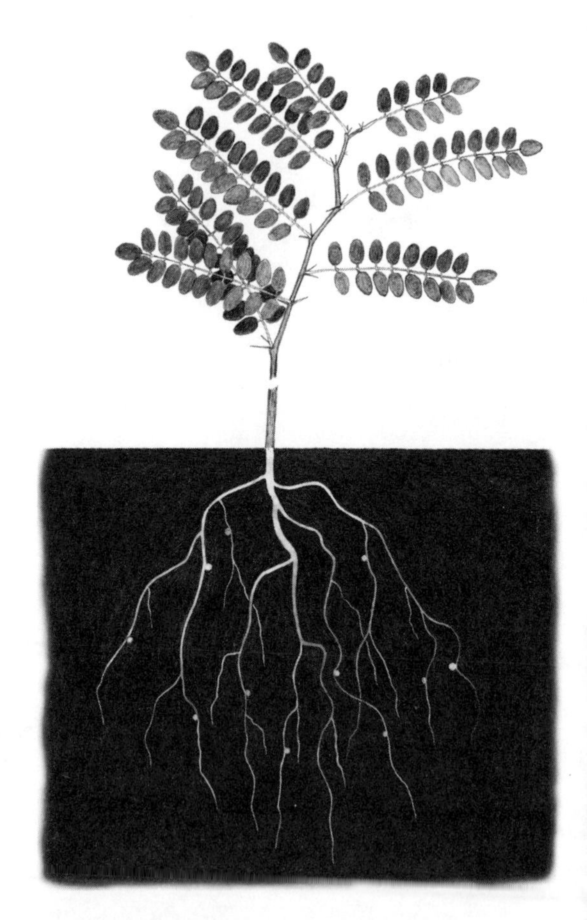

FIG. 38 In microbial nodules on roots of black locust trees, root cells and bacterial cells collaborate to fix nitrogen gas from the atmosphere.

The well-known symbiotic association between nitrogen-fixing bacteria and root nodule cells of green plants was first described in 1888. However, not until almost a hundred years later were bacteria and archaea, not only with nitrogen-fixing ability but also with other talents, shown to colonize tissues aboveground. These microbes not only fix nitrogen but also produce hormones that stimulate tree growth and wound repair. These bacterial and archaeal partners of plants, referred to as endophytes, have been found in all land plants from mosses to trees. The number of species of green plants known to host endophytes is now at least 7,000 (Jung et al. 2020; Griffin and Carson 2018).

In leaves and branches aboveground, these endophytic bacteria and archaea keep company not only with tree cells but also with endophytic fungi. The nitrogen-fixing bacteria that inhabit root nodules enter root tissues through the thin walls of delicate and extremely numerous root hairs. The endophytes aboveground probably continually colonize tree tissues through obvious entry points such as stomata on leaves and lenticels on twigs (Hardoim et al. 2015). While we continue to discover new species of endophytes, we are only beginning to appreciate the extent of nutrient and chemical exchange between microbial cells and tree cells (Santoyo et al. 2016). How these portals into the interiors of leaves and stems—stomata and lenticels—influence the lives of trees is considered in chapter 2 (figs. 39 and 40).

The growth as well as the health of plants is enhanced whenever trees host a combination of endophytic fungi and bacteria. Even when faced with the stressful conditions of drought and high salinity, these symbiotic microbes interact synergistically to improve not only their own lives but also the lives of their hosts (Hashem et al. 2016; Marasco et al. 2012) (LeafScape, 39–41).

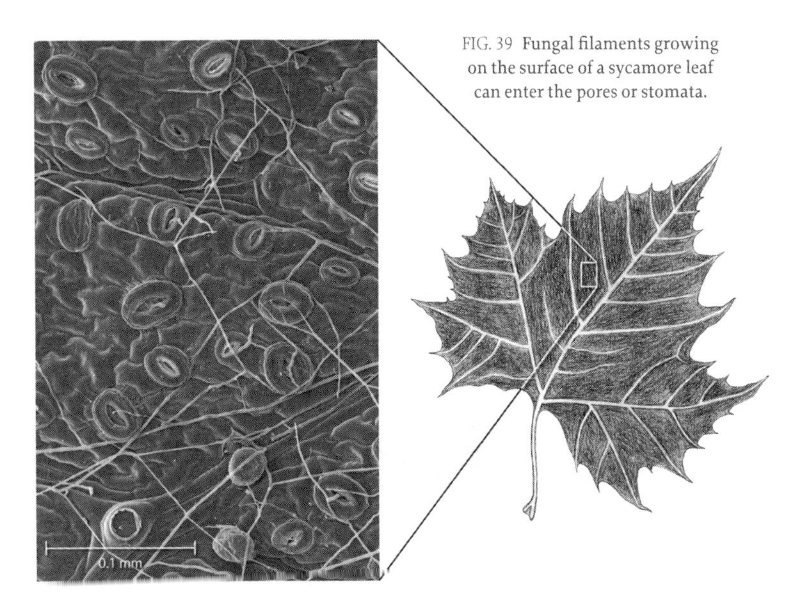

FIG. 39 Fungal filaments growing on the surface of a sycamore leaf can enter the pores or stomata.

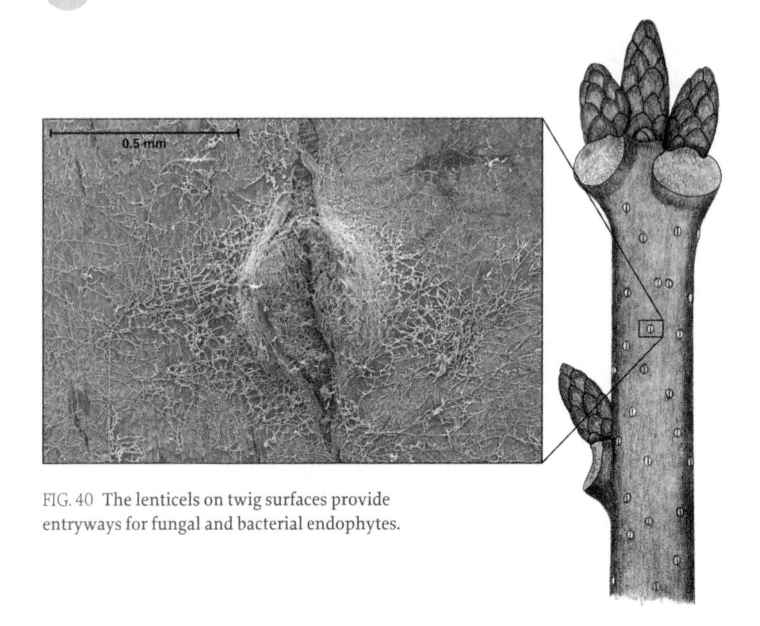

FIG. 40 The lenticels on twig surfaces provide entryways for fungal and bacterial endophytes.

The sugars produced in tree leaves in conjunction with the same plant hormones that orchestrate the tree's response to its visitors aboveground also oversee its interaction with the microbes that take up residence in and around its roots. The sugars and the hormone salicylic acid that roots produce govern which microbes consistently associate as endophytes within root cells and which communities of microbes are consistently found in the soil adjacent to roots, in the soil zone referred to as the rhizosphere (Haney and Ausubel 2015).

Tree Roots and Fungal Partners Exchange Gifts Belowground

We are just beginning to understand some of the many unexpected events that occur in the soil under a tree. The dark world of the soil supports a rich variety of life forms that have established complex and mutually beneficial interactions with each other as well as with trees. What few glimpses we have had into this dark world reveal a web of relationships just as complex as those we encounter in the treetops.

In a bucket of soil scooped from beneath a tree in the forest, a phenomenal number of life forms can be found—billions of bacteria, millions of single-celled animals called protozoa, and thousands of fungi, mites, insects, and other tiny arthropods, as well as hundreds of snails, slugs, and worms. Most of these are decomposers that recycle dead plant and animal matter that has fallen to the ground; during decomposition, these creatures return a long list of mineral nutrients to the soil, water, and air. These nutrients can once again begin their journey through the tree community as they are taken up by tree roots.

The main functions of the familiar large and branching roots of trees are to anchor the tree in the earth, store carbohydrates, produce certain organic compounds, and transport nutrients and water from below the ground surface. The fine, delicate roots that are less than a millimeter in diameter and that lie within the top 10 cm of soil, however, are the roots that absorb nutrients and water from the surrounding soil and leaf litter. These fine roots arise from lateral roots and repeatedly proliferate and branch to form a phenomenal 1,000 roots per cubic centimeter of soil, boasting an impressive surface area of at least 6 square centimeters (Lyford 1980).

These fine roots establish countless connections with their underground fungal allies. Behind the shapely, colorful mushrooms under trees lies a remarkable tale of subterranean partnership between fungi and trees. In addition to contributing to the recycling of nutrients for trees, some fungi of the soil have developed more intimate relationships with trees. The fine roots of trees intertwine with long strands of underground fungi, and these roots together with endophytic fungi form what are called mycorrhizae. Trees and fungi share and benefit from their special symbiotic arrangement. The strands of fungi transfer water and minerals from the soil to the tree, and the tree shares its sugary sap with the fungi. By combining their efforts, the roots and fungi vastly increase the effective surface area for absorption of nutrients from the surrounding soil. Mycorrhizal fungi not only provide nutrients for their green hosts but also help protect them from pathogenic fungi. Trees and fungi depend on each other for their well-being, offering gifts that are equally rewarding to each other (Lowenfels 2017) (figs. 41 and 42).

Tree roots also secrete a hormone to attract the soil fungi that establish mycorrhizal relationships with them. However, this root hormone was first

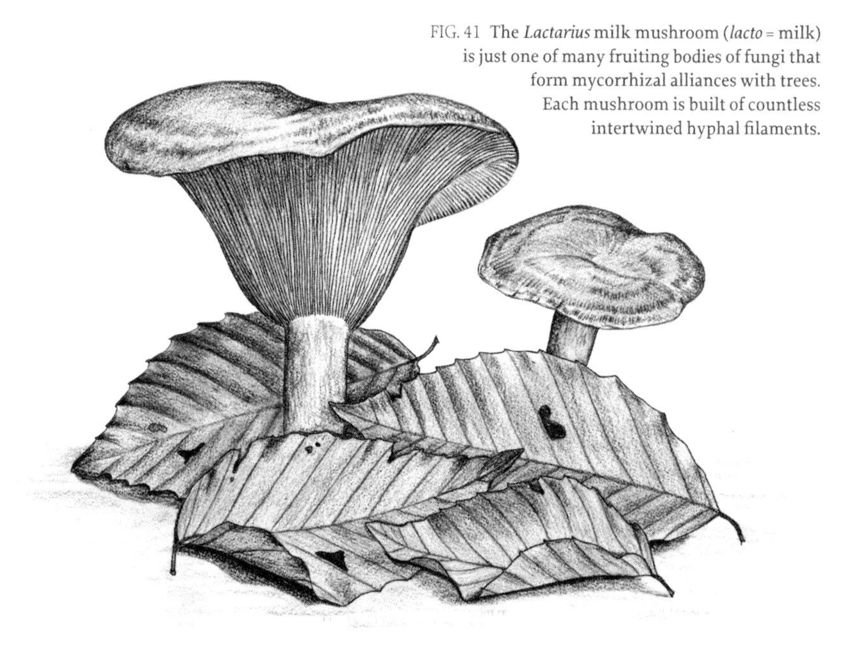

FIG. 41 The *Lactarius* milk mushroom (*lacto* = milk) is just one of many fruiting bodies of fungi that form mycorrhizal alliances with trees. Each mushroom is built of countless intertwined hyphal filaments.

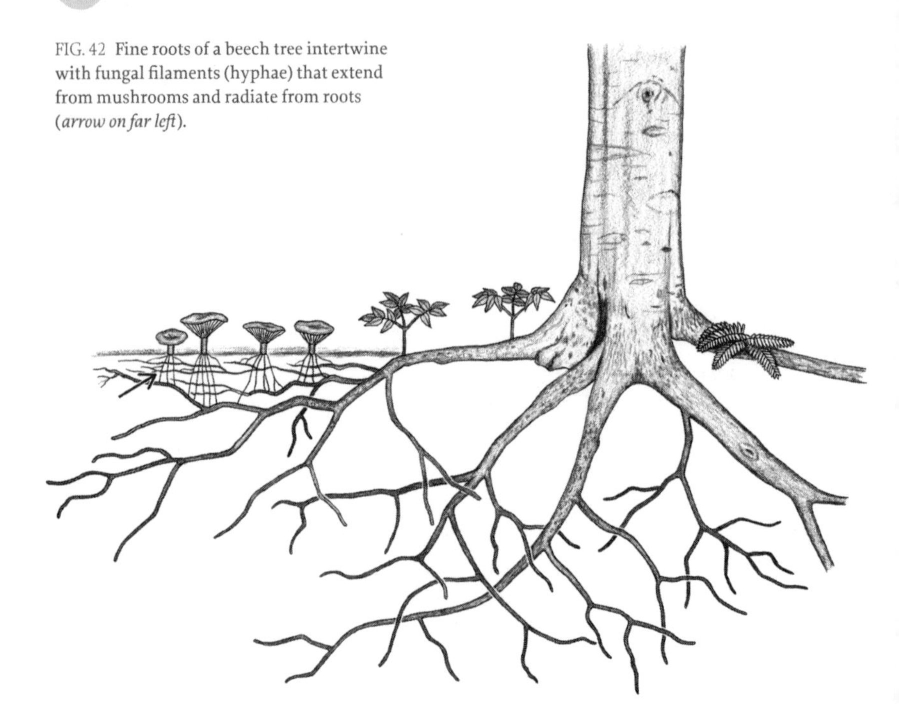

FIG. 42 Fine roots of a beech tree intertwine with fungal filaments (hyphae) that extend from mushrooms and radiate from roots (*arrow on far left*).

discovered not from its mycorrhizal association but from its exploitation by the root parasite witchweed in the genus *Striga*. *Striga* uses this hormone to track down the roots of its host, and the hormone was named strigolactone after its association with this parasitic flowering plant (Conn et al. 2015).

Fungi respond to this hormone secreted from tree roots and soon enshroud and link those roots with a network of their filaments (figs. 43 and 44; RootScape, 25). Fungi that develop mycorrhizae with roots on one tree can also join with the roots of other nearby trees, creating a complex web of interactions beneath the forest floor. Plant physiologists have shown that labeled chemicals can be transferred from one tree to another via the vast network of filaments formed by underground fungi, connecting trees throughout a forest. Even in the shadows of towering ancient trees, tiny seedlings that have joined the network of roots and fungi can grow and share in the abundant nutrients that are constantly being exchanged underground by their elders in the forest (Simard 2021).

At the end of summer, after the fungi have stored up enough sugars produced by the tree, they send strands from the roots to the soil surface and use the stored sugars to produce mushrooms. And what a colorful display the mushrooms can create. Red, orange, blue, green, violet, yellow, and brown mushrooms can carpet the forest floor in late summer and autumn. The mushrooms that pop up under trees after a rainy spell are made of thousands of intertwined strands of fungi that assemble into ornate forms. Before the rain, the strands of fungi were scattered among the soil and roots, actively decomposing fallen branches and leaves. In

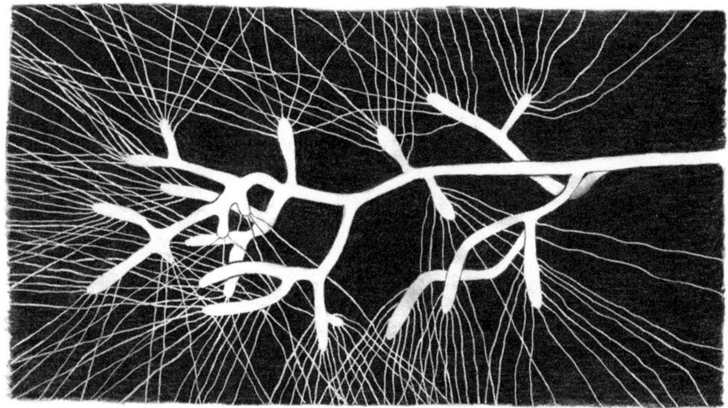

Above: FIG. 43 The filaments of hyphae embrace the cells of each root and extend upward and outward through the surrounding soil.

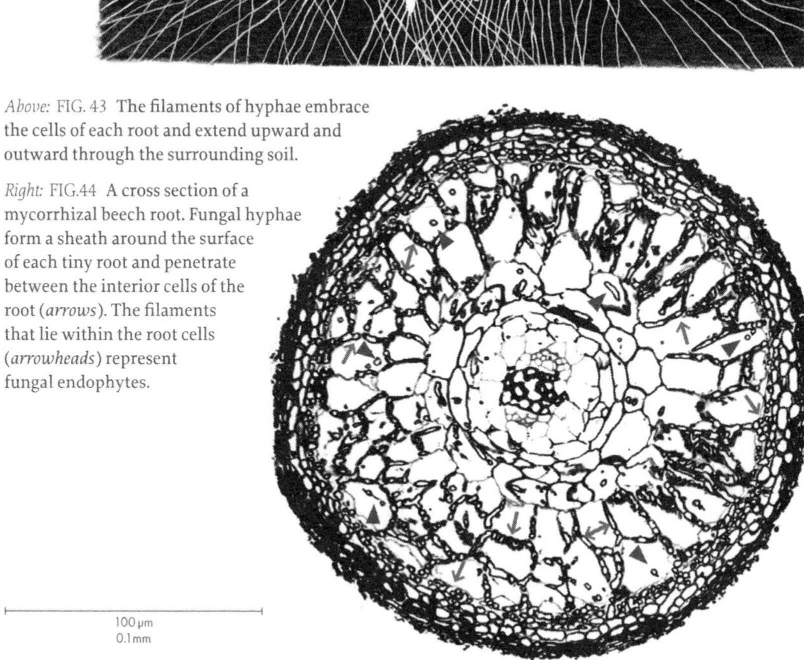

Right: FIG.44 A cross section of a mycorrhizal beech root. Fungal hyphae form a sheath around the surface of each tiny root and penetrate between the interior cells of the root (*arrows*). The filaments that lie within the root cells (*arrowheads*) represent fungal endophytes.

some astonishing and still mysterious way, individual strands of mycorrhizal fungi come together, grow, and create the unique forms of mushrooms aboveground.

From these beautifully sculpted mushrooms come the sexual spores that are carried off by wind or by the innumerable fungus beetles of forests to start new fungi and mushrooms elsewhere. Many species of fungus beetles live on fungi, feed on fungi, and spread fungal spores. These fungus beetles have long-standing, committed associations with fungi and are estimated to number over 100,000 species. Many of the spores they leave along their trails will sprout and form new relationships with trees, ensuring that trees, beetles, and fungi all benefit from their alliances (fig. 45; RootScape, 6, 8). Pleasing fungus beetles represent just one of many beetle families that have formed alliances with fungi.

— PLEASING FUNGUS BEETLES, FAMILY EROTYLIDAE (*erotyl* = darling) (3,500 species; 80 species NA; 2–22 mm).

In the forests of the Pacific Northwest, fungi called truffles form mycorrhizal partnerships with the roots of spruce and hemlock trees. Unlike other mushrooms, truffles live and set spores entirely underground where neither wind nor most fungus beetles can reach them. But in the spruce and hemlock forests, flying squirrels have taken over the job of spreading the truffle spores. The noses of flying squirrels are "tuned in" to the pungent scent of the truffles that lie beneath

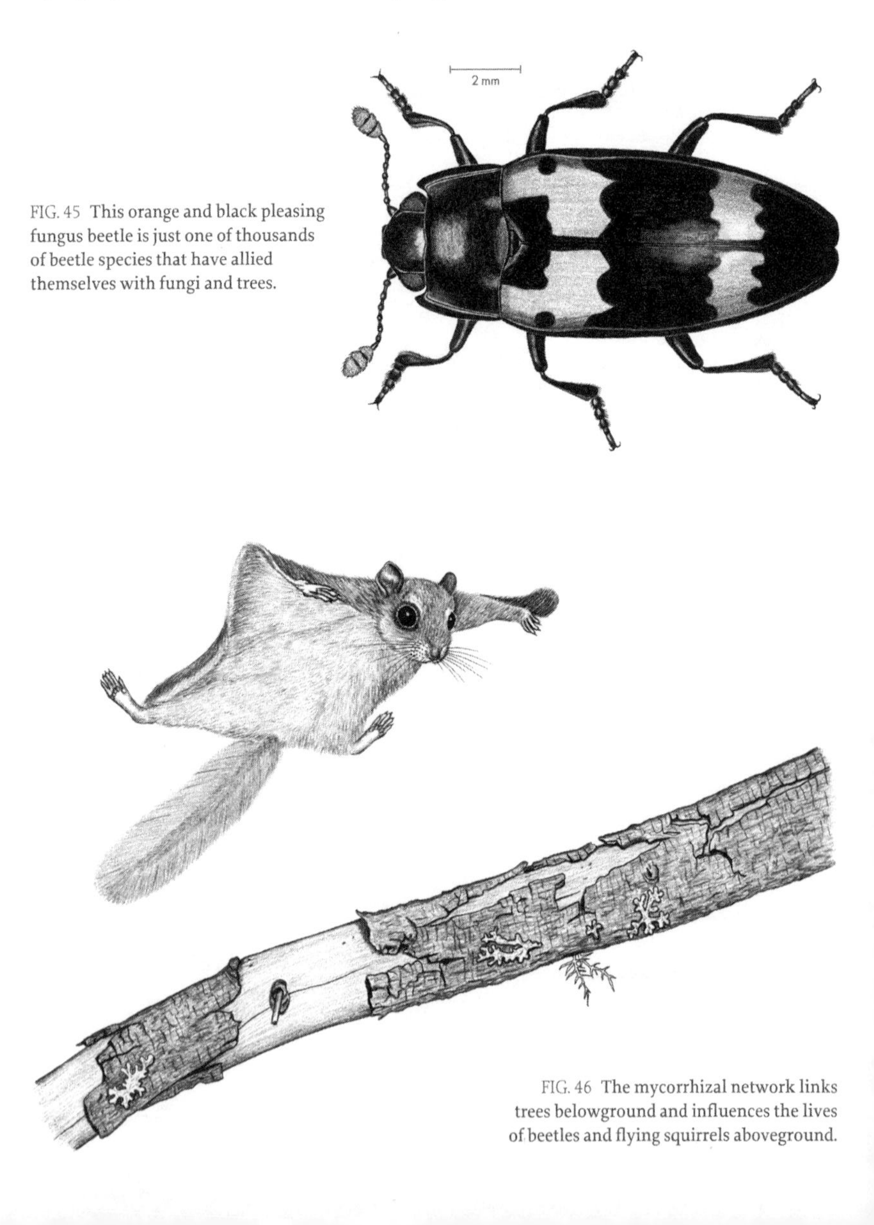

FIG. 45 This orange and black pleasing fungus beetle is just one of thousands of beetle species that have allied themselves with fungi and trees.

FIG. 46 The mycorrhizal network links trees belowground and influences the lives of beetles and flying squirrels aboveground.

the forest floor. They dig up the buried fungi, eagerly devour them, and scatter the spores in their droppings (fig. 46).

Microbial Allies: Enhancing a Tree's Defenses

We often forget that trees have immune systems like mice and humans. They readily distinguish self from nonself and use their immune response to choose the company they keep. To establish their initial contacts with tree cells, symbiotic fungi and bacteria must first confront the defenses that every tree presents to any creature it recognizes as foreign. The symbiotic partner of the tree first surmounts the defense challenges posed by activation of its host's immune response. The symbiotic microbes cope with these challenges and modulate them so a compromise is reached between the symbionts and the host; neither partner gains the upper hand in their relationship. However, this initial encounter leaves the host's defenses in a heightened state of alert for any later encounters with pathogens and intruders (Bastias et al. 2017; Jung et al. 2012). The mycorrhizal and endophytic company of trees helps trees in their battles with insects and pathogens, enhancing the trees' immune response as well as providing their hosts with additional nutrients such as phosphates and nitrates (Gan et al. 2017; Santoyo et al. 2016) (fig. 47; LeafScape, 39–42).

Fungal endophytes and mycorrhizae not only prime the immune systems of their host plants but also enhance the plants' production of secondary defensive compounds that provide another layer of protection against foreign invaders and insect herbivores (Shrivastava et al. 2015; Arnold et al. 2003). This enhanced

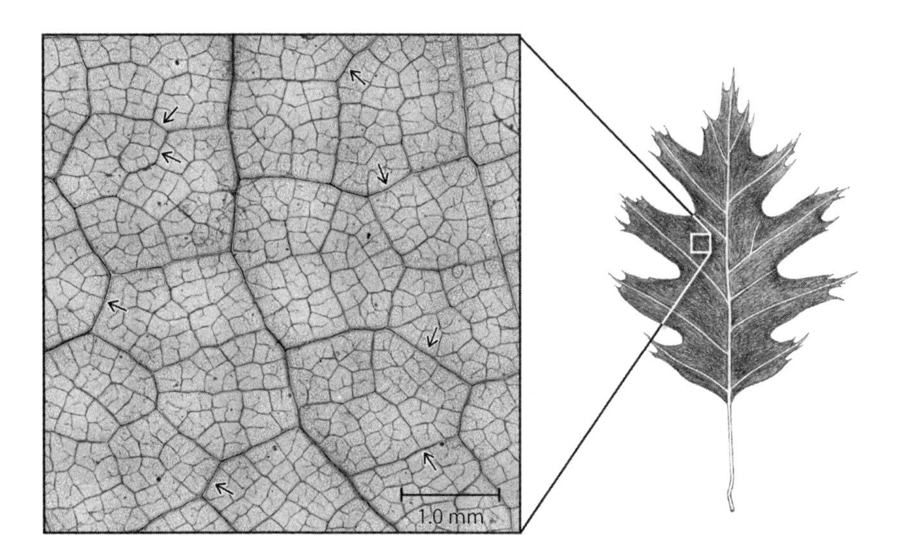

FIG. 47 This oak leaf tissue was cleared and labeled with a stain (chlorazol black E) that binds to chitin molecules, which all fungal filaments contain. The fungal endophytes stand out as the darkest strands within the leaf's vascular system (*arrows*).

ability of the host plant to respond to future foreign encounters occurs without any genetic alterations to the host; however, this alerted immune response is remarkably passed on to future generations of host plants through a stable modification of the parental DNA (Luna et al. 2012). Fungal or bacterial endophytes acting alone exert a positive influence on the growth and health of their host plants (Berg et al. 2017; Hashem et al. 2016; Vega 2008). The combined efforts of both fungal and bacterial endophytes working together in their hosts provide even more formidable challenges for hostile herbivores and potent pathogens.

By producing some of the same chemicals as cells of their host tree, bacterial and fungal endophytes effectively communicate not only with tree cells but also with their environment. Bacterial endophytes can produce key plant hormones—auxins, gibberellins, and cytokinins—and they can indirectly control levels of the hormone ethylene by converting the precursor for ethylene into alternative compounds. Endophytic bacteria isolated from leaves and stems of pistachio trees not only were able to produce several plant hormones but were also proficient at inhibiting the activity of bacterial pathogens (Etminani and Harighi 2018). By producing and responding to some of the same chemical compounds such as plant hormones, endophytes have untold influences on the growth and defense of their host trees. The production of plant secondary metabolites and plant hormones is therefore not the exclusive province of plants and bacteria. Fungi—both endophytes and pathogens—insects, and nematodes also seem to have the ability to produce the same hormones and secondary metabolites that are so

FIG. 48 The hyphae of morel mushrooms not only actively decompose plant litter for trees but also assist trees with their mycorrhizal activities.

crucial for plant development, physiology, and immune responses (Andreas et al. 2020; Dermastia 2019; Giron et al. 2016; Erb at al. 2012; Etminani and Harighi 2018; Chanclud and Morel 2016). Having their own sources of plant hormones that trees universally recognize, all these companions of trees can manipulate tree cells by communicating in a common chemical language (fig. 48).

Tree hosts can get help in countering whatever pathogens come their way from some unexpected sources. Insects may not only transport pathogenic fungi but can also carry endophytic fungi that suppress colonization of the host by microbial pathogens (Feldman et al. 2008). Interactions among diverse members of the tree community are understandably complex and are still poorly understood. From this complexity, however, a harmonious balance among countless members of the tree community seems to emerge.

Fungi as Decomposers and Predators That Supply Nutrients to Trees

As branches, twigs, and leaves fall to the ground, trees provide a constant source of raw materials for fungi and their fellow decomposers. However, decomposing plant material is not the only source of nutrients that fungi provide to their host trees. Fungi can switch back and forth between being decomposers and being predators. As decomposers that feed on decaying leaves and wood, fungi survive on a diet rich in carbon and carbohydrates but poor in nitrogen and proteins. When the opportunity arises to feed on small soil animals such as nematodes or rotifers, many species of fungi drastically switch form to become carnivores that consume the protein- and nitrogen-rich tissues of these invertebrates that share their soil habitat. Somehow the presence of these animals induces the fungi to produce special structures that efficiently trap them. Once the fungus has confined a nematode in one of its traps, the fungus sends a hypha into the body of its victim, digesting the internal organs of the creature and procuring a meal rich in nitrogen-containing proteins to supplement its usual bland, nitrogen-poor fare of decaying leaves and wood (fig. 50).

As they functionally switch from decomposers to carnivores, fungi switch in several other ways as well. A given fungus species can come in two morphological forms—a sexual form (teleomorph, *teleo* = complete; *morph* = form) that produces sexual spores in fruiting bodies or mushrooms, and an asexual form (anamorph, *ana* = without; *morph* = form) that produces spores called conidia (*conidio* = dust) by equal divisions of cells in microscopic stalks called conidiophores.

The two different forms can be so different in appearance that until the recent use of DNA sequencing facilitated comparisons of hereditary material in anamorphs and teleomorphs, the anamorphs were all lumped together in a group known as the imperfect fungi whose sexual forms were a mystery. The insect-destroying or entomopathogenic (*entomo* = insect; *patho* = disease; *-gen* = producing) fungi such as *Metarhizium* and *Beauveria* were always considered to be imperfect fungi that lacked sexual mushroom forms. Among the over 8,000 fungal endophytes that have been recently described are these two genera of fungi, which have been employed as insect biocontrol agents (Vega 2008). These insect-destroying fungi

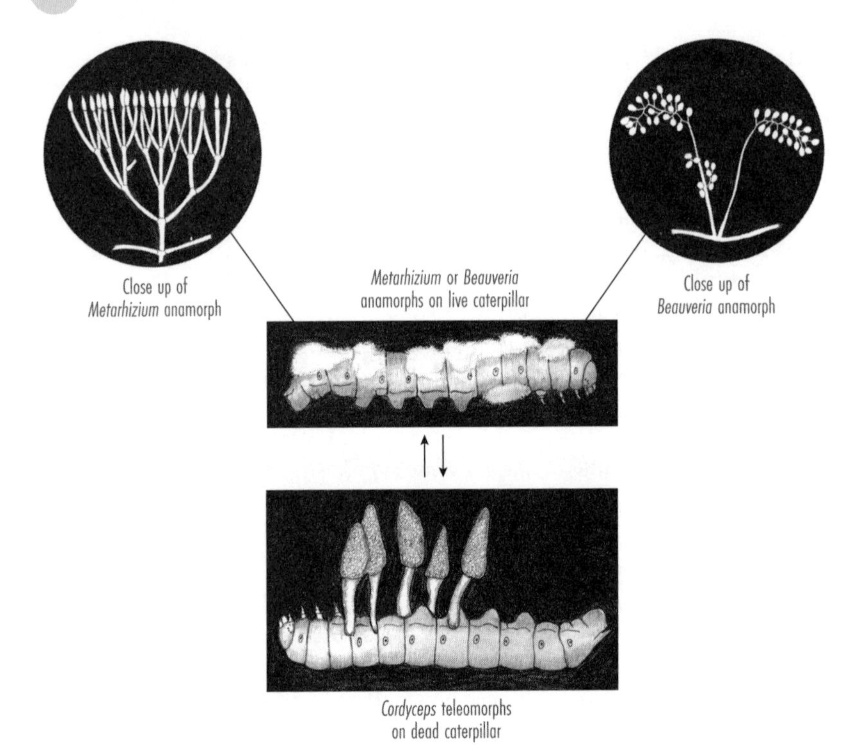

FIG. 49 Many endophytic fungi of trees are anamorphic forms that can produce fruiting bodies or mushrooms (teleomorphs) on the bodies of insects when environmental conditions are appropriate. Anamorphs (*top*) and teleomorphs (*bottom*) reversibly transform according to environmental conditions.

can be free living in soil and on plant surfaces, but they are common anamorphs that have also colonized the canopies of trees (fig. 49).

As observed for a tree's mycorrhizal partners belowground, tracking of nutrient movement with labeled carbon and nitrogen has demonstrated that this aboveground fungal partnership of trees, *Metarhizium*, and *Beauveria* involves exchanging carbon from the host plant's photosynthesis efforts and nitrogen from the insect victims of these fungal endophytes (Behie et al. 2017). After extracting nutrients from the bodies of their victims, the fungi move to the exterior cuticles of the insect mummies and coat them with a thick frosting of fungal filaments and spores.

Now we know that when environmental conditions are suitable, *Metarhizium*, *Beauveria*, and related genera of fungi can transform from masses of intertwining filaments to club-shaped mushrooms that sprout from dead insects lying just beneath the soil's surface. The mushrooms belong to the same free-living teleomorph genus, *Cordyceps*—familiar medicinal mushrooms that have been avidly sought for centuries.

Arthrobotrys and *Dactylella* are the nematode-destroying counterparts of the insect-destroying fungi. Like *Metarhizium* and *Beauveria*, they were also classified

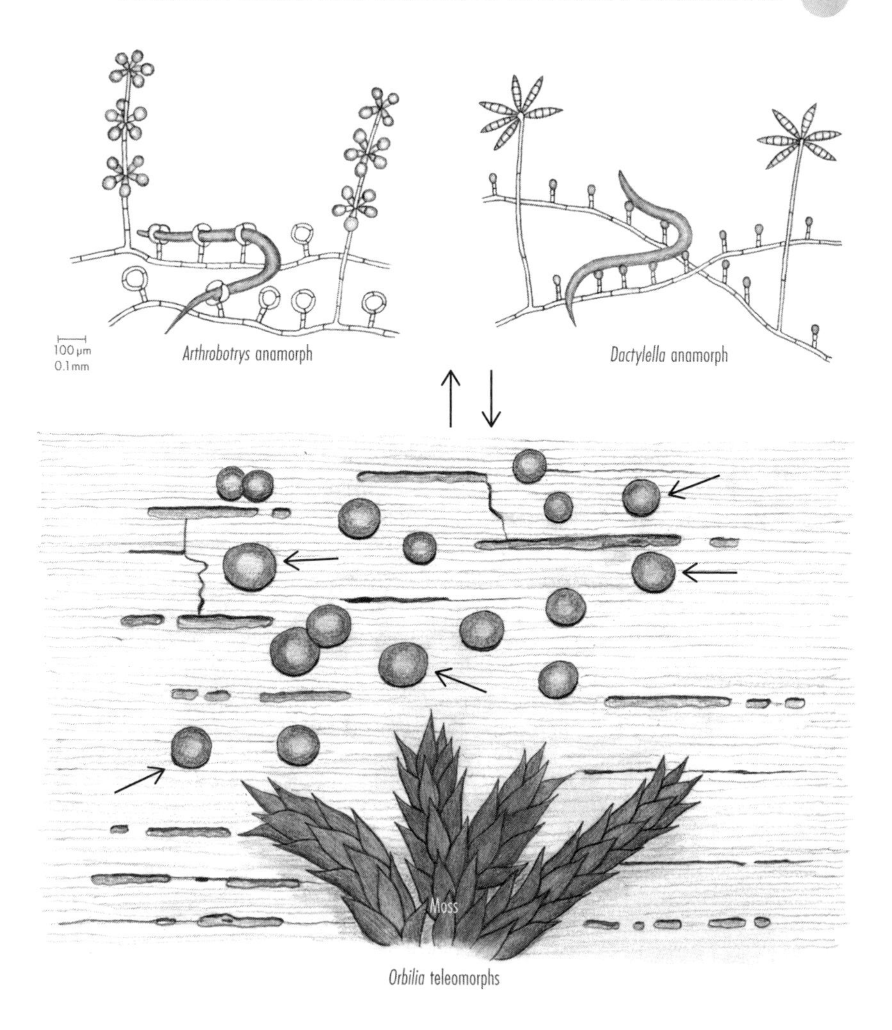

FIG. 50 Top: The anamorphic fungi *Arthrobotrys* (*left*) and *Dactylella* (*right*) are predatory fungi that trap and digest nematodes. Bottom: *Orbilia* are the tiny disk-shaped mushrooms (*teleomorph forms, arrows*) of the nematode-devouring anamorphic fungi. Here *Orbilia* is growing as a decomposer on the surface of a decaying log along with mosses. Anamorphs and teleomorphs reversibly transform according to environmental conditions.

as imperfect fungi. When *Arthrobotrys* and *Dactylella* have exhausted their supply of nematodes, they revert to their sexual mushroom forms. Several anamorph genera such as *Arthrobotrys* and *Dactylella* all have the mushroom genus *Orbilia* as their teleomorph. *Orbilia* is a ubiquitous tiny saucer-shaped mushroom only about half a millimeter across. Hundreds of these colorful orange mushrooms can cover surfaces of damp, rotting tree trunks and branches. From outward appearances, there is no hint that these mushrooms can switch occupations from decomposing tree trunks to hunting nematodes (fig. 50).

One fungus can devour nematodes one day and then decompose a tree trunk a few days later. The nematode-devouring anamorphs *Arthrobotrys* and *Dactylella* switch not only their forms but also their functions to become tiny mushrooms that sprout on dead trees, fallen branches, and logs. These mushrooms are *Orbilia* teleomorphs that now contribute to the essential processes of decomposing and recycling under trees.

Recycling is a community effort. In the litter and soil beneath trees, these countless fungi contribute their share to nutrient recycling along with countless other soil creatures that assist with decomposing. Everyone and everything underground and aboveground ends up as nourishment for the decomposers. Decomposers—large and small—in turn eventually liberate the nutrients that all trees and all companions of trees need for survival. Chapter 6 is devoted not only to the countless animals that share the job of recycling with the fungi but also to the many predators, root feeders, and parasites that round out the community beneath trees.

But in the next chapter the herbivores of trees take center stage. Not only do the insects that feed on leaves and twigs challenge the immune defenses of trees, but they also offer gratifying quarries for the trees' allies of vertebrate and insect predators. The more diverse insects that act as parasitoid allies of trees find an even greater diversity of hosts among a tree's herbivores. Many herbivores are certainly too small to be acceptable meals for many birds, but no herbivores are too small or too large to satisfy the needs of the innumerable species of insect parasitoids and predators.

OUT ON A LIMB

LIVING ON LEAVES, BUDS, AND TWIGS

2

TREES OFFER A BROAD, ASSORTED MENU TO HERBIVORES

All parts of a tree—from the moistest to the driest, the softest to the hardest—provide nourishment for insect herbivores large and small, some with far-ranging tastes, some with fastidious tastes. Under the broad category of herbivores are the leaf chewers, the sap suckers, the gall makers, those that eat living wood, and those that eat dead wood. Then there are those that prefer their dead wood well seasoned with fungi—some eating both dead wood and fungi but some being strictly fungivores.

For the many insect herbivores that harbor microbial partners, their performance as efficient herbivores can be imparted by attributes of their microbial endophytes rather than their own inherent abilities. When deprived of their microbial partners, these insects survive poorly, leave fewer progeny, and consume less plant tissue (Sugio et al. 2014; Hosokawa et al. 2007).

Since the diets of herbivores are so wide ranging, almost every tree has its share of herbivores. Just as farming, medicine, and teaching have their own specialties, the profession of herbivory has many specialties, and some herbivore species have developed special tools for their trade. Some herbivores are so finicky that they will feed on only one tissue of only one tree species. Some even require special preparation of the tissue before consumption. However, being too rigid about one's diet can present problems for survival in an ever-changing world.

Then there are those such as the caterpillars of the white-marked tussock moth (LeafScape, 27), whose tastes are so far ranging that they have been recorded consuming leaves on at least 50 species of both deciduous and coniferous trees. The larvae of one invasive and extremely adaptable moth have been recorded feeding on 78 native species, both evergreen and deciduous, and in the laboratory, this moth's larvae find at least 500 plant species acceptable as food plants. No wonder these invaders from the Old World have been so destructive to the trees in their New World home (Frost 1959).

Some tree species are clearly more popular and tastier than others. Oaks—of all the trees—are by far the top choice of insect herbivores; at least 1,000 different

FIG. 51 Oaks are by far the most popular tree companions for insects.

FIG. 52 Among the cone-bearing trees, pines are the top choice as companions for insects..

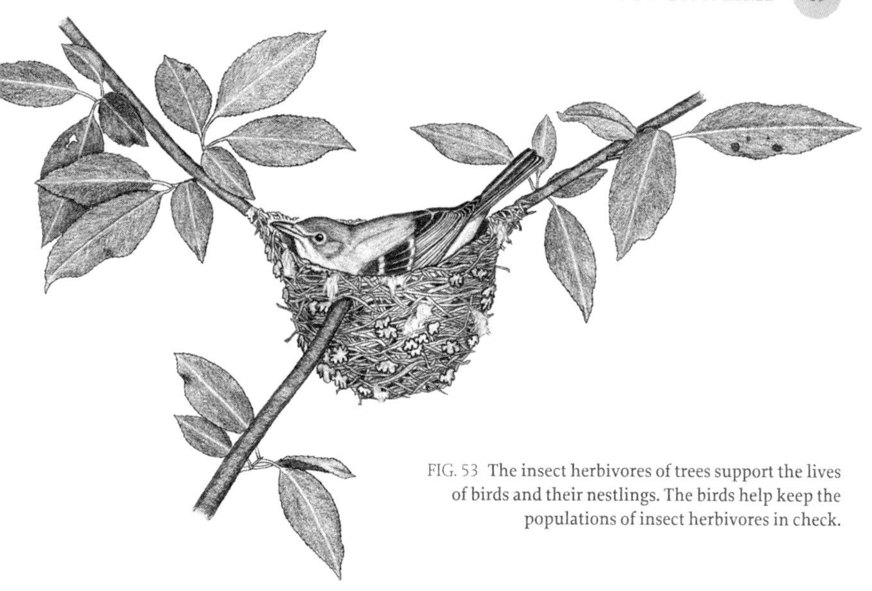

FIG. 53 The insect herbivores of trees support the lives of birds and their nestlings. The birds help keep the populations of insect herbivores in check.

insect species depend on oaks for their livelihood (fig. 51). Apple trees are known to host at least 400 species of insects. Among conifers, pines support about 150 herbivore species (fig. 52). Almost all insects, however, avoid the maidenhair tree (*Ginkgo*) and the tree of heaven (*Ailanthus*); both trees are known for the abundant repellent secondary metabolites in their leaves. These chemicals are antifeedants that repel but do not directly kill insects. However, insect life is so enterprising that among the over one million insect species, at least two discussed in later pages have established lives as herbivores on the unsavory *Ailanthus*, or stink tree (Wakie et al. 2019; Frost 1959). Most tree species, however, support many species of insect herbivores and the birds whose lives depend on them (fig. 53).

A CLOSER LOOK AT LEAVES AND TWIGS: STOMATA AND LENTICELS

How do the anatomy and physiology of a tree influence its inhabitants? The structure of trees and their physiology are intertwined with the structure and physiology of their inhabitants. The tree hosts have lived with their companions for so many years that each has made multiple adjustments to its form and function to accommodate the demands of its partner. Living together can involve conflict, and sometimes one partner gains the upper hand. Living together can also involve cooperation and compromise, with each partner benefiting from the cohabitation.

A sea of air surrounds each tree. For photosynthesis and production of sugars, the tree takes up carbon dioxide from the air and expels oxygen. In the reverse of photosynthesis, respiration uses oxygen to convert a tree's sugar to energy, water, and carbon dioxide. To enter and leave a tree's respiratory and circulatory systems, carbon dioxide, oxygen, and water vapor must pass through tiny passageways

FIG. 54 This close-up of an oak leaf's lower surface reveals the high density of passageways or stomata between the cells inside the leaf and the air outside the leaf.

between the surrounding air and the interior cells of tree leaves. These passageways—stomata (*stoma* = mouth)—might be microscopic but they are very numerous, and through these tiny pores move vast volumes of air and water vapor (fig. 54).

During the warmer days of the year, water evaporates from the many stomata of leaves. As the water vaporizes or transpires from the leaf surface, its removal concurrently draws up water within the long conduits of xylem cells of the tree's vascular system (chapter 3). In this way water continually moves upward through the tree—from the roots, along the trunk, into the branches and twigs, and finally into the leaves. On a hot summer day, transpiration from the leaves of a large tree can draw about 400 liters of water through its xylem channels, from root tips to treetop. Meanwhile the stomata of the tree's respiratory system exchange oxygen and carbon dioxide—in one direction for photosynthesis and in the opposite direction for respiration. The stomata are regularly distributed over the surfaces of all leaves, with an average density of 300 per square millimeter. On a single leaf the size of an oak leaf, approximately five million stomata are found.

Cutting a transverse slice through a leaf gives us an idea of what insects encounter as they construct their homes in or on leaves and consume provisions provided by leaves (fig. 55). On both their ventral and dorsal surfaces, leaves are covered with a monolayer of epidermal cells. Interspersed among the epidermal cells on the ventral surface are the leaf pores or stomata, which permit the exchange of gases and water vapor with the environment. Each stoma is lined by two guard cells that expand to open the stoma by taking in water and contract to close the stoma by expelling water. If the tree is not stressed by water shortage, the stomata

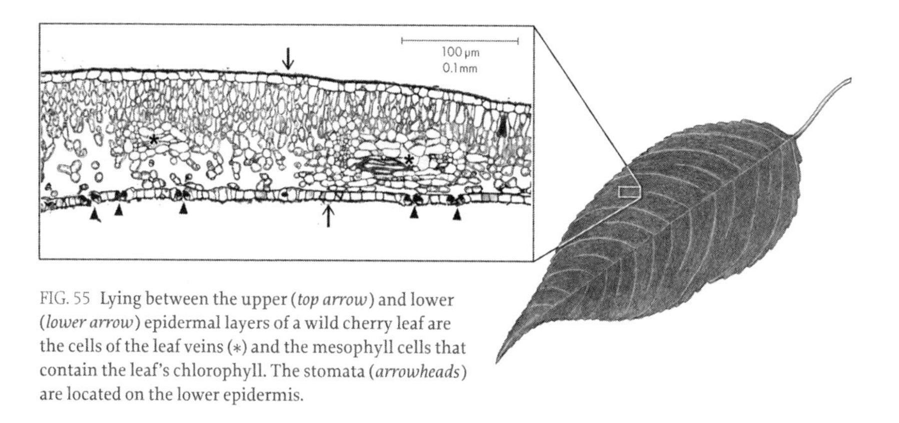

FIG. 55 Lying between the upper (*top arrow*) and lower (*lower arrow*) epidermal layers of a wild cherry leaf are the cells of the leaf veins (*) and the mesophyll cells that contain the leaf's chlorophyll. The stomata (*arrowheads*) are located on the lower epidermis.

remain open for water and gas exchange; however, if the tree roots sense that they are being parched, they communicate this condition to the leaves far above by releasing the plant hormone abscisic acid (Kuromori et al. 2018). Guard cells of the leaf epidermis respond to the abscisic acid signal by closing to conserve leaf water. Guard cells of the stomata stand out among the many surrounding epidermal cells as the only cells in these epidermal monolayers with green chloroplasts.

The cells that fill the space between the two epidermal layers are the tender green mesophyll cells (*meso* = middle; *phyll* = leaf) that carry out most of the leaf's photosynthesis, and the tougher cells that make up the leaf veins or vascular system. A network of veins represents the leaf's vascular system, which is specialized for transport of water and nutrients. Leaf-feeding insects are presented with this buffet of cells from which to choose. Some feed only on leaf sap and not on leaf cells. Nutrient-rich, more tender mesophyll cells are always popular with leaf eaters. Leaf miners avoid eating the waterproof epidermal cells that serve as the floors and ceilings for their homes. The tougher cells of the larger leaf veins are not the choicest cells for ingestion and digestion, so leaf veins often represent physical barriers for the movement and grazing of insects that feed on leaf tissue. Some avoid the leaf veins, leaving just a skeleton of veins behind as they dine on the more tender parts. Some voraciously devour entire leaves—including veins and even the petioles.

After leaf fall and after all stomata have disappeared—until leaf buds open again—a tree's photosynthesis comes to a halt. But the tree continues to respire, using the reserves of organic molecules generated by photosynthesis, as do all living creatures, exchanging oxygen with these organic molecules and finally releasing carbon dioxide to obtain their energy. Throughout the summer, trees use energy from the sun to produce reserves of carbohydrate molecules. With the onset of winter, these reserves are stored in special nutrient storage cells or parenchyma cells. Some reserves will be used to maintain a low level of respiration throughout the winter, and some will remain to power the awakening of the tree in the spring

With its leaves and their stomata gone, the tree resorts to the myriad structures on its twig surfaces known as lenticels (fig. 56). These pores on the surface offer openings to internal tissues through which cells below the twig's surface soak up oxygen from the environment and expel carbon dioxide as the end product of respiration.

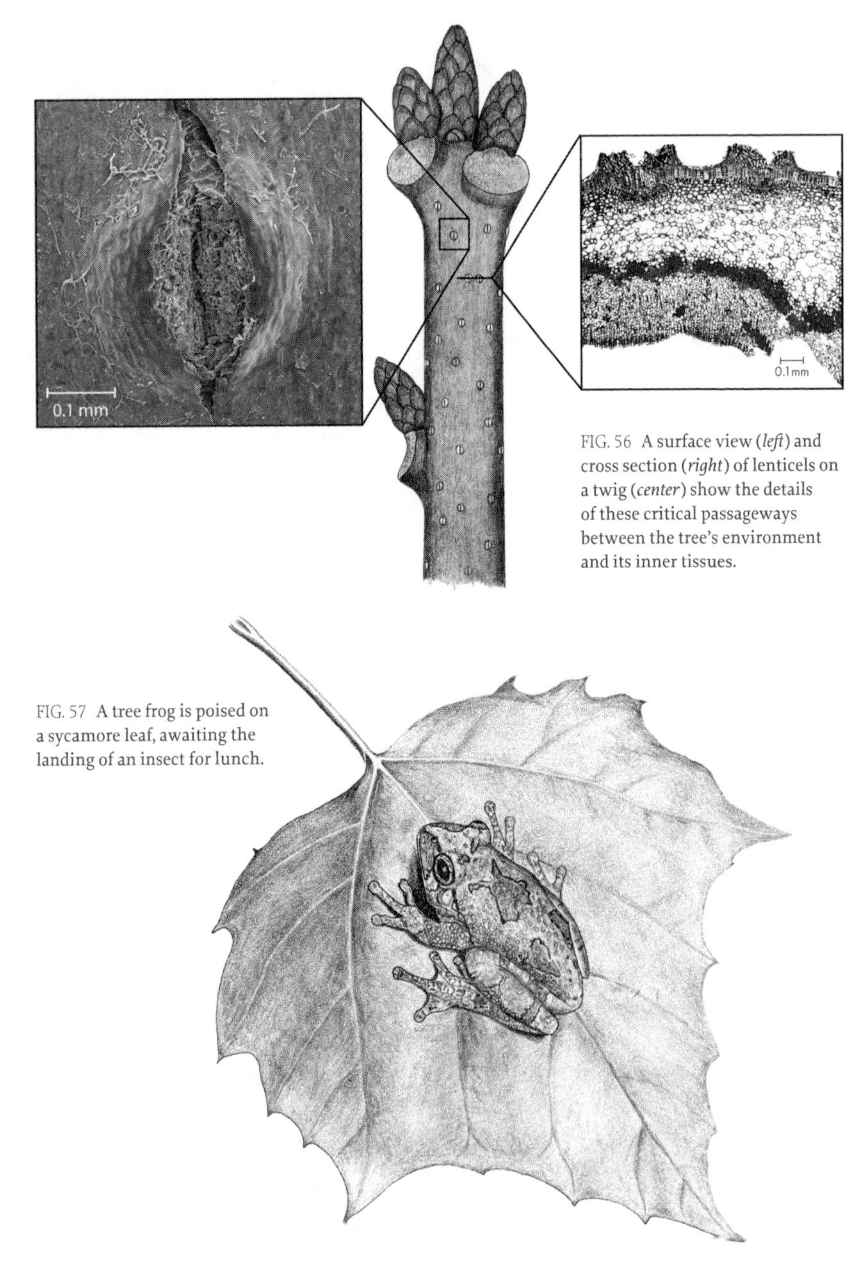

FIG. 56 A surface view (*left*) and cross section (*right*) of lenticels on a twig (*center*) show the details of these critical passageways between the tree's environment and its inner tissues.

FIG. 57 A tree frog is poised on a sycamore leaf, awaiting the landing of an insect for lunch.

Just as carbon dioxide emanating from vertebrate bodies attracts and guides mosquitoes and ticks to their prey's vascular system, carbon dioxide from lenticels and leaf stomata may be the cue that tells bacteria and fungal hyphae which portals offer access to the tree's vascular system. These microbes can move in through these portals to establish themselves as endophytes of the tree (chapter 1).

Although the surfaces of leaves and twigs are relatively smooth and their areas relatively small to human eyes, they offer a vast territory with complex topography measuring thousands of square millimeters where creatures that rarely exceed a few millimeters in length can roam, rest, and dine (fig. 57).

THE MANY WAYS TO PREPARE AND EAT LEAVES

Most leaf feeders are the caterpillars of moths and butterflies (Fürstenberg-Hägg et al. 2013). As caterpillars grow and chomp on leaves, their digestive tracts move into high gear. Unlike the digestive tracts of other insects where the ingested food passes slowly through convoluted intestines, a caterpillar's digestive tract stretches straight from mandibles to anus, feverishly processing whatever leaf tissue passes its way. As caterpillars attain their full growth at the peak of their chomping activity, their droppings rain down noisily on the foliage of a tree's undergrowth—thump, thump, pitter-patter. Droppings return nutrients borrowed from the tree and build up the organic matter in the soil.

— MOTHS AND BUTTERFLIES, ORDER LEPIDOPTERA (*lepido* = scale; *ptera* = wings) (175,000 species: 157,000 moths and 18,000 butterflies; 11,700 species NA: 11,000 moths and 700 butterflies; mature larvae 5–100 mm; wingspan 3–150 mm).

Skeletonizers of Leaves, Devourers of Entire Leaves

Having strong, well-sclerotized mandibles rather than the piercing stylets of true bugs, caterpillars and other chomping insects are equipped to chew through a leaf—often many leaves—in their short lifetimes. The smaller and more delicate among these caterpillars leave behind a leaf skeleton as they avoid biting into the hard and less palatable veins that make up the leaf skeleton in the same way that carnivores often leave behind the bones of their prey as they consume the more tender flesh.

Caterpillars of the family Zygaenidae have earned the common name "leaf skeletonizer moths" from their habit of leaving behind leaf skeletons as they feed. We'll learn more about the exceptional lives of these caterpillars and the colorful moths they become in chapter 4 (fig. 200).

As they grow and molt, other caterpillars graduate from being leaf skeletonizers to consuming the entire leaf. However, for caterpillars in the family Bucculatricidae, just the opposite is true: they end their days as leaf skeletonizers after starting off as leaf miners (fig. 126). These transitions in feeding are often accompanied by drastic transformations in the appearance of the caterpillars. Certain caterpillars in the family Notodontidae start out sporting antlers but shed their adornments after a molt or two (fig. 58). Most leaf-chewing insects, like caterpillars, get off to an early start in the spring before the trees' production of many distasteful substances or feeding deterrents is in full swing. Young caterpillars start by taking small bites from leaves. Immediately after the antlered caterpillar of one notodontid moth hatches from its egg on an oak leaf, it begins using each of its jaws like an ice cream dipper to scoop out the tender green portions of the leaf between the tougher and paler veins. Only the leaf skeleton remains in the path of this caterpillar. As the caterpillar grows and molts, it sheds its antlers, and its jaws become stronger. Soon it begins to devour entire leaves as it chews along.

— PROMINENTS, FAMILY NOTODONTIDAE (*noto* = back; *odonti* = teeth) (3,800 species; 140 species NA; larvae ~ 40 mm; wingspan 20–124 mm).

Many caterpillars in this family are known for their outlandish—sometimes fearsome—appearance and behavior. The puss moth caterpillar has a rear end with a pair of long tails and a front end with a colorful head. The caterpillar rears up and waves both head and tails in a threatening pose. Some notodontid caterpillars are social and gregarious, known for their marching processions that share communal webs.

Another member of this moth family chomps away from the leaf's edge toward its midrib. The two spines on each of the eight dorsal humps of the *Nerice bidentata* (*nerice* = nymph; *bidentata* = two teeth) caterpillar happen to mirror the two spines

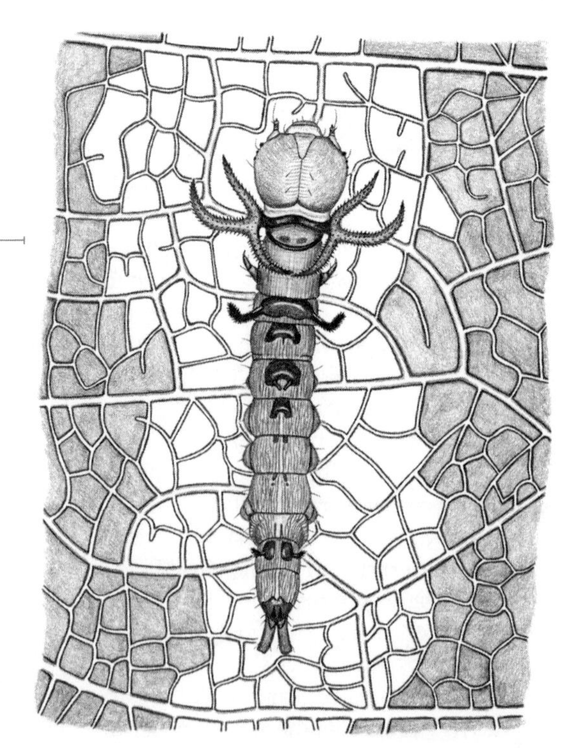

FIG. 58 This newly hatched caterpillar begins larval life eating only the softest tissues of the leaf and leaving the leaf's skeleton, but after a molt or two it sheds its antlers and begins consuming the entire leaf.

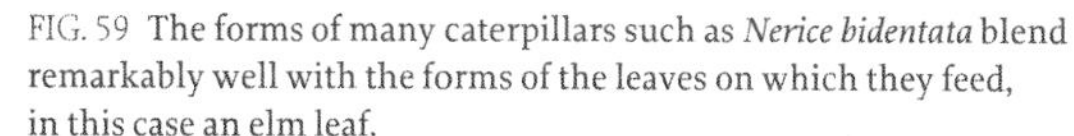

FIG. 59 The forms of many caterpillars such as *Nerice bidentata* blend remarkably well with the forms of the leaves on which they feed, in this case an elm leaf.

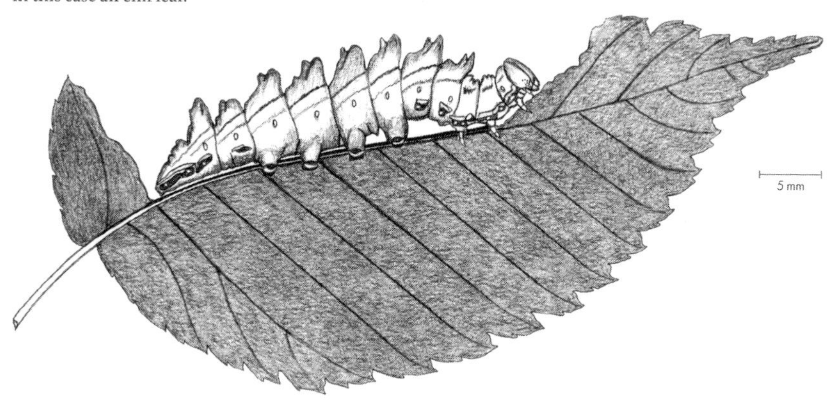

of each tooth on the edge of an elm leaf (fig. 59). Other green caterpillars have brown markings that match the dark blotches, spots, and patches of injured tissue in the middle of aging green leaves in late spring and summer. Many larvae feeding on the edges of tree leaves contort their bodies to align with the curved outlines of the leaf or the holes that they have eaten in the leaf. Rather than adopting a threatening appearance, these caterpillars reduce their visibility and vulnerability by taking on the shapes, colors, and even blemishes of the leaves on which they feed.

The more robust caterpillars dispatch an entire leaf, devouring the rigid leaf skeleton as expeditiously as its tender mesophyll cells. Between meals these insects rest in positions and locations in which they are inconspicuous. Telltale signs of their whereabouts on a tree may be evident; however, locating a hidden, resting insect among the leaves and twigs can be a daunting challenge. Many young caterpillars masquerade as bird droppings, but after their later molts, these same caterpillars emerge from their shed cuticles with very different forms and colors. As mature caterpillars, some adopt the threatening forms of tree snakes, while others blend seamlessly with the twigs or bark that they traverse or with the lichens or leaves on which they feed.

Caterpillar communal living provides safety in numbers. Certain caterpillars of the family Notodontidae congregate and socialize (fig. 60). Tent caterpillars in the family Lasiocampidae not only feed together but also construct a community shelter of silk (fig. 64). Caterpillars of fall webworm moths (*Hyphantria cunea*; *hyphant* = weaver) (fig. 61), which are the larvae of a type of North American tiger moth, fashion silken shelters for themselves and leave trails of communal destruction among the leaves and limbs of the many different tree species on which they feed (fig. 62).

— EREBID MOTHS, FAMILY EREBIDAE (*erebus* = darkness) (25,000 species: 960 species NA, larvae 10–80 mm; wingspan 6–126 mm), TIGER MOTHS, SUBFAMILY

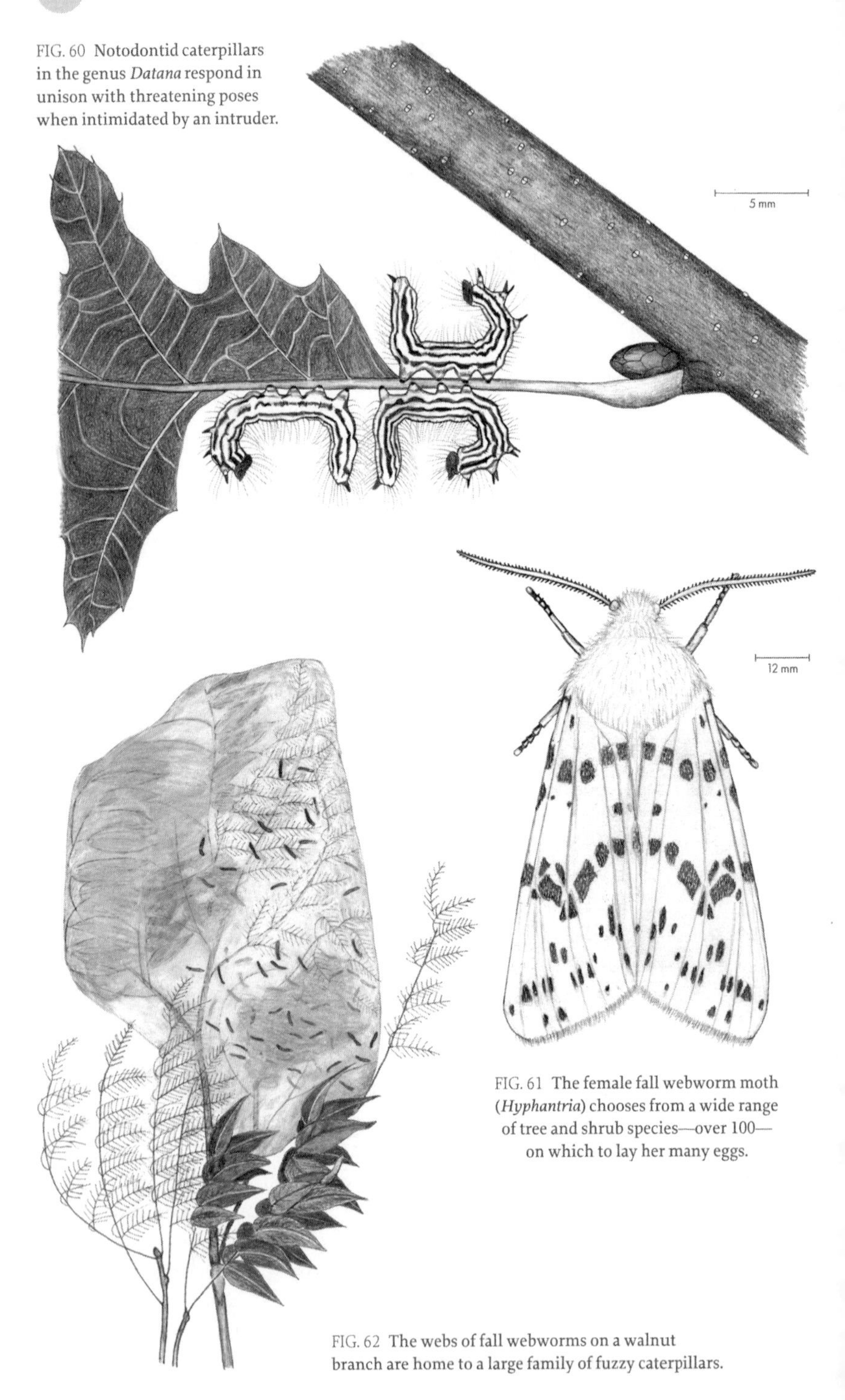

FIG. 60 Notodontid caterpillars in the genus *Datana* respond in unison with threatening poses when intimidated by an intruder.

FIG. 61 The female fall webworm moth (*Hyphantria*) chooses from a wide range of tree and shrub species—over 100—on which to lay her many eggs.

FIG. 62 The webs of fall webworms on a walnut branch are home to a large family of fuzzy caterpillars.

ARCTIINAE (*arctos* = bear, referring to the hairy caterpillars) (11,000 species; 265 species NA; larvae 10–35 mm; wingspan 28–40 mm).

Most insect species that feed on tree leaves are content to fend for themselves as they scatter over the foliage, minimizing their contacts and their competition with each other. However, a few exceptional caterpillar species and sawfly larvae have adopted social behaviors and spend time in each other's company. These insect larvae not only dine together but also, if intimidated, suddenly contort their bodies in unison to threaten and fend off predators and parasitoids. The most common response to danger is for the gregarious caterpillars to suddenly wave their raised heads and tails (fig. 60). The gregarious sawfly larvae also resort to this behavior (fig. 79). Their abrupt and synchronous movements not only prove startling but also make a predator's attack with mandibles or a parasitoid's impalement with an ovipositor almost impossible.

Some of these social caterpillars cooperate in building a communal nest of silk, adding additional layers of silk as the members of the caterpillar herd molt and grow. The tightly woven spring nests of tent caterpillars remain compact and usually confined to the crotches formed by two tree branches (fig. 64); however, the more loosely woven webs of fall webworms continually expand (fig. 62) to cover foliage all along a branch, leaving behind a trail of droppings and decimated leaves. Like the foraging scouts of ant and bee colonies, feeding larvae leave trails of silk marked with chemicals (pheromones) that impart information to their nest mates about which way to crawl to find good sources of leaves. They often form processions through the foliage, trundling single file along the trail left by the leading caterpillar (Costa 1997).

With impressive fidelity, these caterpillars follow the trail laid out by their leader. The French naturalist Jean Henri Fabre, an astute observer of the behavior of insects, tested the strength of the guiding cues used by social caterpillars as they form processions through trees and across landscapes (Fabre 1916). By transplanting the leading caterpillar of a short procession to the rear of the procession, he discovered that the new leader began following the silken trail laid out by the caterpillar taking up the rear position. The entire procession of caterpillars began to move in a circle and could continue circling for days until one caterpillar just happened to get sufficiently out of line to lead the procession forward again.

Caterpillars living communally have adopted this pathfinding method to inform their herd mates about routes leading to the best feeding grounds. But even solitary caterpillars like the spicebush swallowtail (fig. 110) are known to mark their evening trails through the foliage with silk threads. The threads also help them retrace their routes to silk mats where they return to rest during the heat of the day.

— TENT CATERPILLAR MOTHS AND LAPPET MOTHS, FAMILY LASIOCAMPIDAE (*lasio* = shaggy; *campo* = caterpillar) (2,130 species; 35 species NA; larvae 20–80 mm; wingspan 25–75 mm).

Unlike the webworms, which construct their webs around leaves at the ends of limbs, tent caterpillars weave their webs in the forks of tree limbs. The larvae hatch in early spring from the clusters of eggs that mother moths left on twigs the year

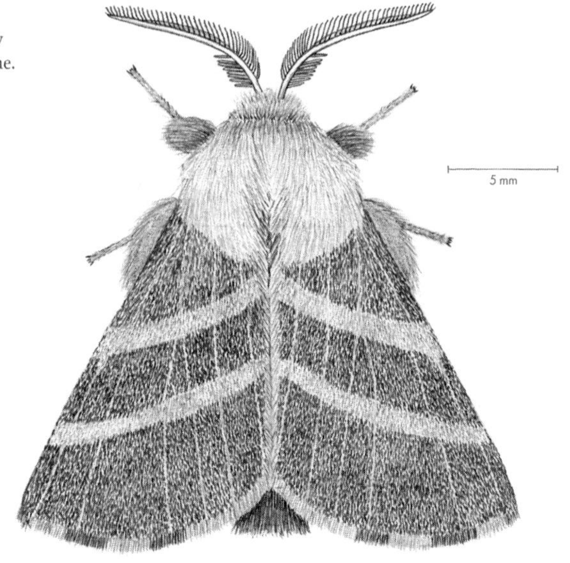

FIG. 63 Tent caterpillar moths almost always choose wild cherry trees as hosts for their social larvae.

before. Note the egg cluster on the lower left branch of the wild cherry tree in figure 64. The newly hatched larvae quickly build a communal "tent" of silk from which they can venture forth to the best feeding grounds along trails that they mark with silk strands for their fellow tent mates.

During their daily routines, songbirds flit from branch to branch, tracking down caterpillars for breakfast, lunch, and dinner. Caterpillars always do their best to evade these hungry birds. Tent caterpillars have fuzzy coats that protect them from most birds, but there are a few undeterred birds like the American cuckoo whose appetites are not daunted by a little fuzziness (fig. 64). The caterpillars turn into adult moths that are just as shaggy and fuzzy as the caterpillars (fig. 63).

— SLUG CATERPILLAR MOTHS, FAMILY LIMACODIDAE (*limax* = slug, snail; *-odes* = form of) (1,800 species; 50 species NA; larvae 10–30 mm; wingspan 15–42 mm). While tent caterpillars are exceptionally fuzzy, many caterpillars of slug moths have excelled at being exceptionally spiny as well as strikingly colorful. Although a few caterpillars in the family are smooth and spineless, a good number have stinging spines that project from ridges, knobs, and bumps that cover much of their body surface (fig. 65). Some of the numerous monkey slug species also masquerade as hairy, spiny spiders. Slug caterpillar larvae have six tiny thoracic legs and crawl along on flat bellies without the typical prolegs and hooked crochets of other caterpillars. They move about more like gastropod slugs than hexapod insects.

Slug caterpillar larvae spend the winter in thin egg-shaped cocoons equipped with circular escape hatches, waiting until spring to transform. While the caterpillars are flashy and flamboyant, most of the moths into which they transform are brown and plain. When resting on a tree, the moths stretch out their forelegs and drape their wings over their bodies like a tent (fig. 66). With reduced or even absent

FIG. 64 Yellow-billed cuckoos are one of the few bird species that can swallow and digest fuzzy tent caterpillars.

FIG. 65 This saddleback caterpillar is fond of oak, apple, maple, and elm leaves but sometimes shows up far from the forest on corn leaves.

mouthparts, limacodid moths are more likely to be spotted at evening sources of artificial light than they are at flowers or rotting fruits.

— INCHWORM MOTHS, GEOMETRID MOTHS, FAMILY GEOMETRIDAE (*geo* = earth; *metro* = measure) (21,000 species; 1,500 species NA; larvae 7.5–25 mm, wingspan 15–50 mm).

The geometrid moths can claim to be one of the largest families of moths, second only to the family Erebidae. The family name refers to a larva's appearing to pace off distances with its measured, methodical "inching" movements along twigs and

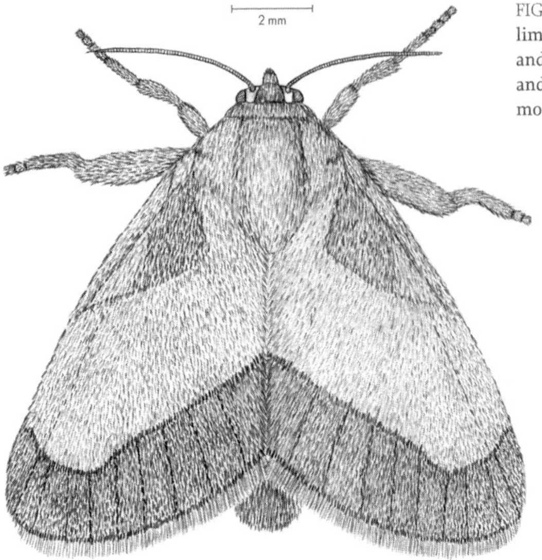

FIG. 66 This slug caterpillar moth, with lime-green scales on its wings and thorax and brown scales on its legs, wing bases, and wing tips, is one of the more colorful moths in its family.

leaves. But the name could just as well refer to some of the splendid geometric patterns of transverse lines—wavy, straight, or jagged—that adorn the wings of so many geometrid moths, such as the tulip-tree beauty shown here (fig. 67; LeafScape, 5).

Walkingsticks and inchworms very successfully disguise themselves as twigs—sometimes brown ones, sometimes green ones—by posing motionlessly for long stretches of time. Inchworms relax their stiff poses and then slink off as only inchworms can slink. The caterpillar begins to slink by first drawing its tail end up to its front end and forming a loop of all its body segments in between. It then extends its whole length forward. Once again, the tail end catches up with the head

FIG. 67 The wings of the tulip-tree beauty moth are embellished with a variety of geometric forms and shapes.

end as the caterpillar methodically "inches" its way along a leaf or a twig (figs. 14 and 68; LeafScape, 29).

— OWLET MOTHS, FAMILY NOCTUIDAE (*noctua* = little owl) (20,000 species; 2,900 species NA; larvae 10–22 mm; wingspan 20–65 mm, most 20–45 mm).

In this family of moths with so many species, members have taken on many different lifestyles and diets—lichens, fungi, green leaves, even dead leaves. In the family Noctuidae as well as some other moth families, some members have

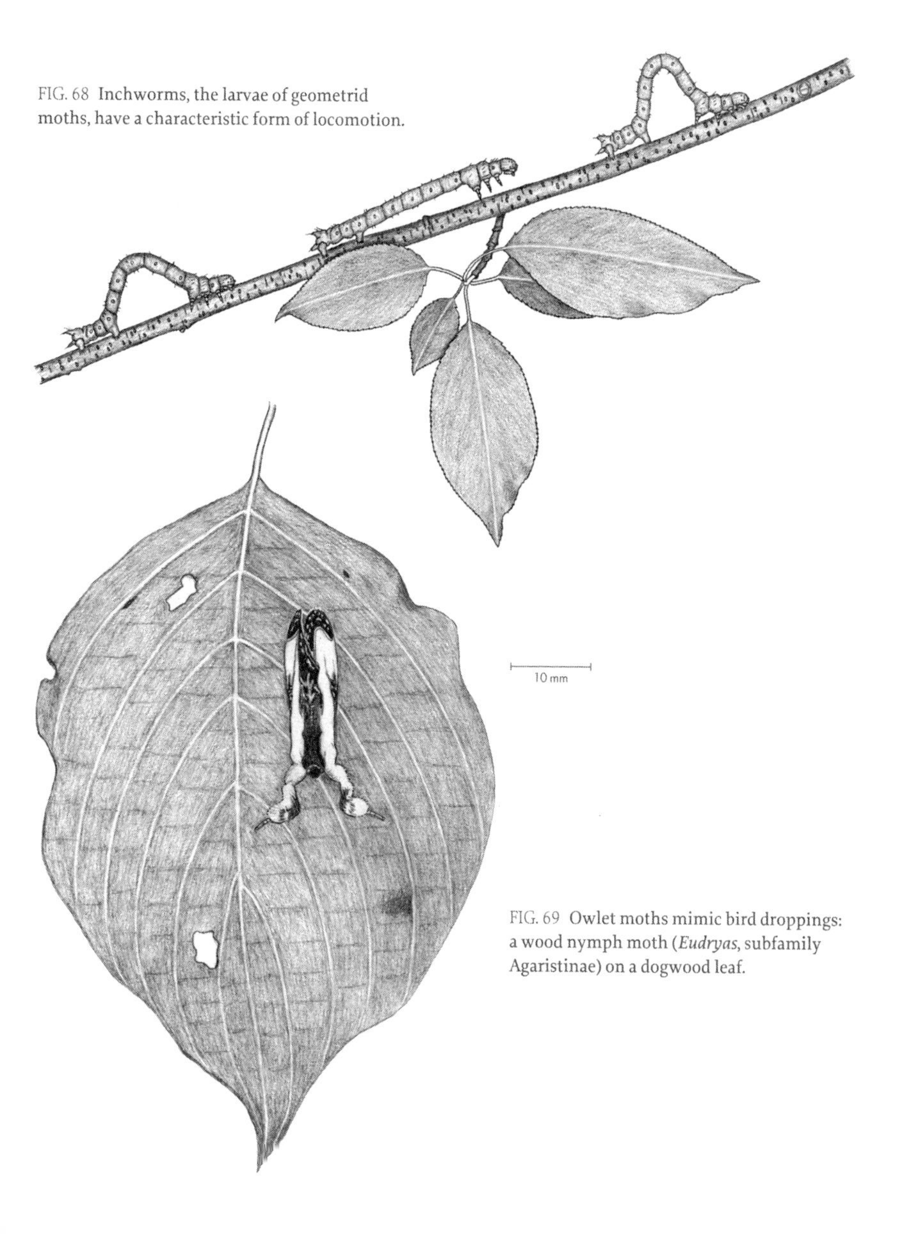

FIG. 68 Inchworms, the larvae of geometrid moths, have a characteristic form of locomotion.

FIG. 69 Owlet moths mimic bird droppings: a wood nymph moth (*Eudryas*, subfamily Agaristinae) on a dogwood leaf.

FIG. 70 A bird-dropping owlet moth, subfamily Acontiinae (*aconti* = dart).

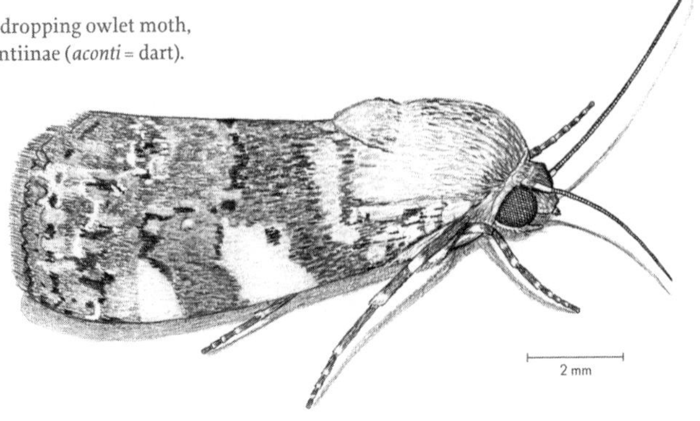

an uncanny resemblance to fresh bird poop as caterpillars and as adults (fig. 12). When many flying moths fold their wings and settle down on a leaf or bark, they suddenly look just like bird droppings. Some moths in this family, such as the wood nymphs (*Eudryas*; *eu* = good; *dryas* = wood; fig. 69), rest on foliage with their antennae hidden and forelegs outspread, looking like droppings that have splattered on the leaf's surface. Many tortricid (figs. 112 and 113), tineid (fig. 202), and gelechiid (fig. 125) moths rest with their black-, brown-, and white-splotched wings close to their bodies; their lengths and widths correspond well with the dimensions of droppings from birds ranging in size from warblers to waxwings. Probably the very best impersonators of bird droppings, however, are the aptly named bird-dropping moths in this family (fig. 70) The birds that so eagerly devour other crawling and flying insects do not show much interest in these moths. Even some tree spiders (crab spiders and bolas spiders, figs. 33 and 35) have discovered the benefits of looking digested and defecated. After all, what self-respecting bird is going to risk mistaking one of its droppings for a moth or spider? By masquerading as something inedible, these creatures have found a way to live in peace with insect-eating birds.

— BAGWORM MOTHS, FAMILY PSYCHIDAE (*psyche* = a butterfly, a soul) (1,350 species; 26 species NA; larvae 8–50 mm; wingspan 12–36 mm).

These caterpillars start building their bags as soon as they hatch from their eggs and feed from the bags throughout spring and summer. The male caterpillars pupate in their bags and emerge about three weeks later as winged moths that set about locating the wingless females. Although the female is sexually mature, she retains a mixture of larval and pupal features. As is the case for females of several other moth species, the female bagworm channels her energy into egg production rather that squandering unnecessary energy on forming adult structures such as wings and antennae. She never undergoes a conventional transformation to a winged moth. Once the female's transition from feeding to egg laying is complete, she emits pheromones to attract males to her bag. Each male has a long, extendible abdomen that he stretches to great lengths, inserting his genitalia through the entrance to the bag and into the female's chamber (fig. 71). Even after mating, the

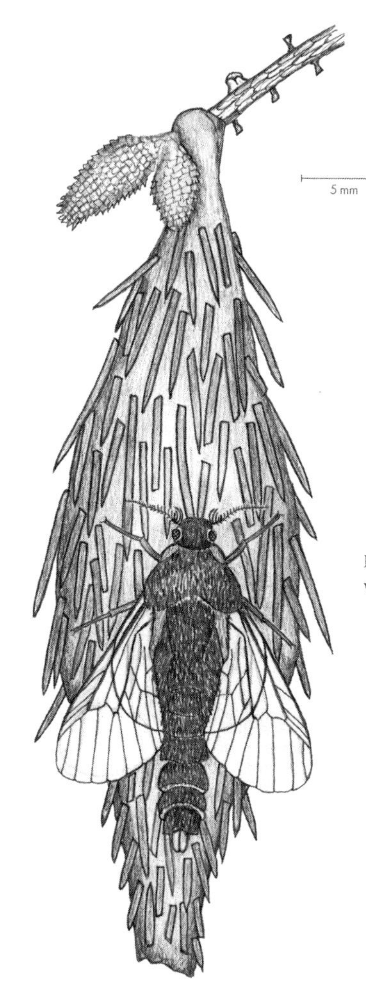

FIG. 71 A male bagworm moth prepares to mate with a wingless female bagworm, which never leaves her cocoon.

female remains in her larval bag, leaving behind about 500 eggs that will weather the upcoming winter and hatch in early spring.

— ROYAL SILKWORM MOTHS, SATURNIIDAE (*Saturnia* = queen of the gods) (2,300 species; 68 species NA; larvae 50–100 mm; wingspan 25–150 mm).
Not all saturniid moths spin silk cocoons as mature larvae, but all saturniids have their last meals as larvae. The adults have no mouthparts with which to feed or drink; they survive long enough on tree nutrients consumed during their larval days to mate and deposit eggs. Saturniid larvae and adults are equally distinctive and handsome. While larvae of rosy maple moths, buck moths, oakworm moths (LeafScape, 3, 4), imperial moths, and royal walnut moths pupate in the soil, our native silkworm moths—cecropia, polyphemus, luna, io, promethea—pupate and overwinter in sturdy, characteristic cocoons (figs. 72 and 73).

The caterpillars of these large, beautiful moths consume many leaves in their lifetimes. If the caterpillar of the polyphemus moth is feeding on maple leaves,

FIG. 72 This male polyphemus moth has just emerged from its cocoon, and its extremely sensitive antennae are waiting to pick up the scent of a female's mating pheromone.

FIG. 73 Promethea silkworm caterpillars often spin their cocoons on sassafras twigs.

it probably consumes 40 or 50 leaves, but if it feeds on the smaller leaves of wild cherry, it can easily chomp through over 100 leaves. All that eating is enough to carry the caterpillar through cocoon spinning, pupation, and adulthood without its ever feeding again.

— TUSSOCK MOTHS, FAMILY EREBIDAE, SUBFAMILY LYMANTRIINAE (*lymantria* = destroyer) (2,500 species; 32 species NA; larvae 20–65 mm; wingspan 15–67 mm). These rather plain moths are better known for their colorful, flamboyant larvae and the prominent dorsal tufts or tussocks of hairs on the first four segments of their abdomens (fig. 74; LeafScape, 27). Instead of being known for ostentatious appearances like their caterpillars, the adult moths of many species are named for their unconventional poses. Some members belong to the genus *Orgyia*, a Greek term referring to their adult resting poses with "outstretched arms." The caterpillars have been recorded feeding on at least 50 tree species—both deciduous and coniferous. The most infamous tussock moth is the forest pest *Lymantria dispar*, introduced to American forests from Europe in 1869 and reported feeding on leaves of over 500 species of trees and herbaceous plants.

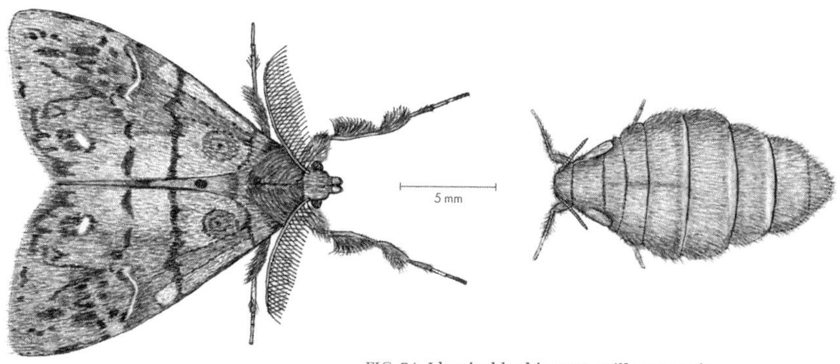

FIG. 74 Identical-looking caterpillars transform at metamorphosis into very different male (*left*) and female (*right*) tussock moths.

After voraciously feeding as larvae, the adult moths never feed. The male moth is preoccupied with finding a female, and with his hefty, imposing antennae, he is well endowed for detecting pheromones that the sedentary female broadcasts from her cocoon. The female has tiny wings and weak legs; she rarely ventures far from her cocoon. She saves her energy, clinging to her cocoon and waiting for a male to appear.

The female moth's life story illustrates the blatant gender disparity among these moths. The nests of cells known as imaginal disks (*imago* = adult), which are destined to form the adult wings in the female, develop normally throughout the life of the caterpillar, but at the time of metamorphosis, these cells undergo massive cell death in female moths only. The same imaginal disk cells in male moths go on to form normal wings with strong flight muscles. The few cells of female wings that survive form tiny, limp wings that can neither expand nor flap. The energy and nutrients salvaged from her dying cells are channeled into forming and maturing the abundant eggs that fill her swollen body. She lays up to 300 eggs in one mass and covers them with a foamy secretion and with scales from her body. The mother moth soon expires, and her eggs wait until spring to hatch.

— ERMINE MOTHS, FAMILY YPONOMEUTIDAE (*yponom* = undermine) (400 species; 35 species NA; larvae 15–30 mm; wingspan 20–30 mm).

Many members of this family have snow-white bodies and wings often marked with specks of black. Their scale patterns resemble the winter coats of weasels or ermines. Some ermine moths, such as the apple ermine moth, start their lives as leaf miners one year and then move to communal webs the following spring to feed until June before pupating. From July to September, most ermine moths lay eggs that overwinter on trees.

One ermine moth decked out in bright orange and black rather than white and black is the *Ailanthus* webworm moth, unforgettable for its stunning wing pattern (fig. 75). It is one of the exceptional insects that have surmounted the notorious defenses of the tree of heaven, or stink tree, whose noxious secondary metabolites make this tree repellent to almost all other insects. And this might be the only moth or butterfly that can survive on that tree.

Each spring this colorful ermine moth migrates to northern latitudes like another orange, black, and white lepidopteran—the celebrated monarch butterfly, whose caterpillars feed on noxious milkweed leaves. Like the monarch, this ermine moth is decked out in colors that warn predators to avoid encounters with it. Neither the *Ailanthus* webworm moth nor its monarch butterfly counterpart can survive the winters of the northern states and Canada at any stage of its life cycle; both must migrate north each spring and south each autumn.

Another butterfly with the same conspicuous warning colors of orange, black, and white is the viceroy. Its caterpillars, however, feed on willow leaves that lack

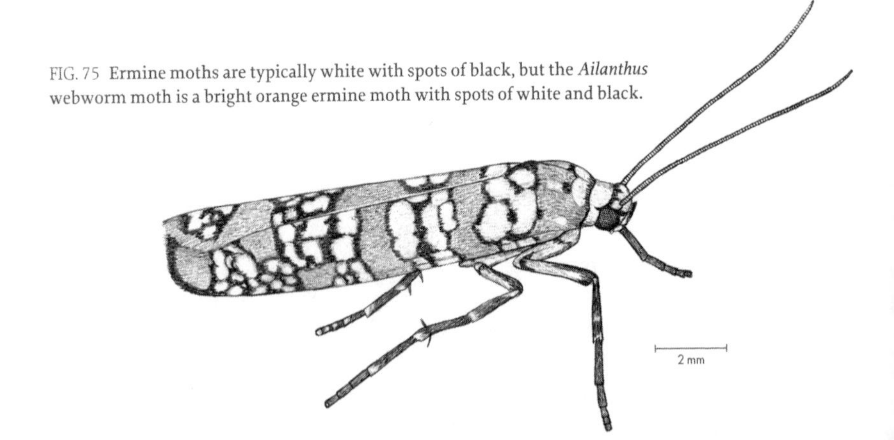

FIG. 75 Ermine moths are typically white with spots of black, but the *Ailanthus* webworm moth is a bright orange ermine moth with spots of white and black.

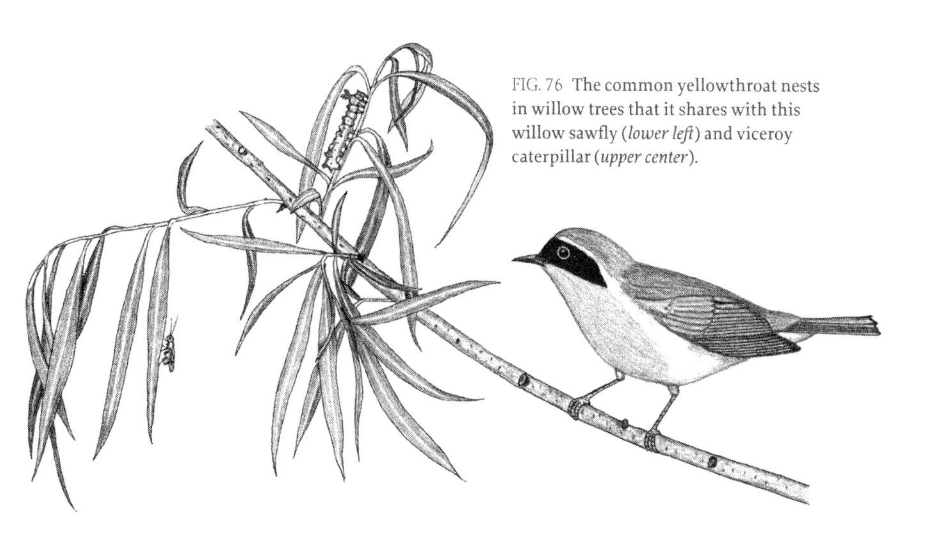

FIG. 76 The common yellowthroat nests in willow trees that it shares with this willow sawfly (*lower left*) and viceroy caterpillar (*upper center*).

the toxins of the milkweed leaves eaten by monarch caterpillars. While viceroy butterflies ward off attacks from birds by mimicking the form and color pattern of unpalatable monarch butterflies, their caterpillars have adopted a different defense. Rather than calling attention to themselves, viceroy caterpillars assume the appearance of bird droppings, dappled in patches of white, brown, and gray like the caterpillar feeding among willow leaves in figure 76. They have little resemblance to the striking caterpillars of monarch butterflies. All caterpillars of the viceroy's close relatives in the same genus, *Limenitis*, the admirals and red-spotted purples, also look like bird droppings and feed on wild cherry in addition to willow leaves.

The monarch, the viceroy, the red-spotted purple, and the admirals are all members of the same butterfly family—Nymphalidae. The butterflies in this largest of butterfly families have at least one feature in common: they all have truncated forelegs ("brushes") and use only their second and third pairs of legs for walking (fig. 77).

— BRUSH-FOOTED BUTTERFLIES, FOUR-FOOTED BUTTERFLIES, FAMILY NYMPHALIDAE (*nympha* = nymph) (6,000 species; 210 species NA; mature larvae 20–60 mm; wingspan 35–90 mm).

Many members of this largest family of butterflies feed on herbaceous plants, but mourning cloak caterpillars choose elms, willows, and poplars; question mark larvae feed on elms; and caterpillars of snout butterflies and hackberry butterflies find only hackberry leaves acceptable.

Caterpillars of many butterflies feed exclusively on grasses and other herbaceous plants; however, caterpillars of the next largest family of butterflies, Lycaenidae, not only feed on herbaceous plants but are also found on leaves of both deciduous

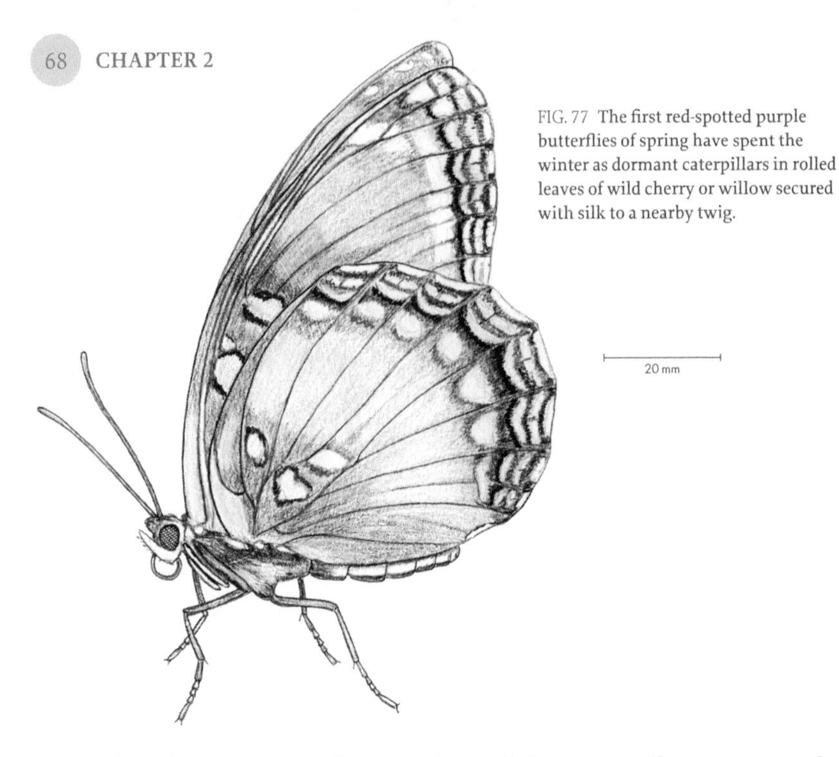

FIG. 77 The first red-spotted purple butterflies of spring have spent the winter as dormant caterpillars in rolled leaves of wild cherry or willow secured with silk to a nearby twig.

trees and conifers. Some members of this family known as elfins are among the few butterflies whose caterpillars feed on pine needles (fig. 78). Its family relatives the harvester, the spring azure, and the red-banded hairstreak—all lycaenids and all with exceptional tastes—are mentioned in chapters 3, 5, and 6.

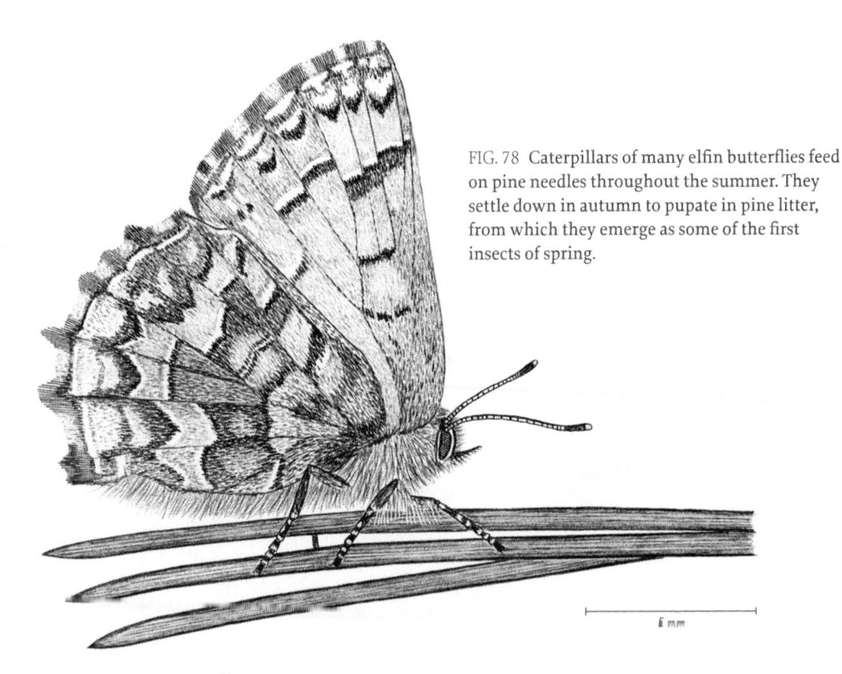

FIG. 78 Caterpillars of many elfin butterflies feed on pine needles throughout the summer. They settle down in autumn to pupate in pine litter, from which they emerge as some of the first insects of spring.

— GOSSAMER-WINGED BUTTERFLIES: HAIRSTREAKS, BLUES, AZURES, ELFINS, HARVESTERS, FAMILY LYCAENIDAE (5,500 species; 142 species NA; larvae 12–20 mm; wingspan < 25 mm).

Caterpillars of lycaenid butterflies have the reputation of being plain, homely, and unadorned. According to the *Field Book of Insects* (Lutz 1948), "lycaenid larvae are short-legged and small-headed." In *Caterpillars of Eastern North America* (2005), David Wagner observed that "while adult lycaenids are surely among the planet's most beautiful animals, their caterpillars are a mundane lot." (LeafScape, 20).

The rather plain and sluggish caterpillars of these tiny, delicate butterflies have developed a variety of partnerships with ants. Many of these lycaenid larvae produce sweet, nutritious secretions like honeydew from special honey glands on their abdomens, which so many ants find irresistible and which prompt ants to fervently defend their caterpillar charges from predators. In addition to their enticing flavor, caterpillars secrete chemicals from their cuticles that mimic the scent of their fellow ants. Some ants have gone to such lengths as adopting these caterpillars as members of their colonies, not only nurturing the caterpillars as though they are their own larvae but even tolerating their brazen devouring of ant larvae (Casacci et al. 2019)

Other insect larvae found on tree leaves eat like caterpillars and look like caterpillars but are more closely related to bees than to butterflies and moths. The adults that these larvae become are known as sawflies (fig. 76, lower left). Sawfly larvae are easily mistaken for moth caterpillars and are as readily devoured by birds as caterpillars are (fig. 79; LeafScape, 17). Both sawfly and moth larvae have three pairs of true legs on their thoraxes; however, almost all caterpillars have five or fewer pairs of false legs or prolegs on their abdomens while almost all sawfly larvae have six or more abdominal pairs of prolegs. Sawfly prolegs are smooth and completely lack the Velcro-like crochets of caterpillar prolegs. Sawfly larvae

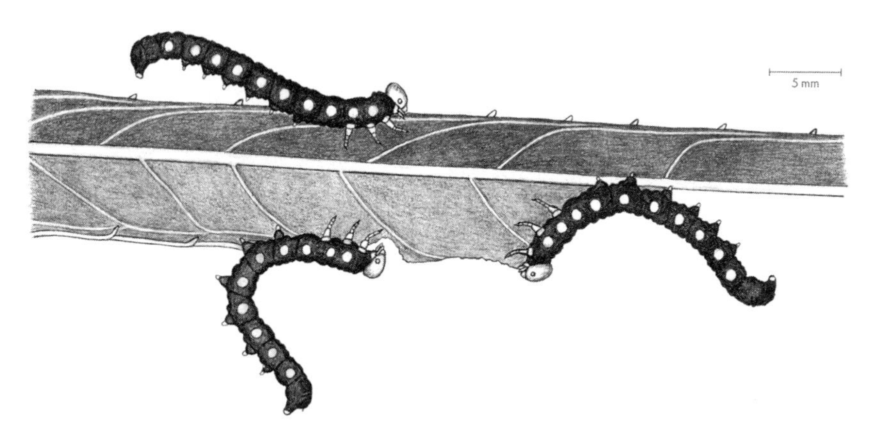

FIG. 79 The hatching of the willow sawfly's eggs will provide ample nourishment for nestlings of many bird species.

are also the only insects chosen as hosts by a small and rare family of parasitic wasps, the Roproniidae, which has only three known species. Sawfly larvae are often gregarious, raising and waving their rear ends in unison to startle and intimidate. This is a pose that caterpillars only rarely adopt (fig. 79).

— SAWFLIES, SUBORDER SYMPHYTA (*symphyo* = growing together) (8,000 species; 1,100 species NA; 2.5–55 mm).

Sawflies are in the same insect order as wasps, bees, and ants—Hymenoptera. Although adult sawflies do not have modified ovipositors or stingers like those of their relatives the bees and wasps, female sawflies deposit their eggs on leaves or twigs with an ovipositor that resembles a tiny saw. Bees, ants, and wasps have conspicuous "waists" separating their abdomens from their thoraxes. Sawflies and their wood-boring relatives the horntails and wood wasps (chapter 4), however, have thoraxes and abdomens that appear to grow together without any constricted region or waist between the two. This obvious feature is conveyed by the name for the suborder of the Hymenoptera to which sawflies belong—Symphyta.

— ARGID SAWFLIES, FAMILY ARGIDAE (*argo* = shining, bright) (800 species; 60 species NA; 8–15 mm).

Argid sawflies are often black or a pleasing combination of black and red; they stand out from other adult sawflies in having antennae with the fewest number of sections (fig. 80). Although the antennae of the adults contain only three sections, an exceptionally long terminal section that takes the form of a tuning fork in male sawflies compensates for the simplicity of their antennae. They are also exceptional in their broad choices of host plants. In addition to being hosted by most deciduous and cone-bearing trees, the larvae can feed on plants that most insects avoid—ferns, abrasive horsetails, and even poison ivy (LeafScape, 17).

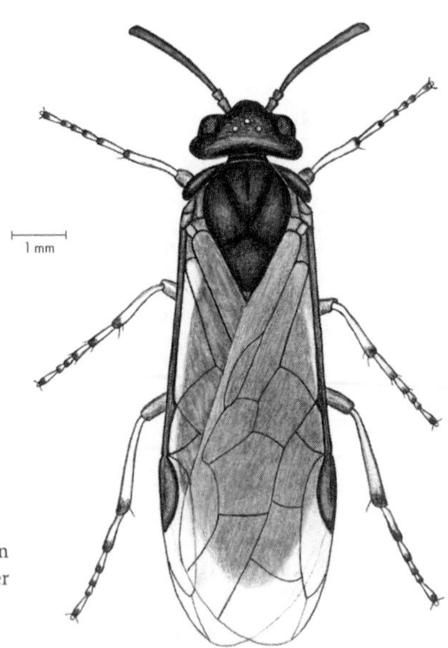

FIG. 80 While some species of argid sawflies can strip birch and willow trees of their leaves, other species have been employed to control weeds such as purslane and poison ivy.

— CLUB-HORNED SAWFLIES, FAMILY CIMBICIDAE (*cimbia* = girdle, referring to the twig-girdling habit of some adult sawflies) (200 species; 12 species NA; 18–25 mm).

Most sawfly larvae form feeding aggregates and abruptly rear their abdomens in unison to startle any intruders. The more robust and sluggish larvae of cimbicid sawflies, however, consume more leaves and feed alone so they will not quickly exhaust their supply of edible foliage. To compensate for the vulnerability of living a solitary life, these larvae adopt coiled postures while they are resting and secrete offensive fluids from glands along their sides when perturbed. The larvae are covered with a thin layer of waxy flakes that make them appear to be frosted. The simple eyes of the larvae stand out as a pair of conspicuous dark spots that impart an undeniable appeal to their faces (fig. 81).

Adult cimbicids look like bumblebees and in this disguise probably fend off unwanted advances from predators (fig. 82). The adults also use their strong mandibles to girdle twigs to reach sap for themselves and probably also to concentrate nutrients in leaves for their solitary larvae. Between a girdle and the end of the twig, the transport of sugar and other nutrients produced in the leaves is halted. Nutrients instead become concentrated in the leaves being consumed by the larva—certainly a great arrangement for larval nutrition. The girdle interrupts only the transport of sweet nutrients in the phloem channels from the leaves to the branches and roots. Water and mineral nutrients in the xylem

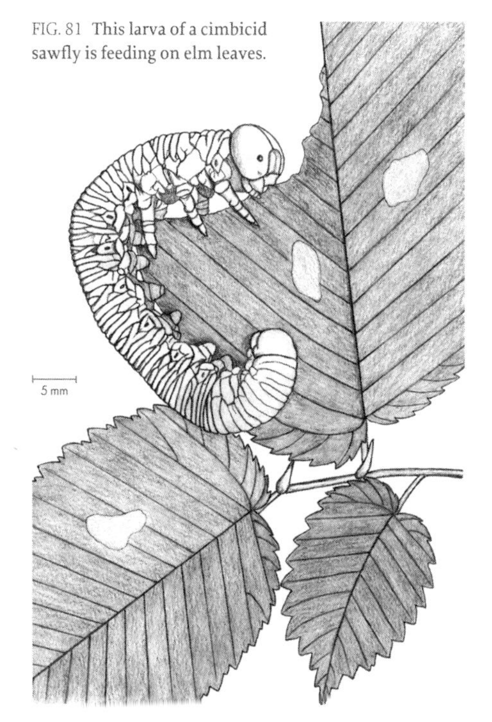

FIG. 81 This larva of a cimbicid sawfly is feeding on elm leaves.

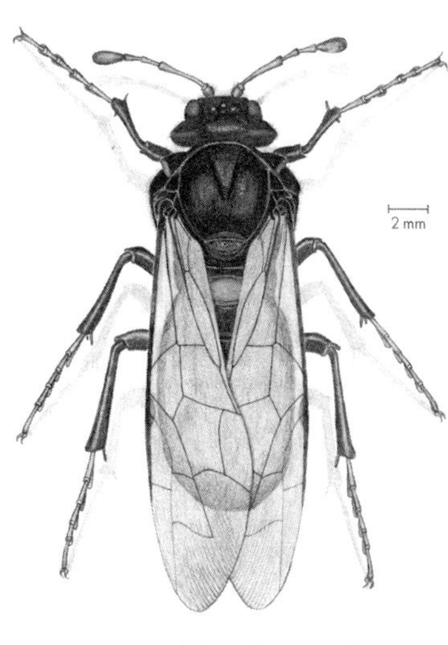

FIG. 82 Adult cimbicid sawflies are some of the largest sawflies and have large jaws for girdling twigs.

channels continue to flow unobstructed from the roots to the leaves where larvae are feeding.

— CONIFER SAWFLIES, FAMILY DIPRIONIDAE (*di* = double; *prion* = saw, referring to the sawlike ovipositor) (140 species; 68 species NA; 6–12 mm).

Trees are certainly not defenseless when threatened by the jaws of caterpillars and other chewers of leaves. For every repellent a tree produces, however, there are certain insects that can break down the repellent metabolites or even use these substances to their own advantage. Conifer sawflies feed on pine needles as larvae and store a variety of noxious metabolites from pine resin in pouches near their mouths. When threatened by a predator, the sawfly larva rears its head and expels a sticky droplet of repellent from its mouth, sometimes even smearing the droplet on its attacker (fig. 83).

Unlike moths, butterflies, and many sawflies that chew leaves only as larvae, beetles have jaws as both larvae and adults. Some beetles like the leaf beetles and weevils feed on leaves as both larvae and adults. Many members of the large beetle family Scarabaeidae are earthbound dung beetles; however, in this diverse family there are two subfamilies, often referred to as chafers, that always begin life feeding on tender roots before switching to a diet of leaves and flowers. Another group of unorthodox chafers are found in decaying hollow hardwood trees (chapter 4; fig. 250). Within this one beetle family, taste preferences range from malodorous dung to fragrant inflorescences.

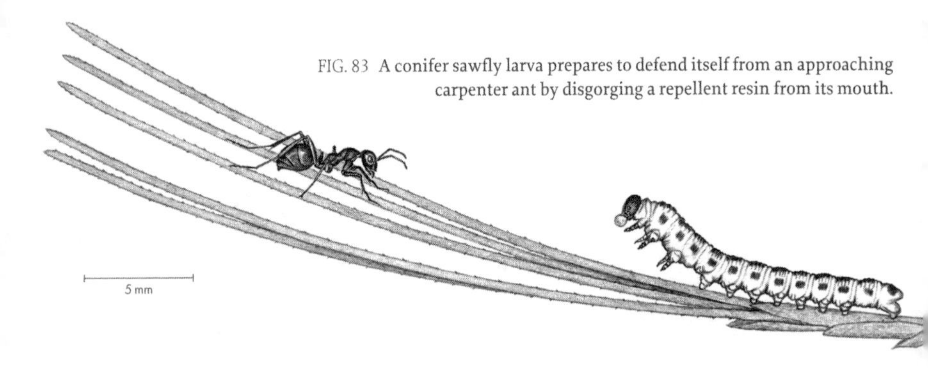

FIG. 83 A conifer sawfly larva prepares to defend itself from an approaching carpenter ant by disgorging a repellent resin from its mouth.

— SCARAB BEETLES, FAMILY SCARABAEIDAE (*scarabeus* = kind of beetle) (30,000 species; 1,700 species NA; 2–160 mm).

— **May Beetles, June Beetles, and Chafers, subfamily Melolonthinae** (*melolonth* = cockchafer) (12,000 species; 650 species NA; 3–35 mm).

— **Shining Leaf Chafers, subfamily Rutelinae** (4,100 species; 1,000 species NA; 10–52 mm).

As these chafer beetles chomp leaves, they also scrape and scratch the leaf surfaces with the armature of spines and spurs that cover their legs. These voracious beetles "chafe" away the upper surfaces of the leaves on which they feed, leaving

FIG. 84 These goldsmith beetles are iridescent gold "chafers" of the cottonweed leaf on which they are feeding.

shredded, wilted leaves in their wake (fig. 84). The most infamous among these beetles are the Japanese beetles. A mob of these beetles can leave the entire foliage of a tree in shreds. The larvae of chafer beetles, however, spend their lives underground (chapter 6; fig. 340; RootScape, 30), grazing on tender tree roots for anywhere from one to four years in darkness.

Since secondary metabolites of trees are so effective in their interactions with others, some animal companions of trees have discovered that they too can benefit from the effects of these metabolites. These animals can sequester these repellent, even toxic, chemicals that the tree host produces in such a manner that they are harmless to them but harmful to their predators. Such animals have a definite advantage in carrying this distasteful protection wherever they go. Among them are beetles in the family Chrysomelidae.

— LEAF BEETLES, FAMILY CHRYSOMELIDAE (*chryso* = gold; *meleos* = useless) (40,000 species; 1,900 species NA; 1–20 mm).

Like their counterparts the leaf bugs (family Miridae, chapter 3) in the order Hemiptera, these leaf beetles are members of a large family, the Chrysomelidae, in the order Coleoptera (fig. 85). Many leaf beetles and many leaf bugs are often brightly colored and have a reputation for providing unpleasant encounters. The leaf-mining beetle, the black locust leaf bug, and the locust-feeding sawfly have not only independently adopted the leaves of black locust trees as their habitat but are also independently adorned with similar orange and black patterns. One

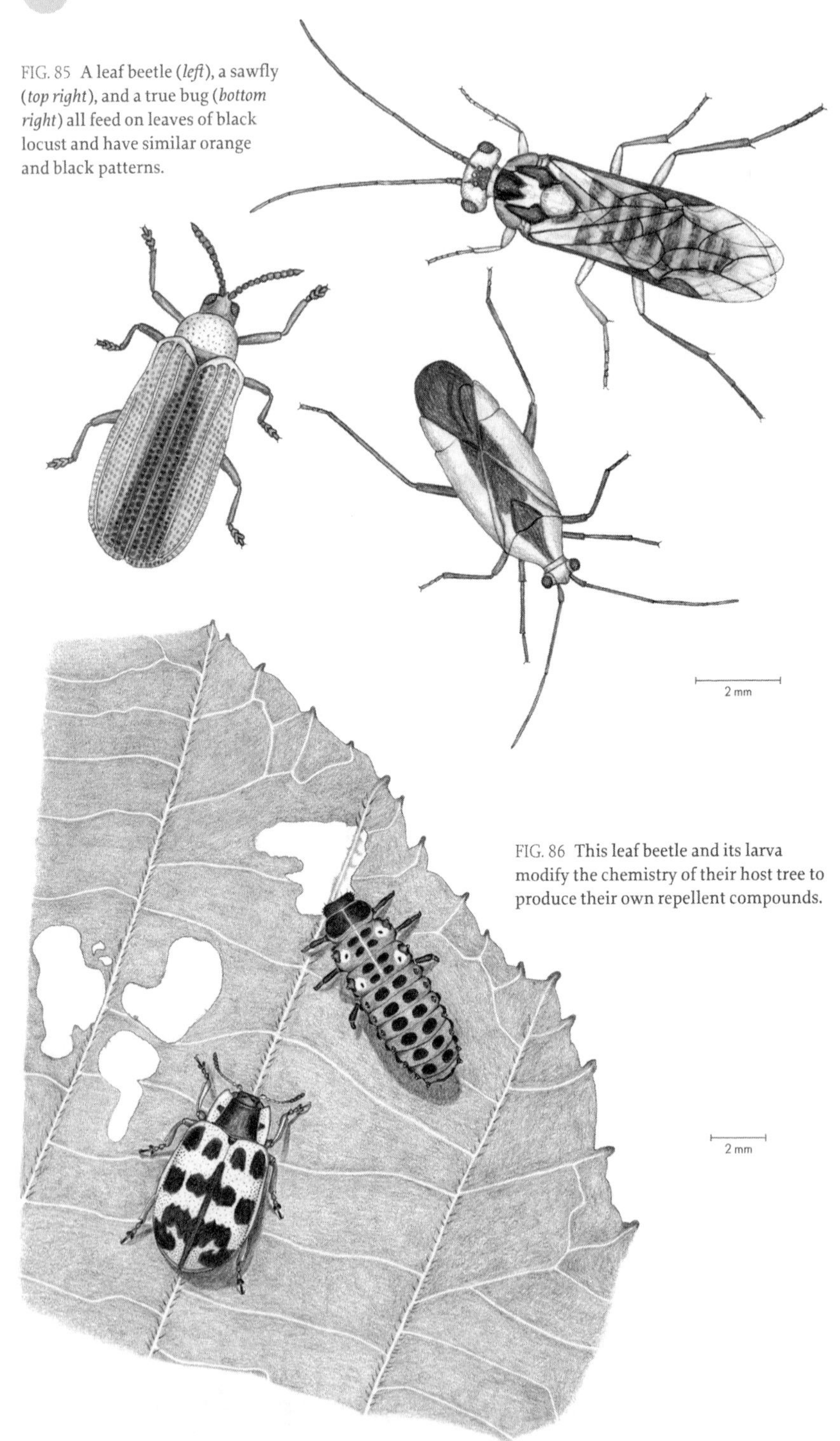

FIG. 85 A leaf beetle (*left*), a sawfly (*top right*), and a true bug (*bottom right*) all feed on leaves of black locust and have similar orange and black patterns.

FIG. 86 This leaf beetle and its larva modify the chemistry of their host tree to produce their own repellent compounds.

must take a close look to determine which is bug, which is beetle, and which is sawfly.

Both the larvae and adults of many leaf beetles that feed on members of the willow family (willows, cottonwoods, aspens) use chemicals from their host tree's leaves to produce their own repellent chemicals, which they store in special glands on the surface of their bodies. Whenever the beetles are disturbed, small droplets of repellent ooze from each gland (fig. 86).

— WALKINGSTICKS, ORDER PHASMIDA OR PHASMATODEA (*phasma* = phantom) (3,000 species; 32 species NA; 15–300 mm).

Walkingsticks pose motionlessly as twigs—usually as green ones in their youth and most often as brown ones as they age (fig. 87; LeafScape, 6). In between poses, walkingsticks move on to find new leaves to chomp. Mimicking twigs seems to be sufficient defense for walkingsticks in our temperate forests. However, in tropical forests where they are most numerous and have more predators, in addition to their camouflage, walkingsticks have come to rely on other highly effective defenses: flashing bright colors to startle intruders, secreting defensive repellents, or adopting threatening poses and swinging legs equipped with sharp, penetrating spines.

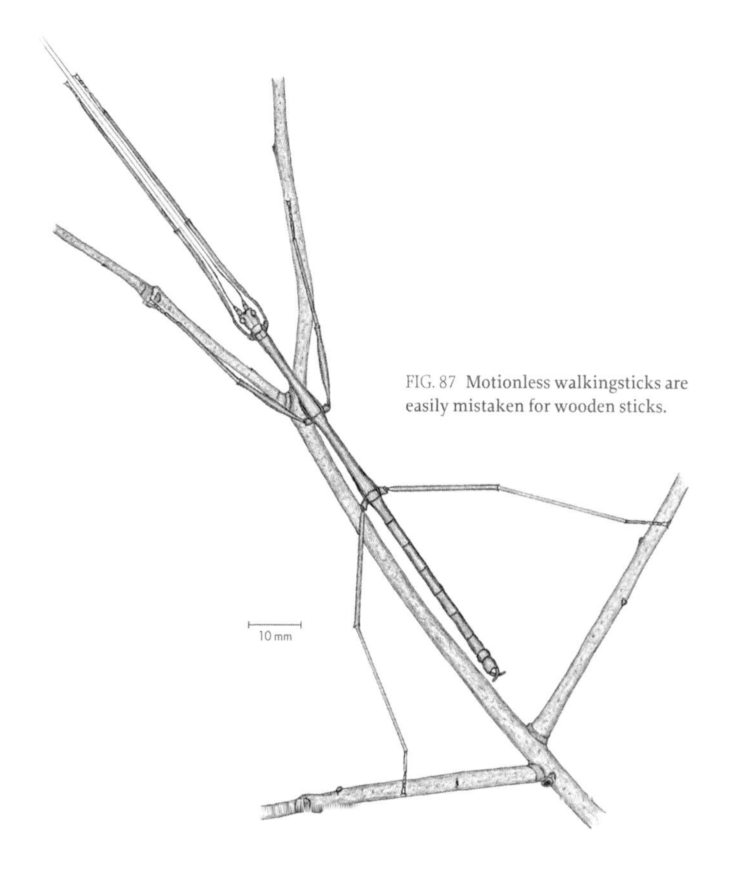

FIG. 87 Motionless walkingsticks are easily mistaken for wooden sticks.

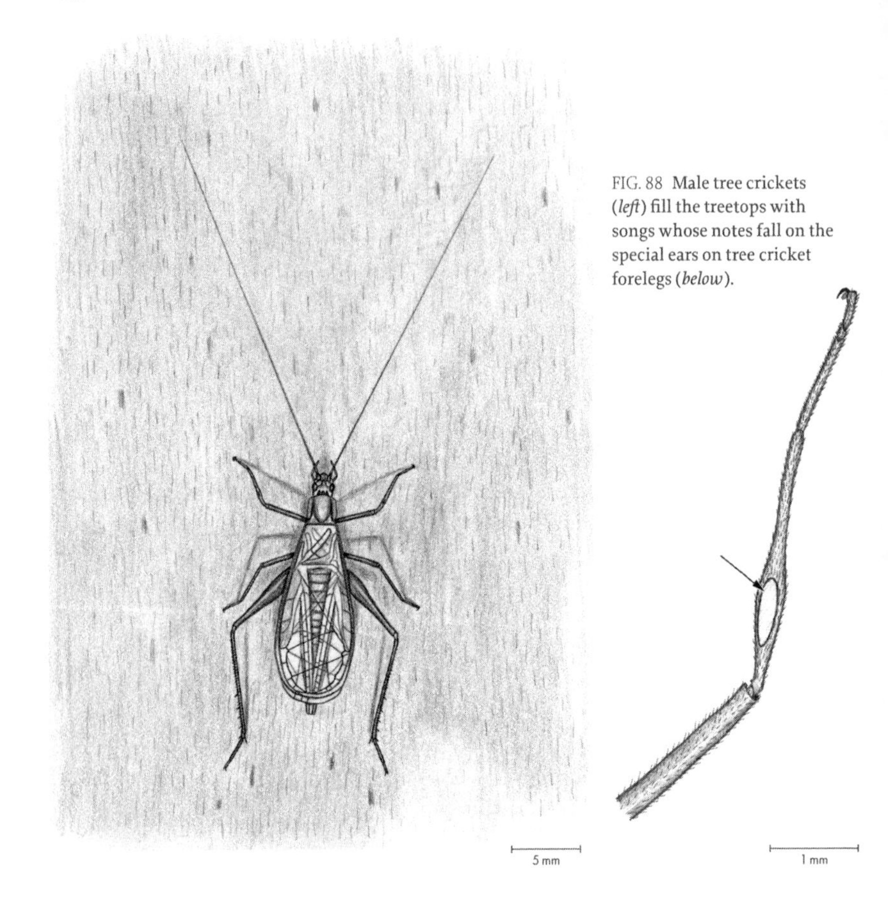

FIG. 88 Male tree crickets (*left*) fill the treetops with songs whose notes fall on the special ears on tree cricket forelegs (*below*).

— TREE CRICKETS, FAMILY GRYLLIDAE (*gryll* = cricket) (900 species; 115 species NA; 3–50 mm), SUBFAMILY OECANTHINAE (*oec* = dwelling; *anthi* = flower) (172 species; 21 species NA; ~ 13 mm).

While the more numerous and more familiar cricket species in the family Gryllidae are ground dwellers, members of this one subfamily spend their entire lives feeding among tree leaves.

You can locate these members of the cricket family, the tree crickets, by listening for their chirps as they perch on bark or on the tops of leaves, chirping and munching most of the day. On hot summer evenings, you can hear the high-pitched chirping of male tree crickets in the treetops. To produce their chirps, they raise their wings and then rub them vigorously together. The tempo of chirping closely follows the rise and fall of the temperature. As summer progresses into autumn and evening temperatures drop, the boisterous chirping of the males fades into feeble rattles. Just as cricket chirping does not arise from a cricket's throat, a cricket does not hear with typical ears. Cricket ears are not found on their heads as expected but just below their knees on their front legs (fig. 88).

Female tree crickets are attracted to the males not only by the romantic sound of their chirping but also by the scent of a pheromone released from a gland that males expose when they raise their wings and commence chirping (LeafScape, 12). The females appreciate males for their odors as well as their chirps. As the female approaches a singing male to sip from the gland that emits the enticing pheromone, she also assumes the position in which she can mate with the male.

— KATYDIDS, FAMILY TETTIGONIIDAE (*tettigonia* = leafhopper) (6,400 species; 245 species NA; 5–130 mm).

Katydids join in the music of the treetops on summer and autumn evenings. The males use their two wings to fiddle out their unmistakable song of "katydid." And every now and then they fiddle out "katy" and "katy didn't" to round out their repertoire. The ears of female katydids, like those of female tree crickets, one on each front leg just below the knee, are tuned to these familiar refrains.

However, not all of the thousands of katydid species sing this same tune. The one and only genus of katydids that fiddles out the iconic lyrics of the katydid song is the aptly named *Pterophylla* (*ptero* = winged; *phylla* = leaf). Not only do the green shades of their wings match the greens of surrounding foliage, but the veins of their wings also mirror the branching patterns of leaf veins and the texture of the leaf surface (fig. 89).

After being courted by serenading males, female katydids go about their maternal affairs in the treetops. They lay their pancake-shaped eggs in neat rows along the twigs. After the leaves fall, the distinctive katydid eggs are often easy to spot (fig. 90).

FIG. 89 The exceptional camouflage of *Pterophylla* katydids reflects the exceptional match of katydid wing veins with tree leaf veins.

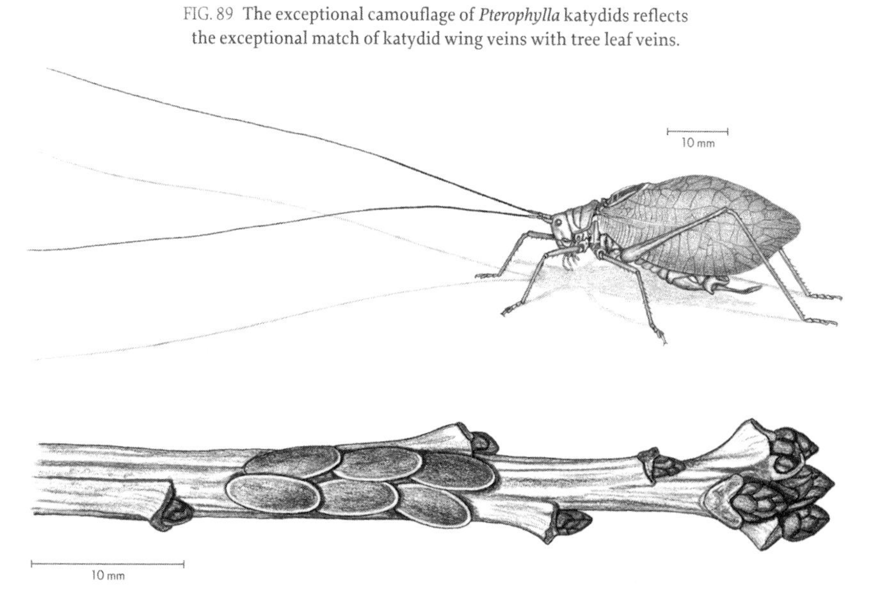

FIG. 90 These katydid eggs will pass the winter on this twig of white ash.

Katydids, tree crickets, and leaf-rolling crickets (fig. 118) have the dubious distinction of being the favorite prey of several sphecid wasps. One group of wasps in the family Sphecidae have a definite preference for tree crickets (LeafScape, 21). These grass-carrying wasps partition hollow twigs or other cavities on a tree with small bundles of grass and then provision these aboveground chambers for their larvae with paralyzed tree crickets.

— GRASS-CARRYING WASPS, FAMILY SPHECIDAE, GENUS *ISODONTIA* (*iso* = equal; *odont* = tooth) (62 species; 6 species NA; 16–20 mm)

Other members of the same family are digger wasps in the genus *Sphex* that, by contrast, provision underground chambers for their larvae.

— DIGGER WASPS, FAMILY SPHECIDAE, GENUS *SPHEX* (*sphex* = wasp) (120 species; 12 species NA; body 10–30 mm).

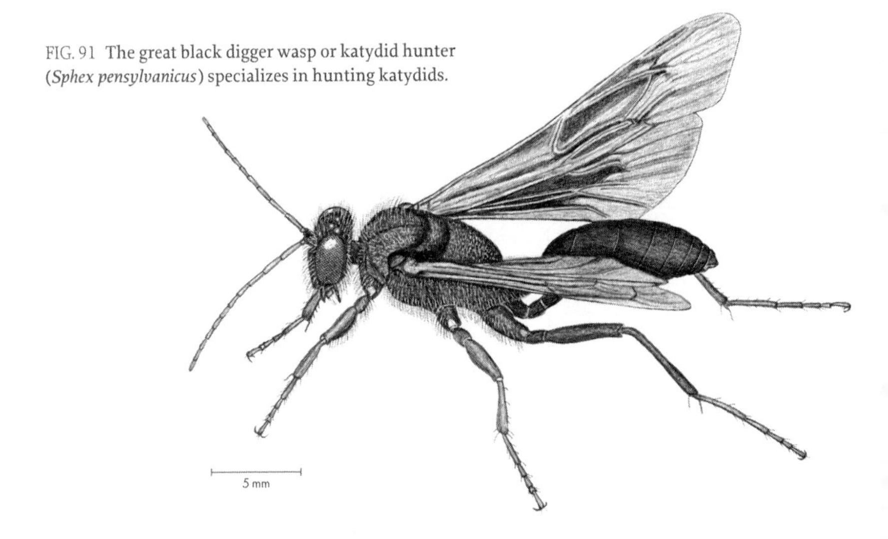

FIG. 91 The great black digger wasp or katydid hunter (*Sphex pensylvanicus*) specializes in hunting katydids.

These handsome wasps are referred to as regal wasps, and the common names bestowed on them are prefaced by the word "great"—great golden digger wasp, great black digger wasp (fig. 91). During the day, these wasps visit flowers for the energy-rich nectar that enables them to maintain their busy activities of excavating underground chambers as nurseries for their larvae and surveying trees for insects with which they will provision these chambers for their carnivorous larvae. These wasps are busy provisioning in early summer well before katydids have attained their full size and have begun singing in the treetops. In each chamber, the wasps usually stock three immature katydids.

How and why did this unwavering tradition of hunting and eating only paralyzed katydids and crickets ever originate? Other digger wasps in this family are equally finicky about their prey choices, and they provision their larval chambers with specific types of arthropod prey—caterpillars, stink bugs, flies, bees, beetles, spiders,

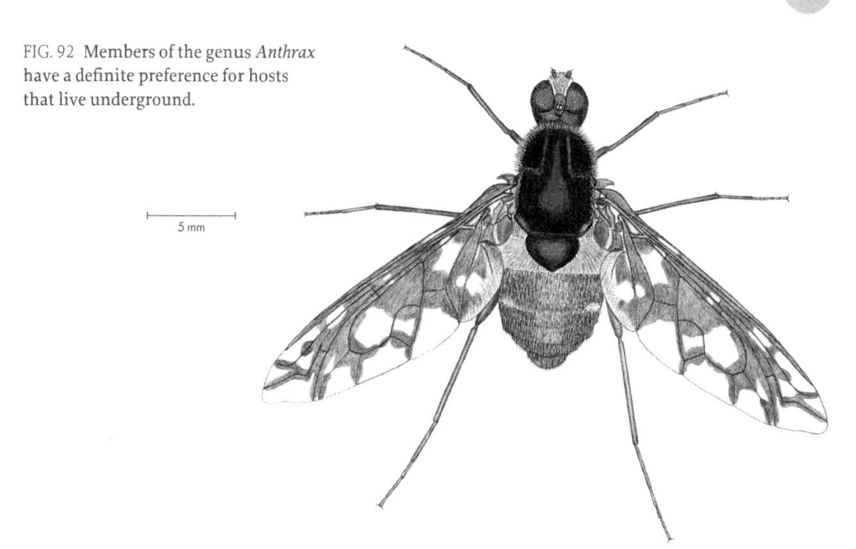

FIG. 92 Members of the genus *Anthrax* have a definite preference for hosts that live underground.

or grasshoppers—but none of these other choices seem acceptable to *Sphex* wasp larvae.

Once the mother wasp has provisioned and sealed the underground chambers so carefully prepared for her larvae, dark female members of the parasitic bee fly family (chapter 5; fig. 271) often appear and hover over the wasp's chambers.

— CHARCOAL BEE FLIES, FAMILY BOMBYLIIDAE, GENUS *ANTHRAX* (*anthrax* = charcoal, coal) (248 species; 37 species NA; 4–20 mm).

Many of these large, dark charcoal bee flies happen to survive as parasites on the larvae of large digger wasps (fig. 92). Other close relatives in this genus choose the larvae of ground beetles and digger bees as their hosts. The mother *Anthrax* simply drops her eggs on the ground and leaves the job of locating growing larval wasps to her newly hatched larvae, which, despite being legless, are endowed with remarkable energy and mobility, invariably locating their wasp hosts beneath the soil surface.

Leaf Trenching

Insects can avoid the unpleasant effects of repellents that some leaves produce and exude. To avoid gumming up their mandibles, caterpillars, katydids, beetles, and sawfly larvae have all been observed disrupting the flow of latex and resin in the secretory canals of many leaves before settling down to a meal on a leaf deprived of its exudate (Dussourd 2017; Dussourd and Denno 1991). Leaves of tropical trees are more frequently endowed with repellent resins and latex than leaves of trees at temperate latitudes.

Insects have resorted to a simple maneuver to avoid the sticky, noxious latex in secretory canals of tree leaves such as mulberry leaves. By chewing a channel across a leaf and severing the secretory canals of leaf veins, insects halt the flow of defensive substances in these canals to the distal end of the leaf. The "trenching" method used by a particular insect on a particular leaf depends on the layout of the secretory canals in that leaf. The secretory material moves primarily along the midribs of leaves in some plants but also along a network of interlacing channels in other leaves. To interrupt the flow of noxious secretions to the distal end of the leaf, the insects chew either through the midrib or across the entire leaf surface traversed by the network of secretory canals. Then they are free to chew on the leaf's far side without encountering secretions that readily gum up their mandibles.

DOMATIA: THE LEAF AS A HOME AND A REFUGE

Each tree also has structures that provide ready-made homes for creatures—holes of all sizes, spaces beneath the bark of a dead limb, crevices in the outer bark, and even depressions or pockets on leaf surfaces called domatia (*doma* = house). Domatia offer refuges into which tiny arthropods can retire for protection from predators or in which equally tiny predators can hide to ambush their prey (Walter and O'Dowd 1992). Domatia are frequent structures on the leaves of more than half the deciduous tree species in North American forests but usually occupy less than 1 percent of the leaf surface area (O'Dowd and Willson 1997).

Leaves of around 2,000 plant species have small depressions or tufts of trichomes (*tricho* = hair) that can be used as refuges in which prey can hide and from which predators can pounce (figs. 93 and 94). Researchers have fabricated artificial domatia by attaching tiny tufts of cotton fibers to the lower surfaces of leaves. The addition of these structures not only attracts small insect predators,

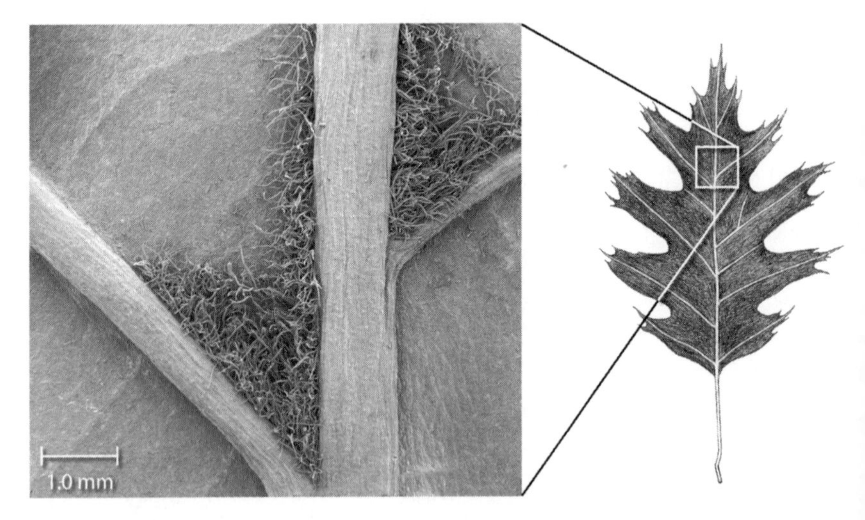

FIG. 93 This image of domatia on an oak leaf was taken with a scanning electron microscope.

FIG. 94 This scanning electron microscope image shows the domatia that lie along the veins of a sycamore leaf.

including predatory mites, thrips, and bugs such as the minute pirate bug (fig. 95), but also diminishes populations of plant-feeding spider mites (chapter 3; fig. 162). The benefits to the host plants from their partnership with these predators show up in plants with more domatia producing more fruits and seeds (Agrawal and Karban 1997).

— FLOWER BUGS, MINUTE PIRATE BUGS, FAMILY ANTHOCORIDAE (*anthus* = flower; *coris* = bug) (600 species; 90 species NA; 1.5–5 mm).

These scrappy little bugs have become known as reputable biocontrol agents. They are almost an order of magnitude smaller than their predatory relatives the assassin bugs and stink bugs (fig. 28) but are no less formidable as predators. One well-known member of this family has been given the fearsome name *Orius insidiosus*, the insidious flower bug or minute pirate bug (fig. 95). *Orius* bugs might not vanquish hefty caterpillars or beetle larvae, but they can readily annihilate aphids, thrips, and spider mites that inhabit the lower surface of a leaf. They can surmount the barriers presented by the best-fortified domatia of tree leaves.

Domatia are just the right size for mites that live on leaves. Mites such as spider mites (fig. 162; LeafScape, 32) can feed on leaves, and other, faster-moving predatory mites feed on spider mites. Most oribatid mites inhabit the soil and leaf litter under trees where they go about their business of decomposing dead leaves and feeding on fungi (chapter 6; fig. 293); however, a few exceptional oribatid mites (fig. 96) live aboveground on surfaces of living leaves where they probably find fungi on which to feed (Walter and Behan-Pelletier 1999) and domatia in which to retreat.

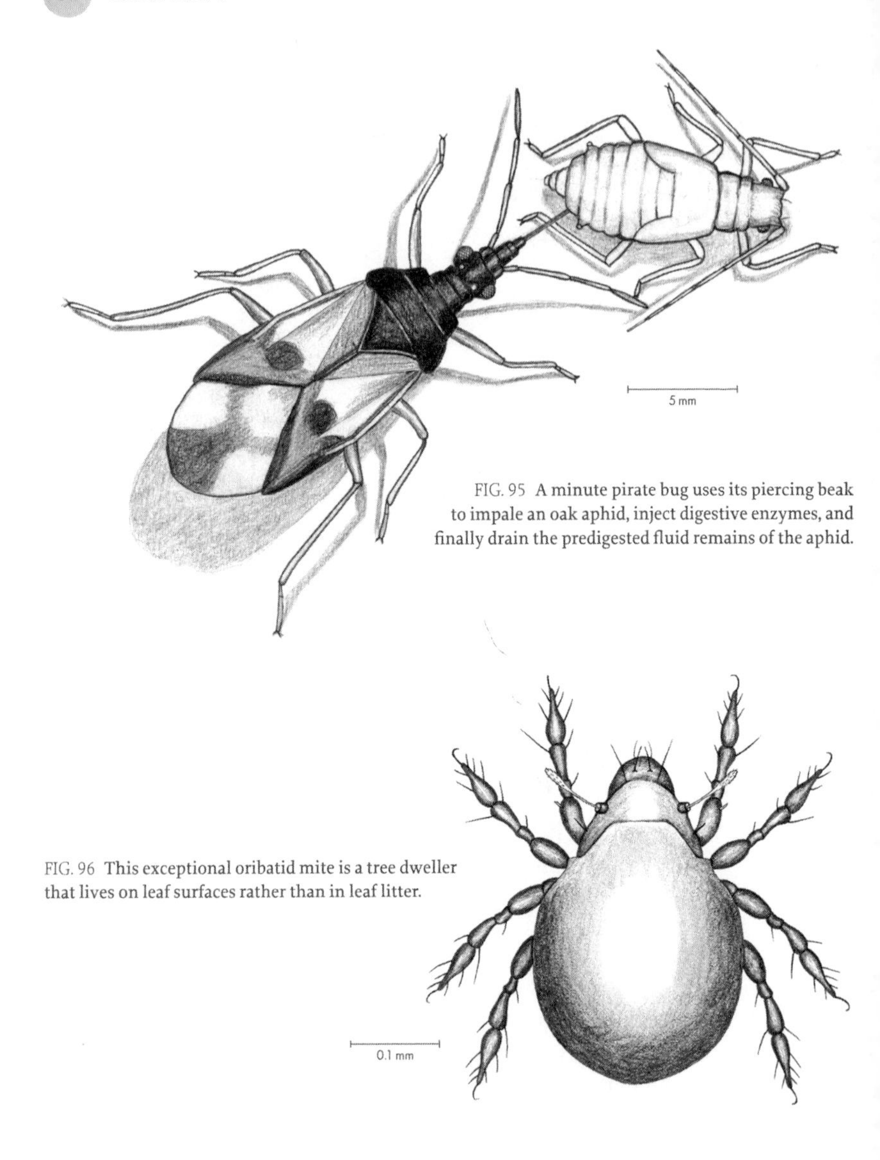

FIG. 95 A minute pirate bug uses its piercing beak to impale an oak aphid, inject digestive enzymes, and finally drain the predigested fluid remains of the aphid.

FIG. 96 This exceptional oribatid mite is a tree dweller that lives on leaf surfaces rather than in leaf litter.

MANIPULATING LEAF FORM AND PHYSIOLOGY: GALL MAKING, LEAF ROLLING, LEAF MINING

Leaves and buds of a tree can be converted to homes and dining areas with some modest remodeling. With jaws and carefully placed silk strands, a two-dimensional leaf can be converted into a three-dimensional leaf roll. And a flat leaf has just enough space between its upper epidermal layer and lower epidermal layer to shelter and sustain the thousands of different insect species—leaf miners—that are small enough and flat enough to nestle inside a leaf's narrowest dimension. Leaf feeders manipulate leaves in a variety of ways to reduce their visibility, maneuvering leaves to construct chambers in which they can feed completely unobserved. Like birds that maneuver plant materials in specific ways during their construction of nests, these insects mold leaves and buds into forms that are specific for their species. The nests these insects create shelter them and their provisions of food during the vulnerable days before they grow wings and are capable of rapid escape. Gall makers go even further than leaf rollers and leaf miners in their manipulation of plant tissues.

Gall Makers

Gall makers do not mechanically manipulate tree tissues to construct their feeding chambers but secrete substances that induce the cells of the tree to form the chambers for them. A gall is made entirely of tissue contributed by the tree as a cooperative effort between the tree and one or more insects or mites. Thousands of different galls are formed by thousands of different mites and insects. Exactly how galls form is a mystery. Simply the maternal act of egg laying or the activity of larval feeding induces the plant to form an elaborate chamber around the gall maker or gall makers. These simple acts manipulate the expression of hormones and genes of the tree host in ways the tree normally never does. A gall-forming insect induces plant tissues to form structures that appear only in response to the presence of that specific insect (Giron et al. 2016; Tooker and DeMoraes 2011).

Insects feeding on a plant prompt it to defend itself with the release of jasmonic acid, cytokinin hormones, and secondary metabolites (Brütting et al. 2017; Bari and Jones 2009). As part of their counterdefense, however, insects and their symbiotic microbes can manipulate levels of plant hormones to mold the tree's physiology and defense responses for their own purposes. These companions of trees can produce the same plant hormones that influence the growth and physiology of adjacent tree cells (Andreas et al. 2020; Takei et al. 2015). Insects can also transfer plant hormones of their own making that not only prolong the lives of the leaves on which they feed but also increase the nourishment these leaves provide (Brütting et al. 2018; Zhang et al. 2017). Insects that are confined to feeding in leaf mines or plant galls stimulate the tree cells that surround them to grow in number and to increase their nutrient content. The hormones produced by insects or by their microbial symbionts when applied on specific tree cells at specific times in the development of a tree's tissues can evoke novel features of leaves, stems, and roots not expressed during a tree's normal development. Tree cells that are exposed to these extraneous hormones reveal their latent abilities to form the unusual, often elaborate and ornate structures of tree galls. The countless malleable cells in tree buds offer many possibilities for the creative stimuli of gall makers (fig. 97).

Certain arthropods have discovered that they can manipulate tree hormones to unmask the hidden talents of a tree's cells and tissues. The novel arrangement of cells in insect galls creates plant tissues that can range from plain to spectacular on their surfaces; in addition, all galls have central chambers formed by the tree

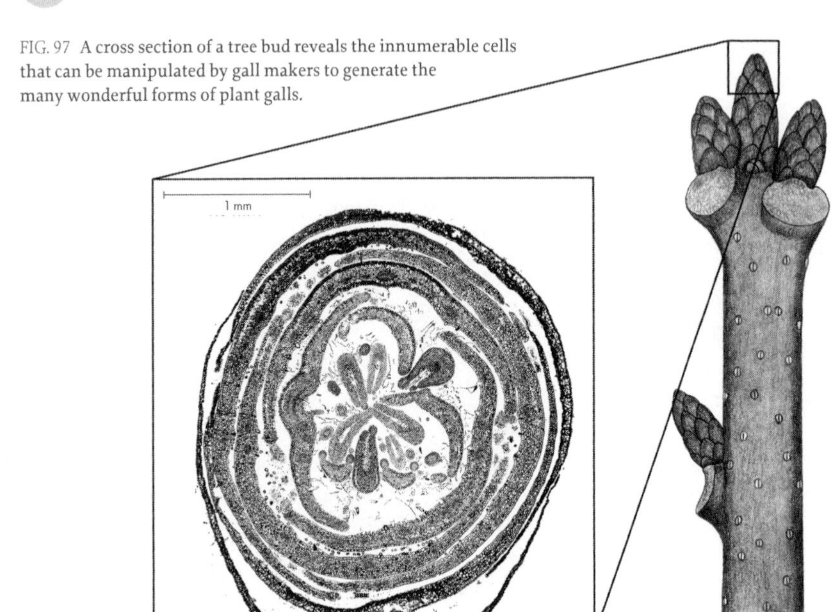

FIG. 97 A cross section of a tree bud reveals the innumerable cells that can be manipulated by gall makers to generate the many wonderful forms of plant galls.

cells in which one or more gall makers spend all or most of their lives. Galls provide both refuges and nourishment for their makers.

The stimulus provided by a gall maker that causes a tree to produce a gall must be very specific, since each gall is unique to a particular insect or mite, even those that start with the same raw materials. Each insect species, even closely related ones, induces a plant structure unmistakably its own. Whatever inducing agent a gall maker uses brings forth latent morphogenetic (*morpho* = form; *genesis* = creation of) features never expressed by the tree alone, demonstrating that only a fraction of what a tree is capable of expressing is actually expressed unless it is prompted to reveal more of its vast morphogenetic potential. Besides being unique, some of these galls are very ornate and lovely, both outside and inside (fig. 98; LeafScape, 10).

The plant galls themselves are better known than their makers. The insects and mites that induce galls are tiny, and related species of these gall makers are so similar in form that members of different species are more readily identified by the forms of their distinctive galls than by their often nondescript appearances.

Ephraim Porter Felt, the state entomologist of New York, published the most comprehensive work on the subject in 1940, titled *Plant Galls and Gall Makers.* Felt's book includes many galls on shrubs and herbaceous plants in addition to those that adorn trees. At the time of the book's publication, the number of known species of gall makers worldwide was 7,930. This figure certainly underrepresents the actual number of species because Australia and Africa were not even included in this count, and new species have certainly been added to this initial list. For

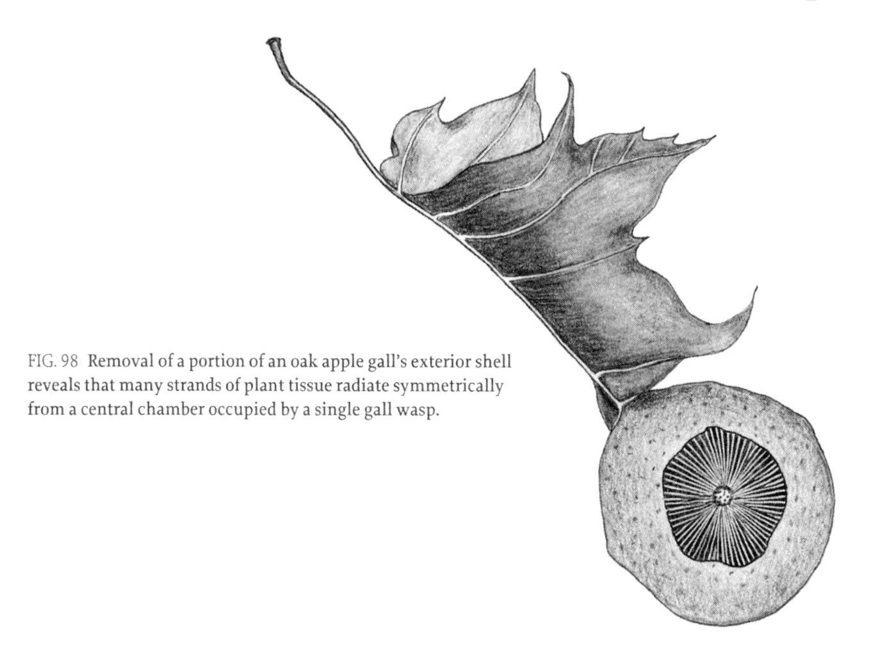

FIG. 98 Removal of a portion of an oak apple gall's exterior shell reveals that many strands of plant tissue radiate symmetrically from a central chamber occupied by a single gall wasp.

North America alone, Felt listed 1,440 species of insect and mite gall makers. The list was dominated by gall wasps numbering 805, gall midges 687, sawflies 200, and eriophyid mites 181. Pale by comparison are the numbers of species of beetles (12), caterpillars (17), and aphids along with a few of their close relatives such as psyllids (80) that are also gall dwellers (Price et al. 1993).

— GALL WASPS, SUPERFAMILY CYNIPOIDEA, FAMILY CYNIPIDAE (*cynip* = kind of insect) (1,300 species, 800 species NA; 1–8 mm).

The estimated 805 species of gall wasps in North America form galls on members of six plant families. Ninety percent of these species, however, live on oak trees (LeafScape, 10, 18, 31). Although most of these oak galls are found on leaves, all parts of the tree are chosen: roots (41), buds (45), twigs and branches (175), flowers (21), acorns (34), and leaves (275). A single larva of these small and stingless wasps usually inhabits the center of each gall, but hundreds of wasp larvae may occupy some galls.

One species of gall wasp is responsible for an amazingly complex structure formed on twigs of white oak trees each spring called the wool-sower gall (figs. 99 and 100). This gall looks like a ball of wool about an inch in diameter and is home to several hundred tiny wasps, each of which develops in one of several hundred pink chambers suspended in a web of gossamer strands. The chambers are evenly spaced throughout the gall and are joined by delicate strands that radiate from each one (fig. 100). The mother of the wasps that emerge from the wool-sower gall laid her eggs on a bud of a white oak tree weeks earlier, perhaps while snow still covered the ground and certainly before new oak leaves had begun to appear.

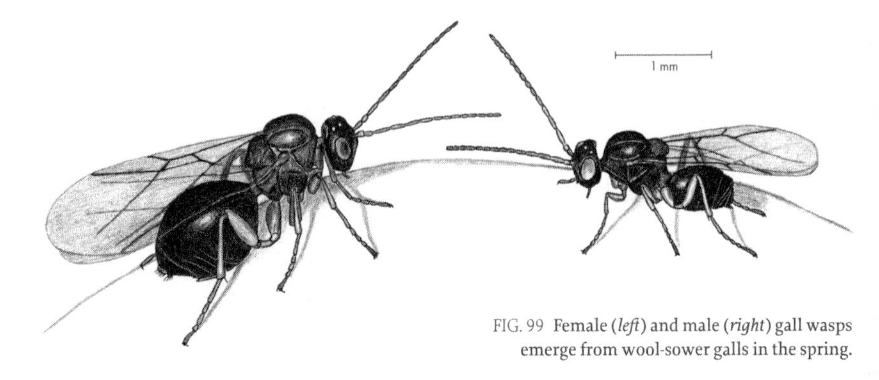

FIG. 99 Female (*left*) and male (*right*) gall wasps emerge from wool-sower galls in the spring.

Since there were no male wasps with whom to mate, her sons and daughters were conceived without a father. Unlike their mother, these sons and daughters mate but produce only daughters.

As you are probably beginning to realize, gall wasps have some peculiar family histories. Each year there are two generations of wasps—a spring generation and an overwintering generation—each forming its own unique gall. Each species of gall wasp produces two different and distinctive galls. All members of the

FIG. 100 Several hundred gall wasps can develop in each ornate wool-sower gall on white oak.

overwintering generation are females, but their offspring are both males and females immaculately conceived, produced without any male parents. After these males and females mate, the cycle begins anew, and only daughters hatch from the fertilized eggs they lay. The daughters look more like their grandmothers than they do their own parents. They also overwinter in galls like those of their grandmothers and not like those of their parents. A year in the life of a gall wasp family is often so complex that the complete family histories of most gall wasps are still unknown. No one knows, to my knowledge, where wasps from the wool-sower gall lay their eggs and in what galls their daughters and mothers live. The wasp that makes the oak hedgehog gall is one of the few gall wasps whose entire life cycle is known (fig. 101).

Because the two generations of wasps in each life cycle look different and live in different galls, many gall wasps have probably been described as separate species when they are really the same species. Learning how the lives of gall wasps are connected with the lives of oak trees is a job for a very patient and careful observer.

— TORYMID WASPS, SUPERFAMILY CHALCIDOIDEA, FAMILY TORYMIDAE (960 species; 600 species NA; 2–13 mm).

Gall wasps have few large predators except for woodpeckers that sometimes try to peck their way into the galls. However, the most worrisome enemies of gall insects are small chalcidoid parasitic wasps, such as torymid wasps, that lay their eggs on the gall maker. Torymid wasps are about the same size as gall wasps, but instead of being shiny black or brown like gall wasps, torymid wasps are most often a shiny, lustrous green or blue (fig. 102; LeafScape, 30). The female torymid

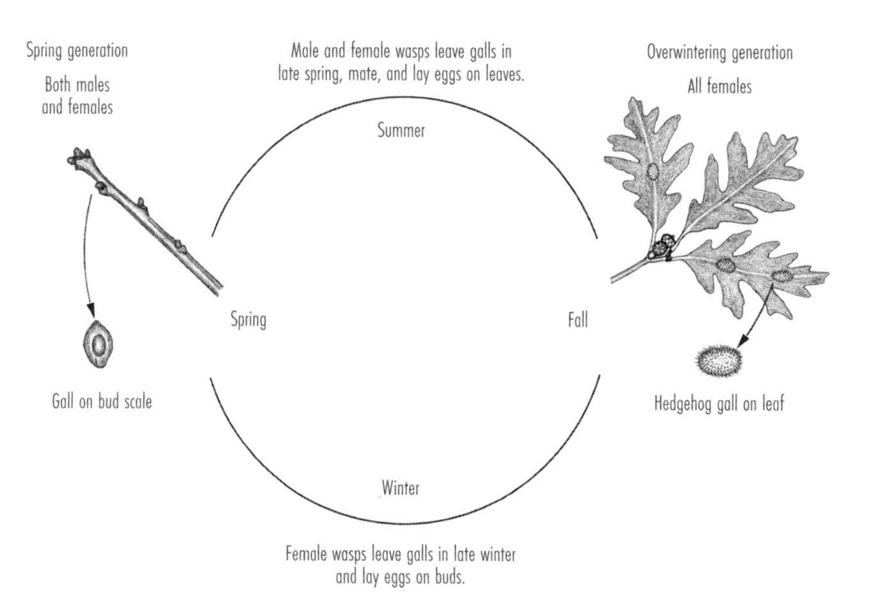

FIG. 101 Tiny gall wasps have complex life cycles with alternating generations.

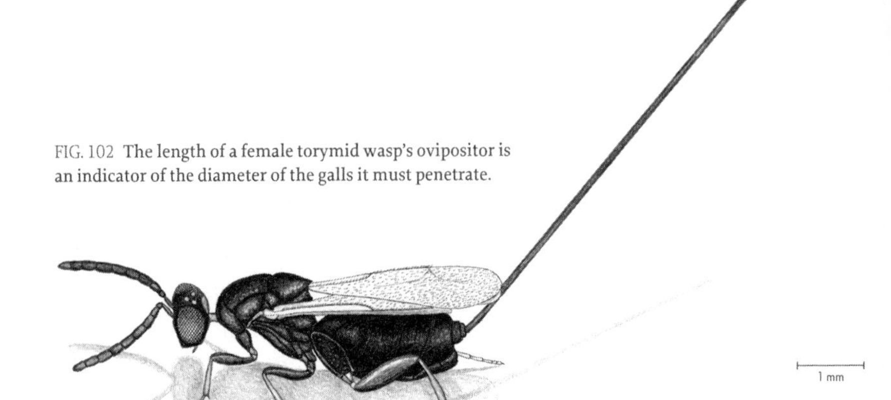

FIG. 102 The length of a female torymid wasp's ovipositor is an indicator of the diameter of the galls it must penetrate.

has a saber-like ovipositor that is often longer than her body. During egg laying this ovipositor can reach even the innermost chambers of a gall. The many fine bristles that cover the ovipositor are sensitive to touch and help guide the ovipositor with great precision toward the larva or larvae that lie unseen within the gall. The eggs of the parasitic torymid wasp hatch into larvae that devour their hosts and take over their galls.

What hatches from a gall can be the gall maker, the parasitic wasp that usurped the gall maker's home, or even another small wasp called an inquiline (*inquilinus* = tenant)—a harmless squatter that may move into the spacious, unoccupied quarters provided by the luxuriant tissue of the gall.

— GALL SAWFLIES, FAMILY TENTHREDINIDAE (*tenthredin* = type of wasp), SUBFAMILY NEMATINAE (*nemato* = thread, apparently referring to their long, threadlike antennae) (1,250 species; 430 species NA; 3–15 mm).

Gall-forming sawflies have a particular fondness for willow trees and are members of the same family as the leaf-mining sawflies (fig. 131). Over 200 species of sawflies in this family induce galls on willows (Nyman et al. 2000), and almost all reside in the Northern Hemisphere. These willow galls take many forms and demonstrate how malleable willow tissues can be under the influence of gall makers: galls on leaf blades, galls on petioles, galls on stems (fig. 103), galls on buds, galls that roll leaves, galls that fold leaves.

— GALL MIDGES, FAMILY CECIDOMYIIDAE (*cecido* = gall; *myia* = fly) (6,000 species described; 1,100 species NA; 1–3 mm).

Cynipid wasps have a definite preference for oak trees, and gall sawflies are certainly partial to willow trees. Galls of gall midges, by contrast, are formed on herbaceous and woody plants of over 70 plant families. Their favorite trees are members of the willow (fig. 103), oak (LeafScape, 14, 37), elm, linden, maple, walnut, magnolia, cypress, and pine families. These midges have developed a symbiotic

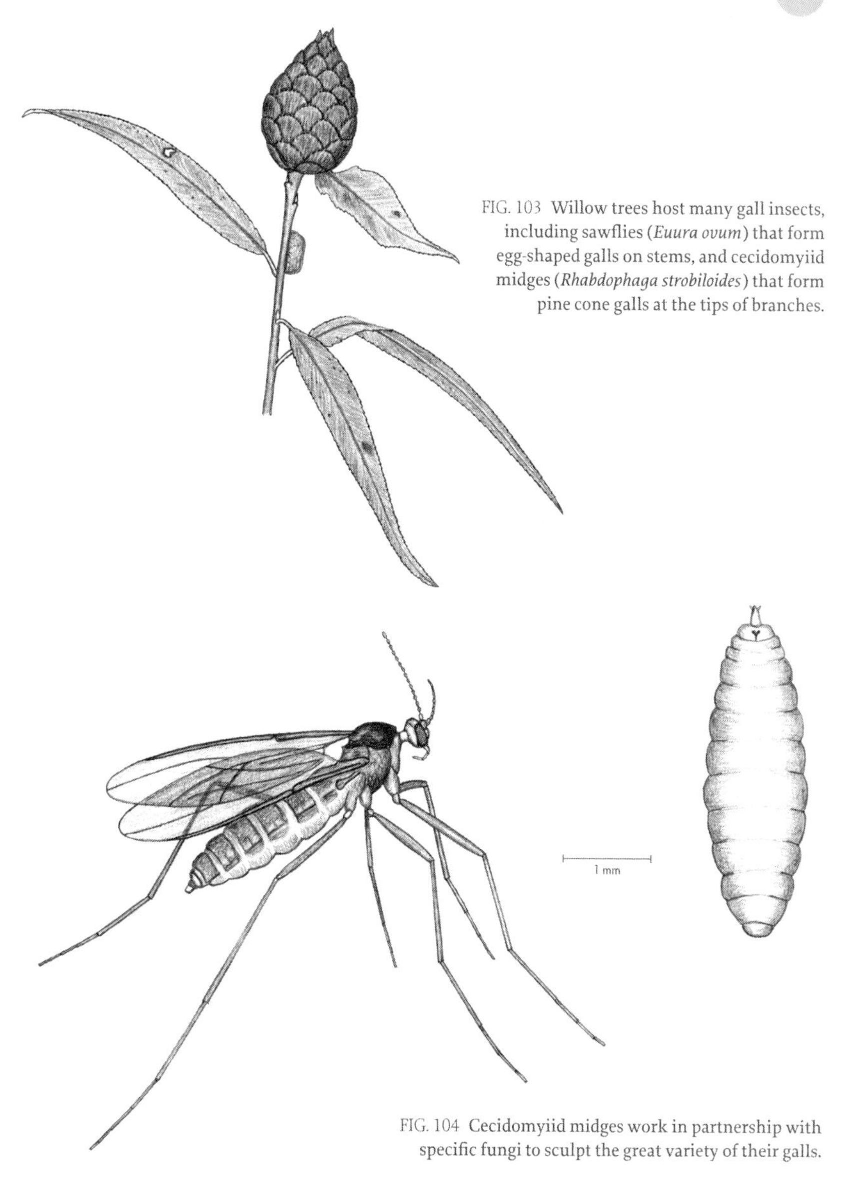

FIG. 103 Willow trees host many gall insects, including sawflies (*Euura ovum*) that form egg-shaped galls on stems, and cecidomyiid midges (*Rhabdophaga strobiloides*) that form pine cone galls at the tips of branches.

FIG. 104 Cecidomyiid midges work in partnership with specific fungi to sculpt the great variety of their galls.

relationship with fungi that has been deemed responsible for driving the great diversification of these tiny flies and their extensive colonization of different plants. A 2015 study conducted in Canada identified an estimated 16,000 species. Based on this great diversity of gall midges in Canada alone, the researchers estimated that 1.8 million species of gall midges inhabit the entire planet. In number of species, gall midges rival—and may even surpass—the highly diverse beetles and parasitic wasps. Thanks to their fruitful partnerships with fungi, gall midges have been able to establish homes on more tree species than any other group of gall makers (Hebert et al. 2016; Joy 2012; Rohfritsch 2008; Borkent and Bissett 1985) (fig. 104).

Such a diverse family of insects as the gall midges has attracted the attention of a correspondingly diverse family of even smaller parasitic wasps, the Platygastridae (fig. 335). To ensure that it obtains the most benefit possible from its small larval host, a platygastrid wasp larva postpones its own development until the host attains its full size.

While the larvae of most flies, gnats, and midges are legless and nondescript, the larvae of gall midges have certain features that stand out. They are spindle shaped and are often an intense orange color. Against this bright background, a microscopic dark, fork-shaped plate referred to as a spatula stands out prominently on the larva's ventral surface, immediately behind its cone-shaped head (fig. 104).

Cecidomyiid larvae have adopted a great range of lifestyles, from gall making and leaf mining to being fungivores under bark or in leaf litter. Some are even predators or parasitoids of aphids, thrips, lace bugs, and other small arthropods. Some of the eccentric members of this family of gall makers are in the genus *Endopsylla* (*endo* = inside; *psylla* = flea) and have specialized as parasitoids of another family of well-known gall makers, discussed in the following section—the psyllids. With a body plan for the family that is simple and conserved, the larvae of gall midges have nevertheless diversified in many occupations.

— APHIDS, FAMILY APHIDIDAE (5,000 species; 1,350 species NA; 1–6 mm), GALL APHIDS (at least 500 species; about 80 species NA).

— JUMPING PLANT LICE, FAMILY PSYLLIDAE (*psylla* = flea) (800 species; 134 species NA; 2–4 mm).

Aphids and a few of their relatives the psyllids often take up residence in galls on tree leaves. As tree tissue forms a communal gall around aphids, psyllids, or mites, a narrow passageway always remains from the interior chambers to the outside. None of these gall makers happen to have the mouthparts with which to chew their way out of galls. Gall wasps, gall sawflies, and gall midges either chew an escape route to the gall's surface before they pupate, or they use sufficiently strong adult mandibles to excavate an exit. Having a gall with a built-in escape route at one end ensures that the delicate aphid and psyllid inhabitants have a struggle-free departure from their summer homes on tree leaves.

Inside their galls where they suck sap from phloem cells, aphids and young psyllids are sheltered from the sun, rain, and predators. The hackberry nipple galls of psyllids and the cottonwood petiole galls of aphids can be extremely numerous and easily spotted (figs. 105 and 106). Another singular and conspicuous aphid gall is the cockscomb gall on elm leaves. Each of these galls on the top surface of a green elm leaf is home to many aphids, which escape through a narrow slit on the leaf's lower surface. Only in late summer or early autumn do adult aphids and adult psyllids leave their galls to spend the winter in some sheltered spot under the tree. The adult psyllids that emerge from the hackberry galls in vast numbers look remarkably like cicadas that have shrunk in size almost tenfold.

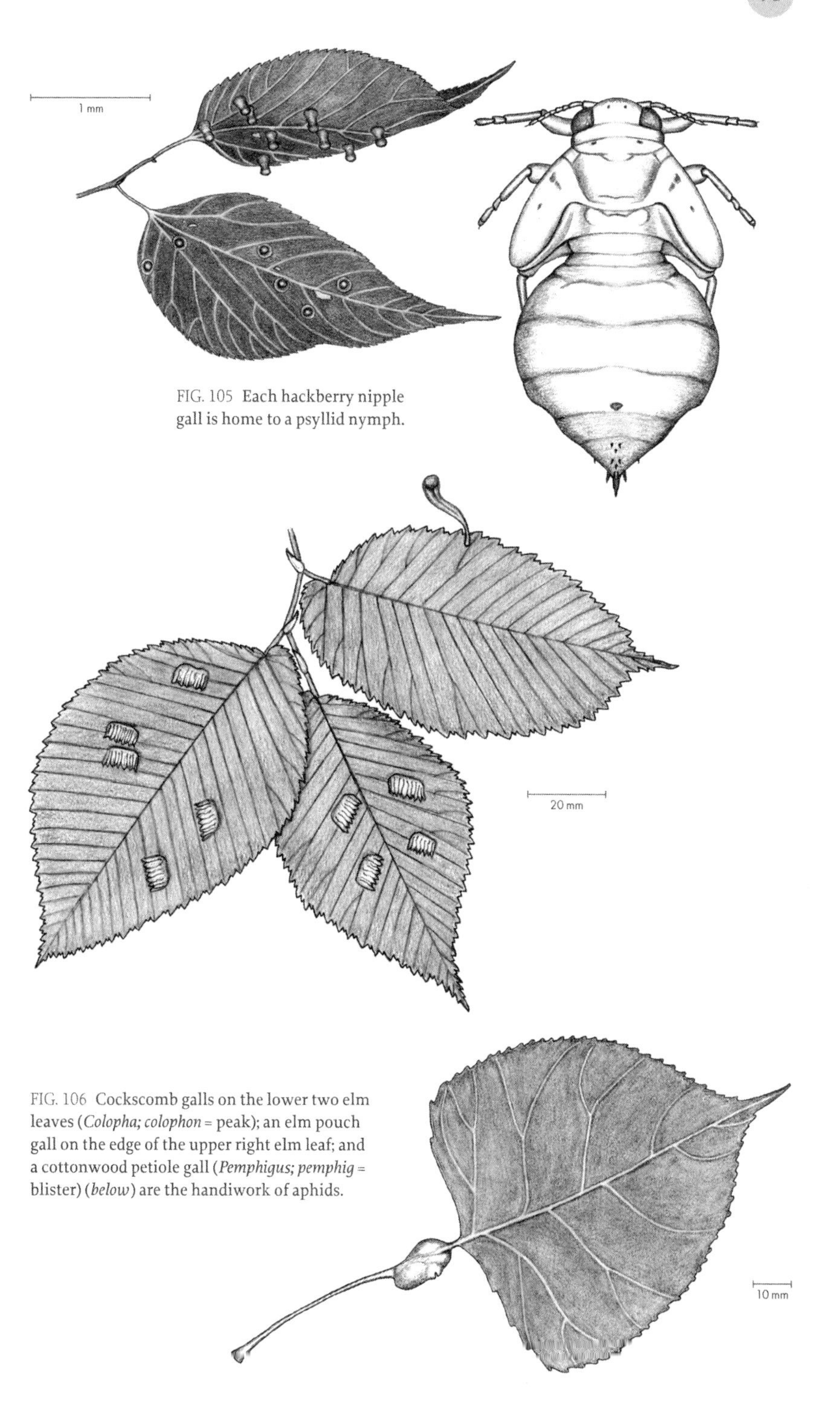

FIG. 105 Each hackberry nipple gall is home to a psyllid nymph.

FIG. 106 Cockscomb galls on the lower two elm leaves (*Colopha; colophon* = peak); an elm pouch gall on the edge of the upper right elm leaf; and a cottonwood petiole gall (*Pemphigus; pemphig* = blister) (*below*) are the handiwork of aphids.

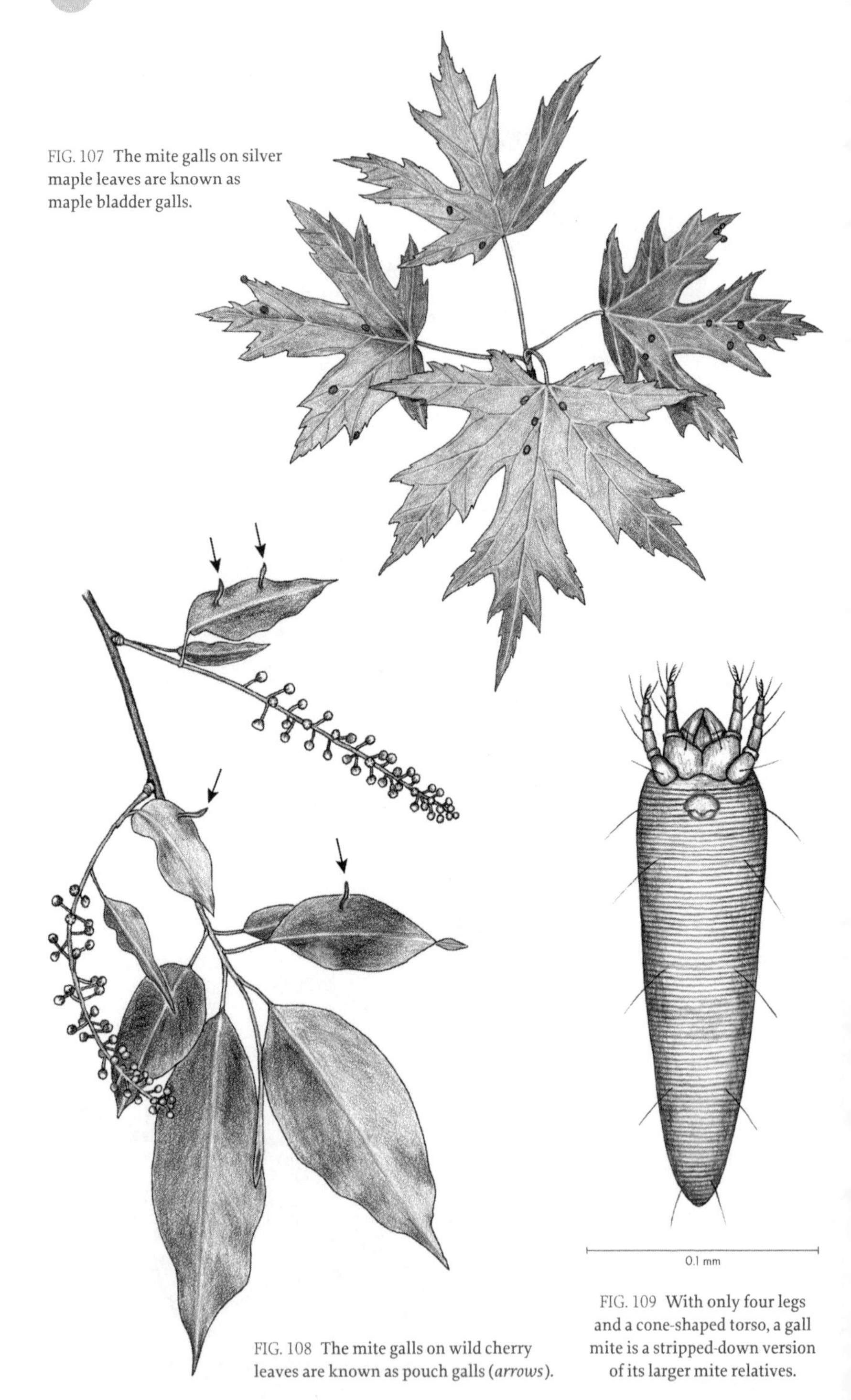

FIG. 107 The mite galls on silver maple leaves are known as maple bladder galls.

FIG. 108 The mite galls on wild cherry leaves are known as pouch galls (*arrows*).

FIG. 109 With only four legs and a cone-shaped torso, a gall mite is a stripped-down version of its larger mite relatives.

Mites can live in just about any environment, so it should not be surprising that a good number of mite species have taken up residence in plant galls.

— GALL MITES, CLASS ARACHNIDA, SUBCLASS ACARI, FAMILY ERIOPHYIDAE (*erio* = wool; *phya* = grow, apparently referring to the woolly interior texture of each gall) (3,400 species; 182 species NA; 0.15–0.35 mm).

The easiest gall mites to find are probably the ones that live in pouch-like galls on wild cherry and maple leaves. Each gall may have a family of 10 or more mites crowded inside, and a single leaf can often be covered with over 100 galls (figs. 107 and 108). The number of gall mites on a single tree can easily add up to several million. Each mite is smaller than the period at the end of this sentence. These gall-dwelling mites are usually so lightweight that breezes and winds carry them from tree to tree. A microscope with exceptional optics reveals that these mites have four short legs and look like long cones with short beaks at their front ends (fig. 109).

Leaf Rollers

Leaf rollers wrap themselves in leaf tissue, creating a shelter where they lie hidden from the eyes of predators and parasitoids. But most leaf rollers take full advantage of what a leaf roll has to offer, creating both a shelter and a leaf-feeding trough for themselves. A leaf roll can consist of several layers of leaf tissue, and the larva usually feeds on the innermost layers. Some caterpillars feed on the ends of the rolled layers, and others extend their bodies from the shelter of their rolls to feed on adjacent leaf tissue. The predominant builders and occupants of edible leaf rolls are caterpillars. Moth caterpillars are certainly the master leaf rollers, but a few butterfly caterpillars have also taken up this occupation.

— SWALLOWTAIL BUTTERFLIES, FAMILY PAPILIONIDAE (*papilio* = butterfly) (560 species; 33 species NA; mature larvae 30–50 mm; wingspan 90–165 mm).

Among the simplest constructions are the curled leaves on which swallowtail caterpillars settle down on mats of silk, remaining motionless and as inconspicuous as possible while songbirds scour the tree for insect meals. The colorful larvae of swallowtail butterflies that feed on sassafras or tulip tree leaves have adopted an incipient form of leaf rolling by resting during the day on mats of silk spread over the upper surfaces of leaves (fig. 110). As the caterpillar spins out its silk mat along the midrib of a sassafras leaf, the silk strands shrink and curl the upper surface of the leaf inward. The rolled leaf cradles the caterpillar so that it is now visible only from above. Whoever looks down on the caterpillar during the day as it pauses from feeding, however, encounters two eyespots on the top of its thorax that peer back threateningly. The caterpillar's real eyes are tiny and located at its far anterior end, while these eyespots on its expanded thorax can easily be mistaken for the eyes of a green tree snake (LeafScape, 2).

— SKIPPER BUTTERFLIES, FAMILY HESPERIIDAE (*hesperi* = evening) (3,500 species; 275 species NA; larvae 20–50 mm; wingspan 25–50 mm).

With their broad heads and narrow first thoracic segments, skipper caterpillars are best known for their conspicuous necks. Most build shelters with leaves of their

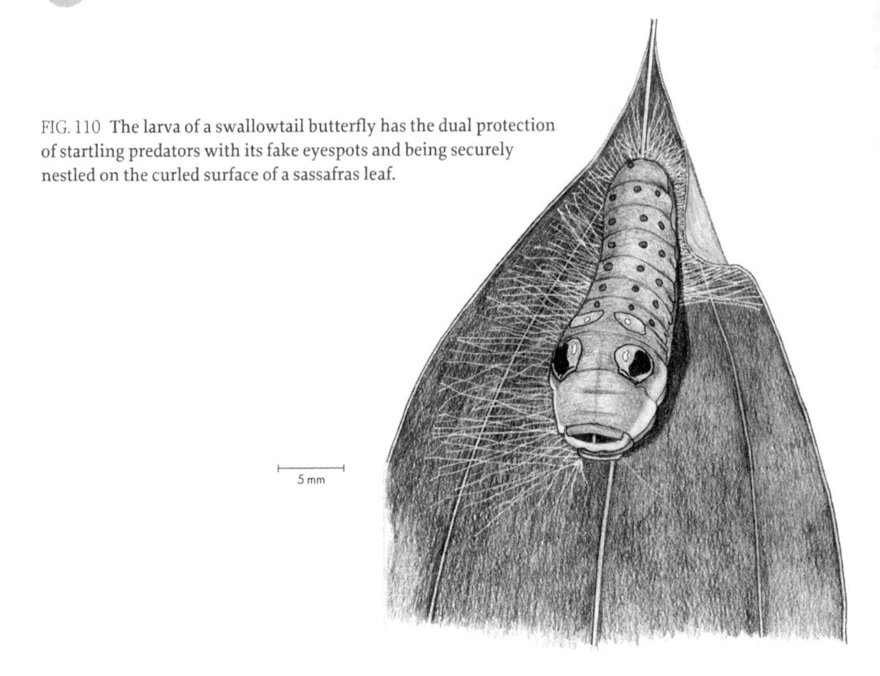

FIG. 110 The larva of a swallowtail butterfly has the dual protection of startling predators with its fake eyespots and being securely nestled on the curled surface of a sassafras leaf.

FIG. 111 The "red-neck" larva of the common silver-spotted skipper shelters within a leaflet of black locust.

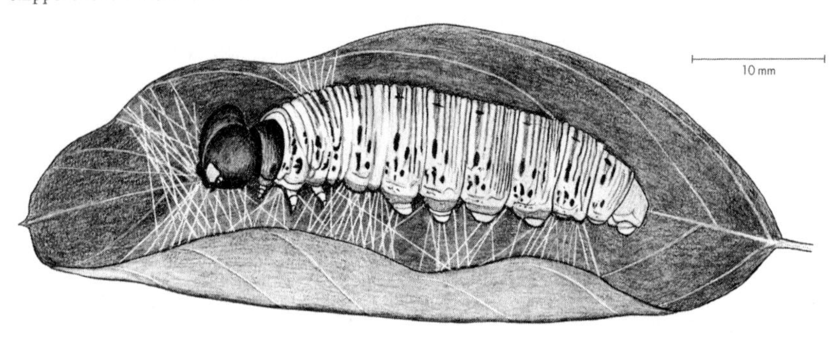

food plants by folding and securing the edges of the leaves around themselves with silk strands (fig. 111). Folded leaves provide secure shelters for growing caterpillars during the summer. In autumn, the leaf shelters fall to the forest floor and settle in the leaf litter, where the mature larvae and pupae within are buffered from the vagaries of weather and predators until time for their metamorphosis in the spring.

Another example of an incipient leaf roll is the shelter adopted by the larva of the pine tube moth, a tortricid moth named *Argyrotaenia pinatubana* (*argyro* = silver; *taenia* = band; *pina* = pine; *tub* = tube; *-ana* = belonging to). The caterpillar fastens a cluster of white pine needles into a tube from which it can reach out and feed on the ends of the needles that make up its protective shelter (fig. 112).

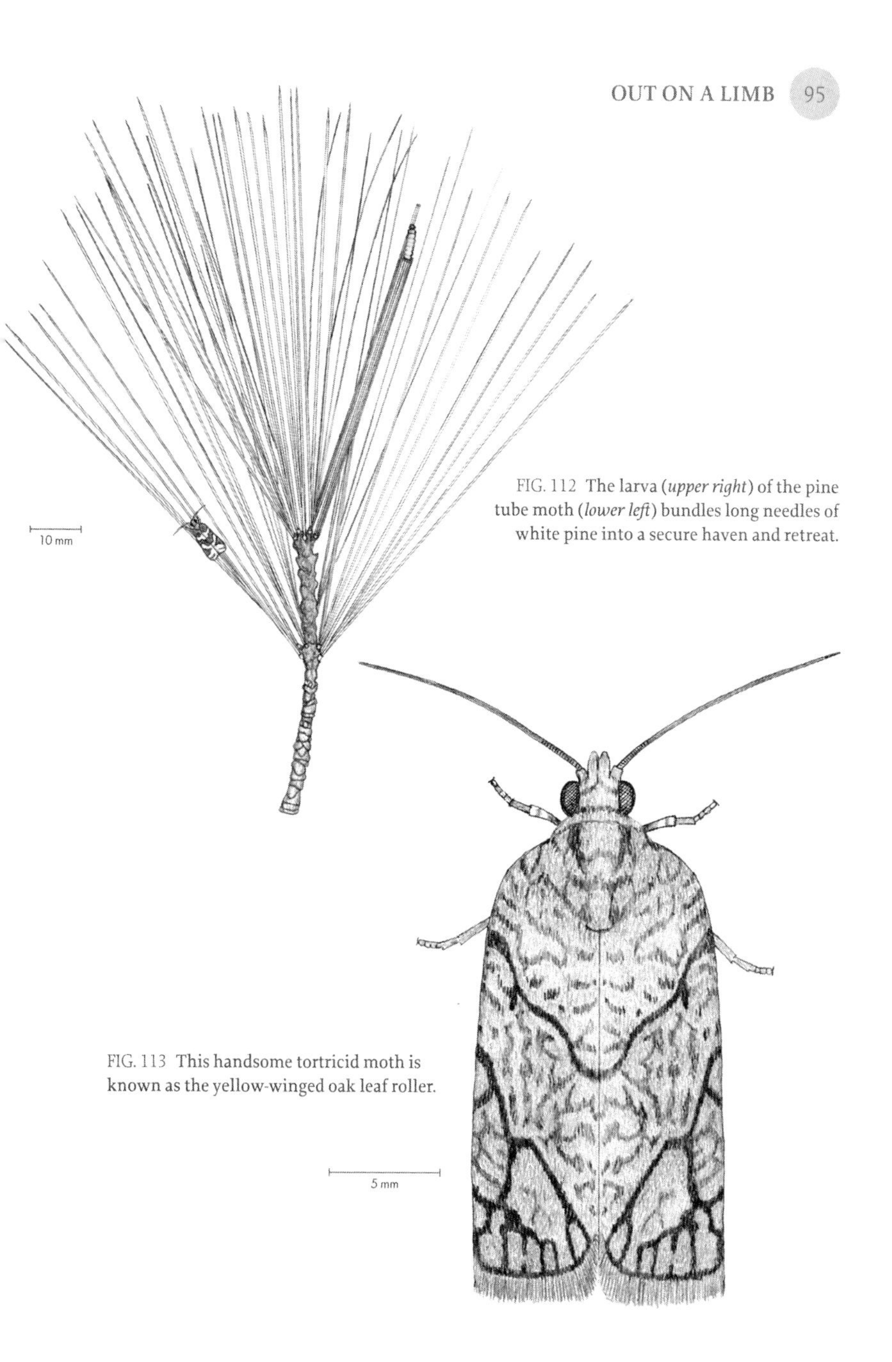

FIG. 112 The larva (*upper right*) of the pine tube moth (*lower left*) bundles long needles of white pine into a secure haven and retreat.

FIG. 113 This handsome tortricid moth is known as the yellow-winged oak leaf roller.

— **LEAF ROLLER MOTHS, FAMILY TORTRICIDAE** (*tortric* = twisted) (10,350 species; 1,400 species NA; larvae 8–25 mm; wingspan 10–33 mm).

Most members of the family of moths to which the pine tube moth belongs are leaf rollers. At rest, the wings of these moths lie folded against the body, rounded at the shoulders and almost squared at the ends, giving these moths bell-shaped silhouettes (fig. 113). The wing patterns of some moths are stunning and lovely, some mimic bird droppings, and some are just plain and dull. Some species such as the codling moth of apple trees are quite infamous pests, but most tortricid moths lead quiet lives of little notoriety (Wilson 1972).

Members of one genus of tortricid moths, *Choristoneura*, are referred to as budworms from their habit of feeding on new needles emerging from conifer buds. Although they do not roll leaves or needles like their other tortricid relatives, they spin silk around nearby needles to create a frail refuge from predators. Warblers that prefer nesting in coniferous forests—Cape May warblers, bay-breasted warblers, Tennessee warblers—also prefer budworms as prey. The budworm populations usually peak in June and July just when the warblers, thrushes, and sparrows of the boreal forest are intensely foraging for their nestlings.

— CRAMBID SNOUT MOTHS, FAMILY CRAMBIDAE (*crambus* = dry, parched) (11,700 species; 860 species NA; larvae 8–40 mm; wingspan 10–35 mm), SUBFAMILY PYRAUSTINAE (*pyra* = fire; *usti* = scorched) (1,400 species; 160 species NA).

While still very small, the caterpillar of the basswood leaf roller makes a cut on one side of a large basswood leaf and then begins rolling itself into the cut edge of the leaf. This green caterpillar with a shiny black head keeps the curled edge from

FIG. 114 The basswood leaf roller moth (*above*) begins life on the edge of a basswood leaf (*right*).

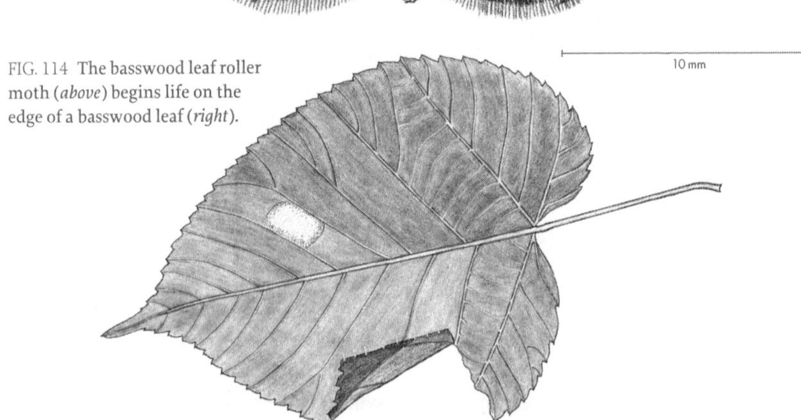

unrolling by securing it with a few silken strands. In the privacy of its leaf roll, the caterpillar feeds until the time comes for it to transform into a moth with the characteristic pose for pyraustine moths. They rest on leaves and bark with their wings spread and folded flat (fig. 114).

Deceptive Leaf Rollers

The olfactory sensors of predators and parasitoids are finely attuned to detecting the odors emitted by damaged plants and by the digested plant fragments of caterpillar droppings. As caterpillars avidly devour leaves, they try to confuse any parasitoids or predators that may be searching in their neighborhood. The caterpillars would leave their droppings in or near their leaf rolls if it were not for their ability to fling their droppings far from their feeding chambers. After each dropping passes through the digestive tract, it is flung far from its origin to lead predators and parasitoids astray. At its anus, the caterpillar has a stiff structure called an anal comb that flings each dropping horizontally before it has a chance to fall vertically. Caterpillars in the families Tortricidae and Gelechiidae, many of which are leaf rollers, have anal combs and anal forks of many shapes (fig. 115). Mother birds use a similar ploy in rearing their young brood; as droppings of baby birds are added to the nest, the parents carry the droppings off to sites many feet from the nest to deceive any vertebrate predators. This trick of birds and caterpillars has proved very effective in misleading predators and parasitoids regardless of their size.

The list of leaf-rolling insect species includes larval members of a small family of sawflies.

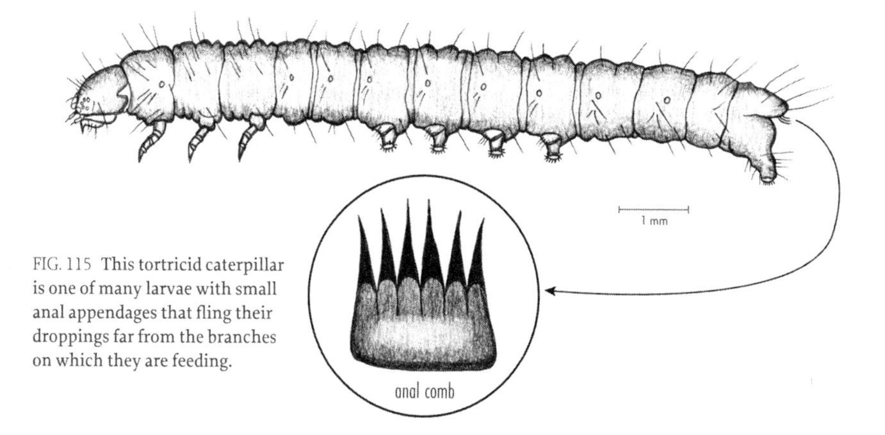

FIG. 115 This tortricid caterpillar is one of many larvae with small anal appendages that fling their droppings far from the branches on which they are feeding.

— LEAF-ROLLING AND WEB-SPINNING SAWFLIES, FAMILY PAMPHILIIDAE

(*pam* = all; *philo* = love of) (300 species; 72 species NA; 3–20 mm). Pamphiliid sawflies stand out among the sawflies in having long, many-faceted antennae, an attribute shared by both adults and larvae (fig. 116). Larvae of these sawflies are silk spinners and create leaf rolls and webs. Web spinners are usually social, while leaf rollers of the family are usually solitary. Two of several unique features of larvae in this family are their exceptionally long, thin antennae of seven segments and their complete lack of prolegs. Most sawfly larvae have short, stubby antennae and at least six abdominal prolegs. Even without these abdominal appendages found on other sawfly larvae, these pamphiliid larvae can still scoot about on their silken mats. In place of the prolegs found on other sawfly larvae, each pamphiliid larva has structures on its last abdominal segment that no other sawfly larva has: a pair of segmented appendages on its lower surface and a single hook on its top surface. Perhaps these unusual structures are the leaf-rolling sawfly equivalents of the anal forks and combs of the dung-slinging leaf-rolling caterpillars. No one has ever checked.

— LEAF-ROLLING WEEVILS, FAMILY ATTELABIDAE (*attelabus* = wingless locust) (2,500 species, 60 species NA; 3–6 mm).

While leaf-rolling caterpillars and sawfly larvae construct their own leaf chambers with larval mandibles and silk, the larvae of leaf-rolling beetles rely on their

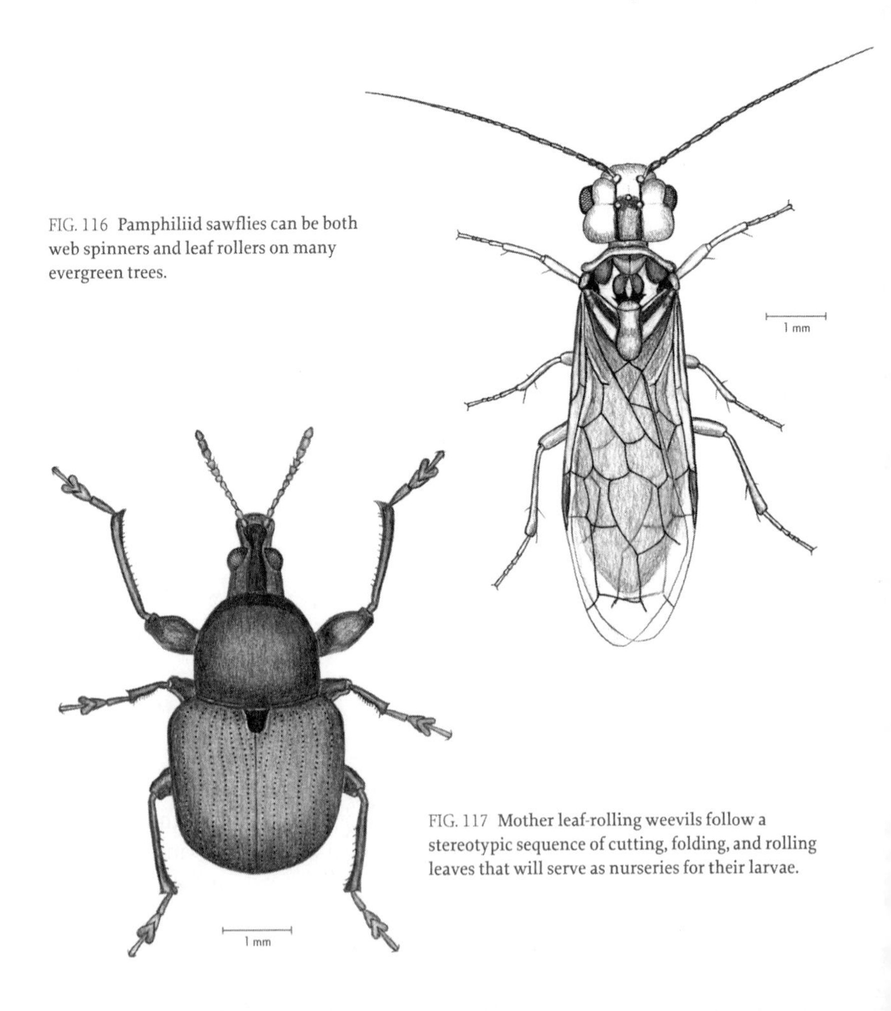

FIG. 116 Pamphiliid sawflies can be both web spinners and leaf rollers on many evergreen trees.

FIG. 117 Mother leaf-rolling weevils follow a stereotypic sequence of cutting, folding, and rolling leaves that will serve as nurseries for their larvae.

mothers to prepare a rolled leaf nursery for them; the mother beetle places her egg in the leaf roll only after she has molded the leaf tissue into a suitable thimble-shaped chamber (fig. 117; LeafScape, 22, 24).

One of America's pioneering entomologists and a founder of the journal *The American Naturalist*, A. S. Packard (1839–1905), described the origami artistry with which the mother leaf-rolling weevil cuts, folds, and molds nurseries for her offspring (Lutz 1948): "When about to lay her eggs, the female begins to eat a slit near the base of the leaf on each side of the midrib, and at straight angles to it, so that the leaf may be folded together. Before beginning to roll up the leaf, she gnaws the stem nearly off, so that after the roll is made and has dried for perhaps a day, it is easily detached by the wind and falls to the ground. When folding the leaf, she tightly rolls it up, neatly tucking in the ends, until a compact, cylindrical solid mass of vegetation is formed [LeafScape, 24]. Before the leaf is entirely rolled, she deposits a single egg, rarely two, in the middle next to the midrib where it lies loosely in a little cavity." Once the compact leaf nursery falls to the ground, the larva finishes consuming the provisions of its cradling leaf roll and pupates in the shelter of the soil.

After devoting so much care and attention to the welfare and future of each of her offspring, a mother leaf-rolling weevil can have her nursery confiscated by another female member of her own beetle family known as the thief weevil (LeafScape, 23). These female weevils wait in the wings while the mother weevil completes her leaf rolling and egg laying. Then the mother thief weevil moves in to devour the egg so carefully nestled in its leaf roll and replaces it with her own egg. These beetle thieves that wait for opportunities to take advantage of the leaf-rolling efforts of others have been given the generic name *Pterocolus* (*ptero* = wings; *-colus* = inhabiting).

Leaf-rolling weevils are found around the world. One particularly noteworthy member of the leaf-rolling weevil family is a brightly colored red and black beetle found only on the island of Madagascar. This singular weevil, known as the giraffe weevil, lives and feeds on only one tree species, the giraffe weevil tree. While the female weevil resembles other female leaf-rolling weevils in appearance and leaf-rolling prowess, the male beetle has an exaggerated neck up to three times as long as others in the family. The male uses his extralong neck to spar with other males during courtship for the attention of female weevils. Madagascar is known for its most unusual vertebrate animals, and the giraffe weevil *Trachelophorus giraffa* (*trachelo* = neck; *phorus* = bearing; *giraffa* = giraffe) is no doubt its most unusual endemic insect.

A few exceptions to the domination of the leaf-rolling habit by leaf-feeding beetles and caterpillars are observed among unrelated, strictly predatory arthropods that never consume any part of their leaf rolls. A silent, wingless cricket known as the Carolina leaf roller retires to its leaf roll after an evening of feeding on aphids and other small insects. Sac spiders in the family Clubionidae remain hidden and tranquil in their leaf rolls during the day until they emerge for a night of hunting.

— LEAF-ROLLING CRICKETS, FAMILY GRYLLACRIDIDAE (*gryllus* = cricket; *acrid* = grasshopper) (600 species; 1 species NA; ~ 50 mm).

Not all crickets that live in trees feed on leaves or sing from the treetops. The wingless cricket called the Carolina leaf roller lives inside leaves that it rolls and sews with its silk (fig. 118). Only this one species of leaf-rolling cricket is found in North America, but it is a common inhabitant of treetops and happens to be a favorite prey of a digger wasp that provisions its underground larval chambers with paralyzed crickets and katydids. This cricket stays inside its roll during the day and then ventures forth at night to feed on aphids and other small insects and spiders. Most crickets are musical leaf chewers, but this one is a silent hunter that is also a silk spinner.

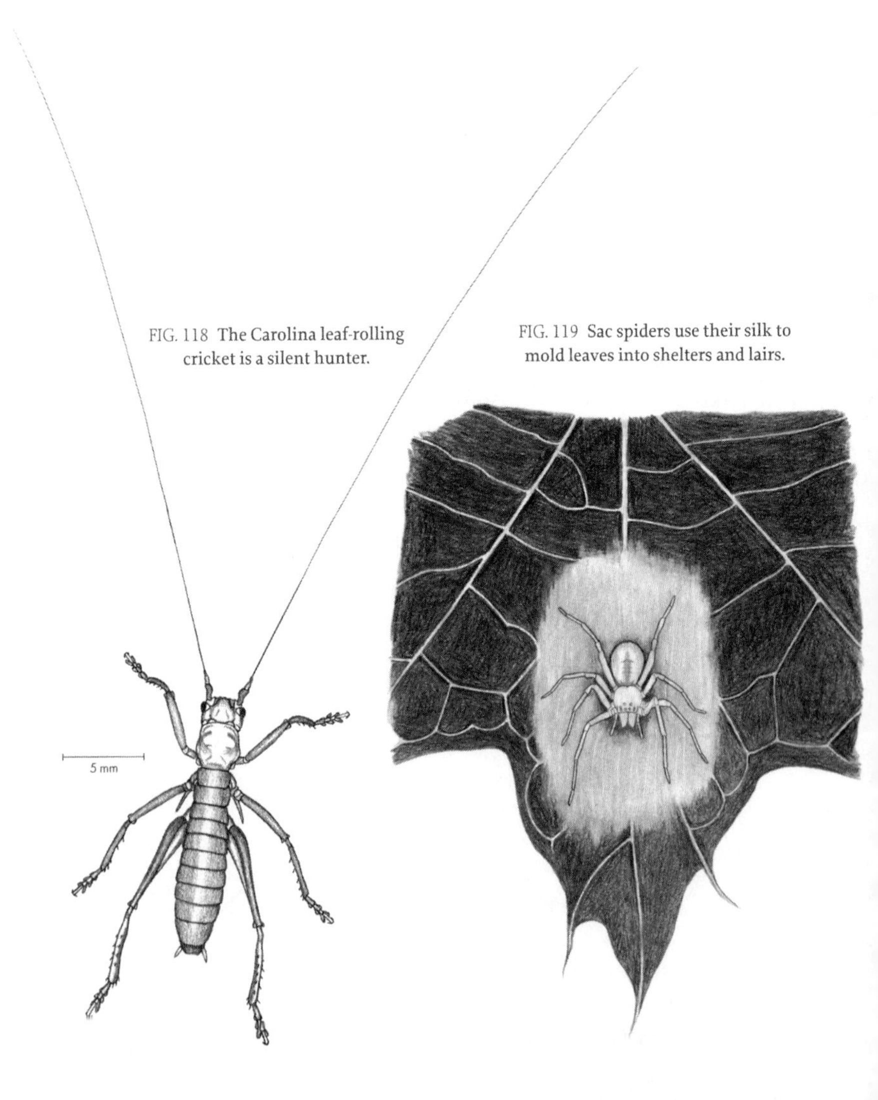

FIG. 118 The Carolina leaf-rolling cricket is a silent hunter.

FIG. 119 Sac spiders use their silk to mold leaves into shelters and lairs.

— SAC SPIDERS, FAMILY CLUBIONIDAE (*clu* = close; *bio* = life) (580 species; 58 species NA; 4–12 mm).

Leaf rolling is an occupation usually reserved for caterpillars, a few beetle and sawfly larvae, and even fewer crickets. However, an exceptional family of spiders, the Clubionidae or sac spiders, use their spinning talents to roll leaves into tubular refuges or sacs in which they can retire during the day. Sac spiders do not build webs and use their silk only for constructing their sac retreats; rather than ensnaring their prey in silken tangles, these long-legged hunters resort to ambush and speed as they set forth each night from their silken sanctums (fig. 119).

Leaf Miners

An estimated 11,000 species of insects have adopted leaf mining. About 1,900 are known in North America and about 1,800 from Europe. All of these moths, flies, beetles, and sawflies have larvae that mine the leaves of trees. In North America at least 50 species of leaf miners live on oak trees alone. The most numerous are caterpillars of moths and maggots of flies (Eiseman 2019; Frost 1959; Needham et al. 1928).

Most leaf-feeding insects chew through all layers of a leaf; however, leaf-mining insects eat only the soft, green mesophyll tissue lying between the waxy layers on the top and bottom of a leaf, leaving these respective waxy layers as a roof and a floor. The miner then lives snugly between the two epidermal layers of the leaf, first as a larva and often as a pupa as well. An adult insect—whether it is a moth, beetle, fly, or sawfly—that spends its larval days in the narrow space between the top and bottom layers of a leaf is destined to be tiny (fig. 120).

The leaf-mining lifestyle is found in about 20 families of moths, but most species of leaf-mining caterpillars and moths are in the four families Nepticulidae, Tischeriidae, Gelechiidae, and Gracillariidae. Among these, the most species of leaf miners reside in the family Gracillariidae (1,850 worldwide).

These moths usually go unnoticed because of their diminutive sizes; whoever takes the time to observe them more closely, however, will discover the truth expressed by H. S. Smith, an early American scholar and admirer of the innumerable species of these moths, which are most often referred to as microlepidoptera: "Many of them are veritable gems of beauty, far exceeding in brilliancy and richness their relatives of larger size."

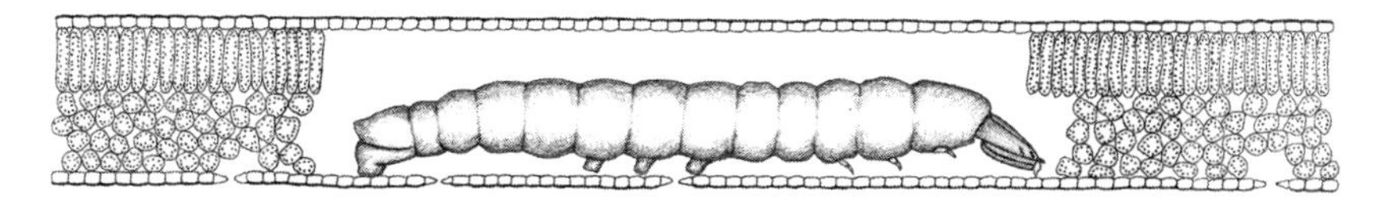

FIG. 120. **Leaf mining larvae are a select group** that live and feed inside leaves rather than on leaves.

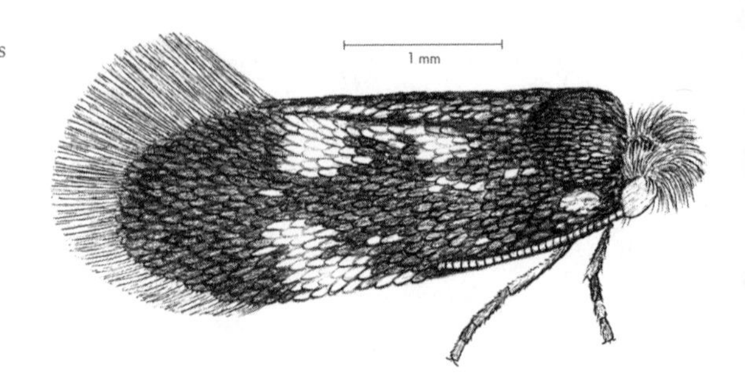

FIG. 121 These tiniest of moths have leaf-mining caterpillars that consume only a fraction of a leaf during their brief larval lives.

— MIDGET MOTHS OR PYGMY MOTHS, FAMILY NEPTICULIDAE (*nepti* = granddaughter; *cula* = little) (819 species; ~ 100 species NA; larvae < 6 mm; wingspan 3–10 mm).

Since the smallest of all moths are found in this family, the two common names for this family—midget moths and pygmy moths—are well-earned epithets. Not surprisingly, the tiny larvae of these moths mine only small portions of their home leaves during their brief lives. Larval development in nepticulid moths is rapid; usually less than two weeks after hatching, larvae leave their mines to spin cocoons and pupate in the soil or in a cleft of bark. The tiny moths that emerge have flamboyant tufts of scales arranged in enlarged caps over their eyes (fig. 121).

— SPARKLING ARCHAIC SUN MOTHS, FAMILY ERIOCRANIIDAE (*erio* = wool; *cranus* = helmet) (26 species restricted to Northern Hemisphere; 13 species NA; larvae < 11 mm; wingspan 6–13 mm).

These tiny moths literally sparkle from patches of wing scales that reflect light and act like tiny mirrors (fig. 122). The adults are some of the first insects to appear on trees in the spring, and their eggs hatch before leaves completely form, harden, and produce tannins and other insect-repellent compounds. The legless larvae of eriocraniids feed between the epidermal layers in their leaf mines and grow to full size in less than two weeks. Then they drop to the ground and spin a cocoon in the soil, where they remain until the following spring. These moths have a novel way of escaping from their cocoons. Other moths lack chewing

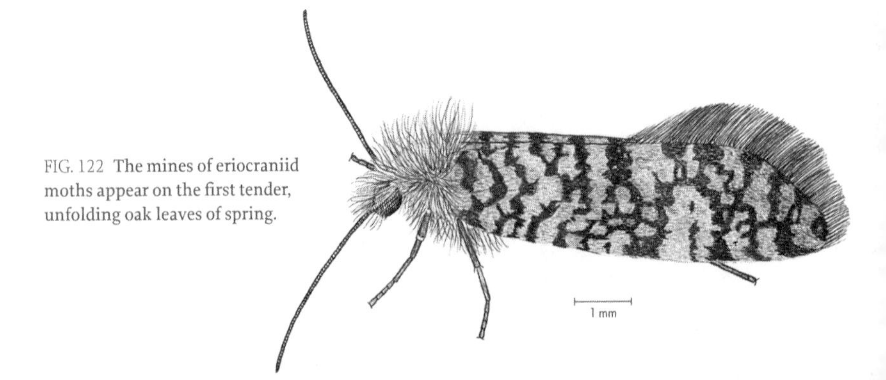

FIG. 122 The mines of eriocraniid moths appear on the first tender, unfolding oak leaves of spring.

appendages to facilitate escape from their cocoons and produce special enzymes to break through them. However, these moths have large jaws to chew through their tough, silken cocoons. They emerge sporting a crown of long, disheveled scales. These moths prefer flying in daylight to flying at night, and in many other ways they are exceptional through all their life stages.

— TRUMPET LEAF MINER MOTHS, FAMILY TISCHERIIDAE (named after German biologist Carl von Tischer) (110 species; 47 species NA; larvae < 6 mm; wingspan 6–11 mm).

As larvae grow, their mines lengthen, expand laterally, and flare out into trumpet shapes. In the center of its mine, the mature larva of the oak blotch miner, *Tischeria quercitella*, constructs a silken nest or nidus (*nidus* = nest) that shelters it while it pupates and transforms into a moth. A tuft of scales projects from the top of the heads of tischeriid moths (fig. 123), and their compound eyes stand out clearly, while the eyes of leaf-mining gracillariid moths are covered by a cap of scales and are often not visible from above (fig. 124).

— LEAF BLOTCH MINER MOTHS, FAMILY GRACILLARIIDAE (*gracil* = slender) (1,850 species; 270 species NA; larvae < 10 mm; wingspan 4–20 mm).

What tortricid moths are to leaf rolling, gracillariid moths are to leaf mining (LeafScape, 15, 16). The Gracillariidae is the largest family of leaf miner moths. *Phyllonorycter lucetiella* (*phyllo* = leaf; *orycter* = digger) constructs a rectangular mine between two veins of a basswood leaf (fig. 124). Remarkably, the newly hatched larva marks out the final boundaries of its mine as it begins eating and then commences to feed from the outer edges of the mine always toward the center, leaving a trail of droppings along the outer edges before it finally pupates in the mine. At an early age, the larva must be a good judge of how much leaf tissue it will consume in its lifetime as it lays out the borders for its future mine.

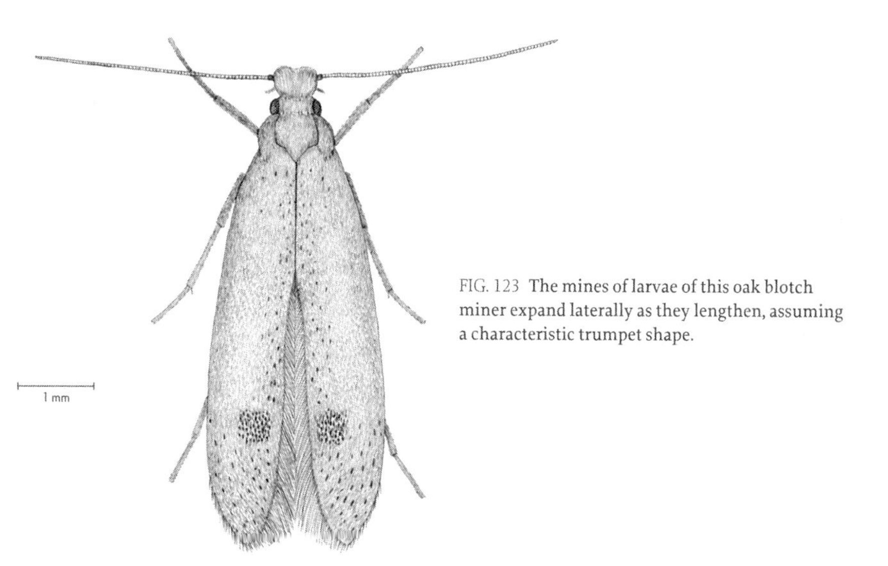

FIG. 123 The mines of larvae of this oak blotch miner expand laterally as they lengthen, assuming a characteristic trumpet shape.

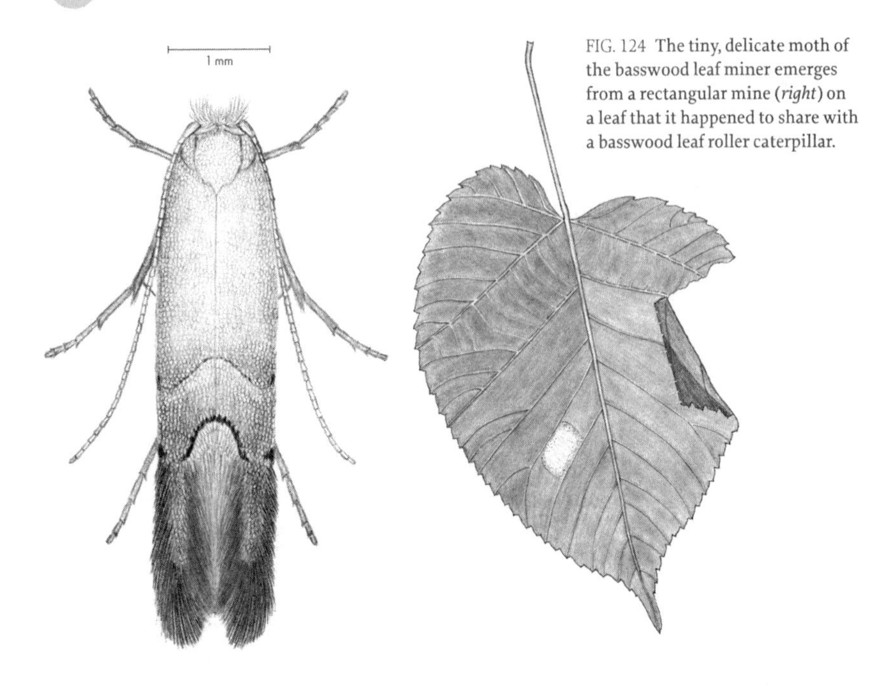

FIG. 124 The tiny, delicate moth of the basswood leaf miner emerges from a rectangular mine (*right*) on a leaf that it happened to share with a basswood leaf roller caterpillar.

— TWIRLER MOTHS, FAMILY GELECHIIDAE (4,600 species; 886 species NA; larvae 4–15 mm; wingspan 4–20 mm).

Members of the family Gelechiidae go by the common name twirler moths (fig. 125). The name is well earned from their habit of spinning in circles on leaf surfaces. At the bases of their long, thin antennae, a pair of what appear to be long, curved horns project upward and backward. These horns are tapered projections

FIG. 125 In addition to their leaf-mining activities, larvae of twirler moths such as this redbud leaf-folder moth can fold or roll leaves.

called labial palps that arise from the moth's labium or lower lip. Gelechiid larvae lead hidden lives in trees. In addition to leaf mining, they have adopted other lifestyles involving seclusion and concealment—leaf folders, leaf rollers, gall makers, and borers of stems, seeds, and roots (chapter 5).

— RIBBED COCOON–MAKING MOTHS, FAMILY BUCCULATRICIDAE (*buccula* = little mouth, probably referring to the larval mouth that excavates the very narrow mine; *-trix* = suffix for a feminine agent) (250 species; 110 species NA; larvae 5–10 mm; wingspan 8–10 mm).

The caterpillars of this family of tiny moths may start out as leaf miners, but after outgrowing the confines of their narrow mines, they end their larval days as leaf skeletonizers (LeafScape, 8). Each species is particular about the trees on which it feeds; each is a specialist on a particular tree, whether birch, oak, apple, or hickory. The larvae are pale green caterpillars with no exceptional spines, spots, or particularly striking attributes, but they leave a trail of leaf skeletons in their wakes. If the branch on which they are feeding is disturbed, the caterpillars release their grip on the leaf and dangle from firm silk threads to evade capture. These nondescript larvae eventually use their silk to construct unmistakable oblong cocoons with long parallel ridges. The moths that emerge from these cocoons have heads covered with rough, ruffled scales, and folded forewings with tufts of raised black scales halfway down their length (fig. 126).

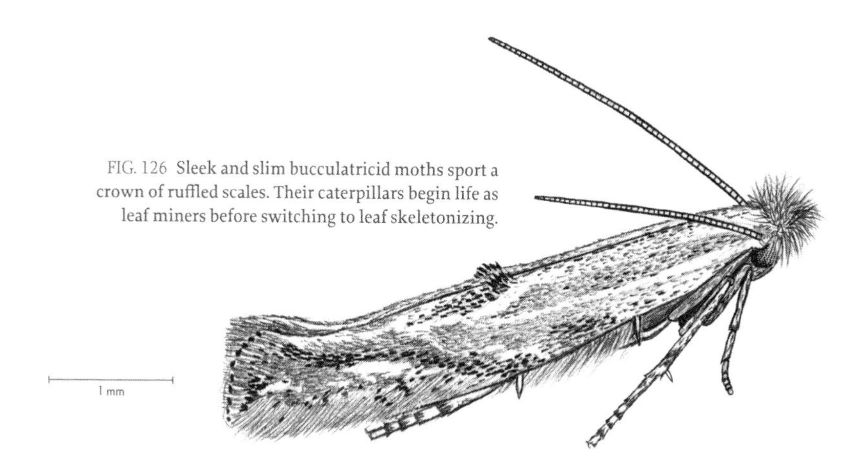

FIG. 126 Sleek and slim bucculatricid moths sport a crown of ruffled scales. Their caterpillars begin life as leaf miners before switching to leaf skeletonizing.

— CASEBEARER MOTHS, FAMILY COLEOPHORIDAE (*coleos* = sheath; *phor* = carry) (1,100 species; 158 species NA; larvae 5–10 mm; wingspan 5–26 mm).

Case-bearing caterpillars also mine leaves, but they live in cases that they construct from small pieces of leaves, and they venture into their mines only to chew on the tender leaf tissues. A tiny hole that they chew in the bottom layer of a leaf is both entrance to and exit from their leaf mine (figs. 127 and 128). Their cases are open at both ends. Like other leaf miners, they are neat about their housekeeping and back into their cases when they need to discard droppings from the rear openings of their cases. Casebearers move about on a leaf as well as from leaf to leaf, mining leaves during the summer. In the autumn they take their cases with them and move to sheltered spots on the bark where they wait out the winter in their cases. As leaves appear the following year, the nearly fully grown caterpillars resume mining new leaves until they retire in late spring to pupate and become moths.

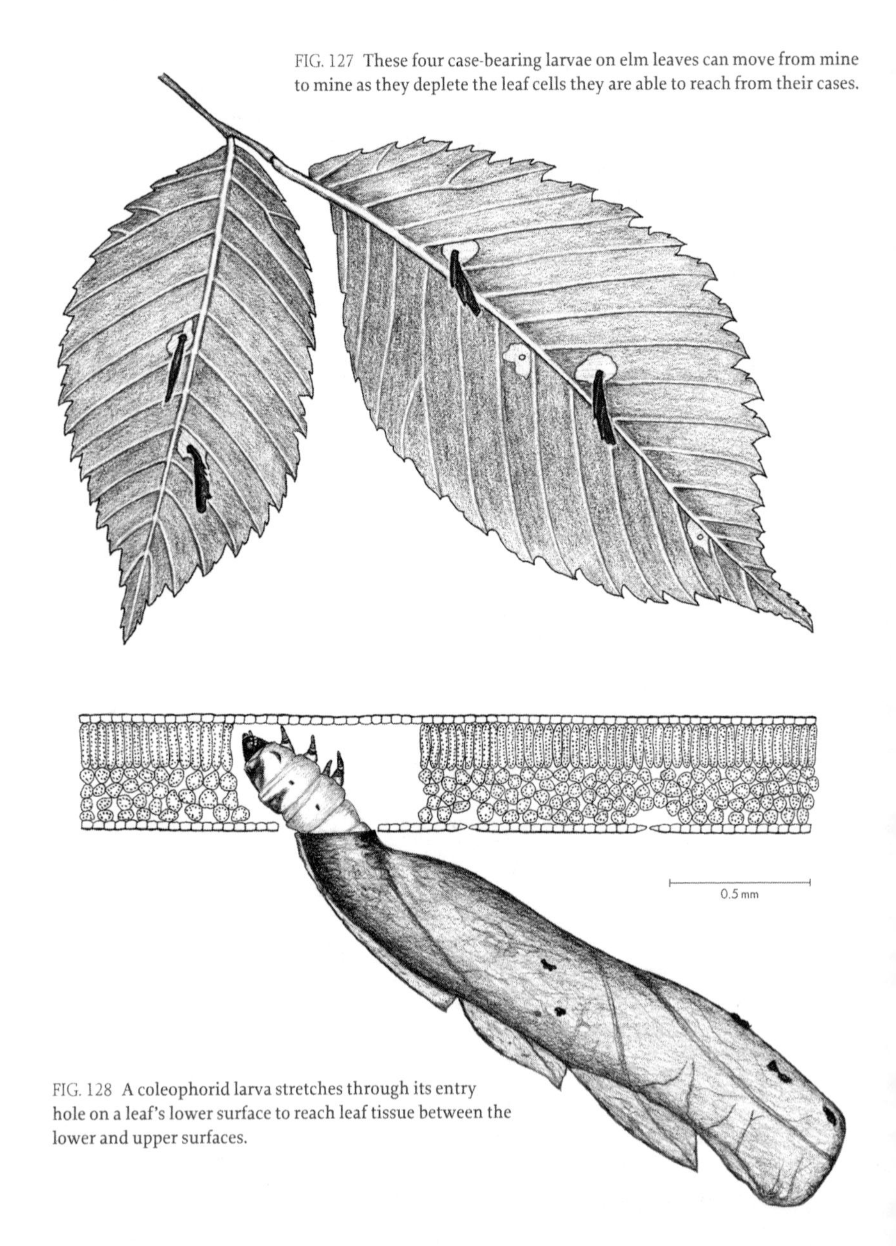

FIG. 127 These four case-bearing larvae on elm leaves can move from mine to mine as they deplete the leaf cells they are able to reach from their cases.

FIG. 128 A coleophorid larva stretches through its entry hole on a leaf's lower surface to reach leaf tissue between the lower and upper surfaces.

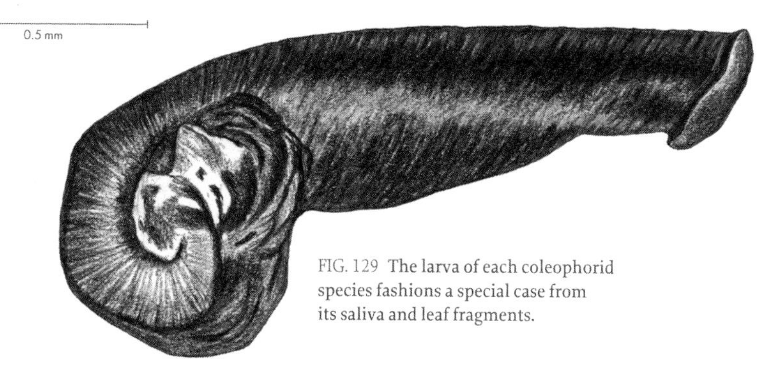

FIG. 129 The larva of each coleophorid species fashions a special case from its saliva and leaf fragments.

Casebearer moths are slender, dull-colored moths with conspicuous fringes on the edges of their wings, but they are difficult to identify as adults. Different species of case-bearing caterpillars look very much alike, but their cases come in very distinctive shapes. Some look like snail shells; others look like pistols or cigars. Casebearer moths and caterpillars are better known for their cases than for any of their other physical attributes (fig. 129).

— LEAF MINER FLIES, FAMILY AGROMYZIDAE (*agro* = field; *my* = fly; *za* = very) (2,860 species; 640 species NA; 1–6 mm).

Agromyzid flies are tiny and have long bristles called vibrissae (*vibrissae* = whiskers) (fig. 130). While almost all leaf-mining flies are in this one family, a few leaf-mining species are scattered among other fly families. These families include Anthomyiidae (root-maggot flies), better known for living in roots and stems

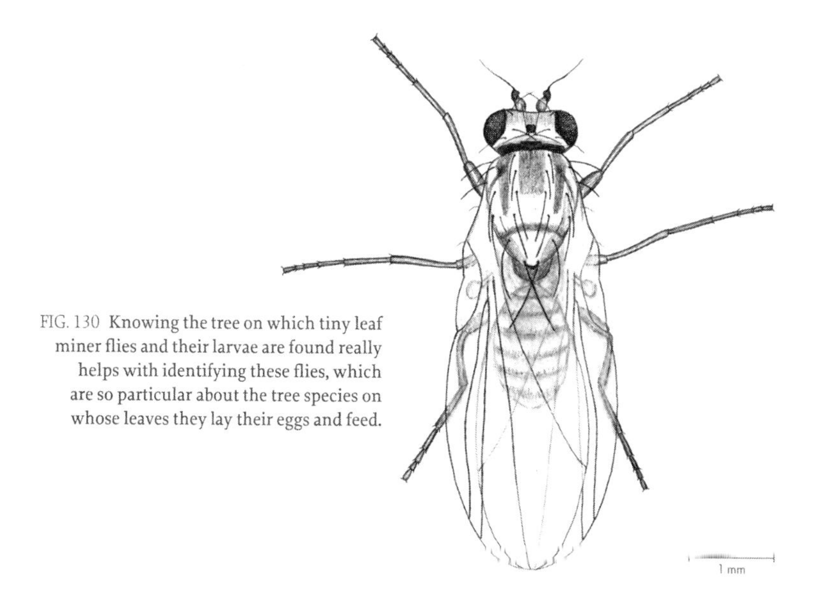

FIG. 130 Knowing the tree on which tiny leaf miner flies and their larvae are found really helps with identifying these flies, which are so particular about the tree species on whose leaves they lay their eggs and feed.

and decaying plants, and Cecidomyiidae (chapter 2), the gall midges, known primarily for their gall-making prowess. While almost all species of the vinegar fly family Drosophilidae feed on rotting plant material, some unconventional larvae in the family mine leaves. The approximately 5,000 species of the fruit fly family Tephritidae (chapter 5) are best known for being fruit feeders, but a small number are gall makers and leaf miners. The flower fly family Syrphidae, whose larvae are better known as aphid predators and tree hole inhabitants (chapter 4), has a few leaf-mining members that have respectively switched from predaceous and detritus-feeding traditions to the herbivorous tradition of leaf mining. In the Tipulidae (chapter 6), a large family whose members inhabit many aquatic, soil, and tree habitats, at least one species in Hawaii has taken up the leaf-mining vocation. With creatures as diverse as insects, it is not surprising that there are always nonconformists.

— LEAF-MINING SAWFLIES, FAMILY TENTHREDINIDAE (*tenthredin* = type of wasp) (7,500 species, 900 species NA; 5–20 mm), SUBFAMILY HETERARTHRINAE (*heter* = different; *arthro* = joint) (150 leaf-mining species, 37 species NA; 5–10 mm).

Most members of this largest family of plant-feeding sawflies feed on leaves (figs. 76, 79–83, 85, 116), but a few produce galls (fig. 103) or bore into stems. The few that mine leaves are in the subfamily Heterarthrinae (fig. 131).

— LEAF-MINING BEETLES. Among the beetles, three of the largest families—the metallic wood borers, the leaf beetles, and the weevils—each have a few members that have taken up leaf mining. As expected for insects that inhabit the cramped spaces between the tops and bottoms of leaves, these leaf-mining members are among the smallest of their families. More information on each of these three beetle families that interact with trees in so many other ways is presented in chapters 4, 5, and 6.

While almost all members of the family Buprestidae (metallic wood borers) carve tunnels under bark (figs. 216 and 217), some of the smaller and flatter members of

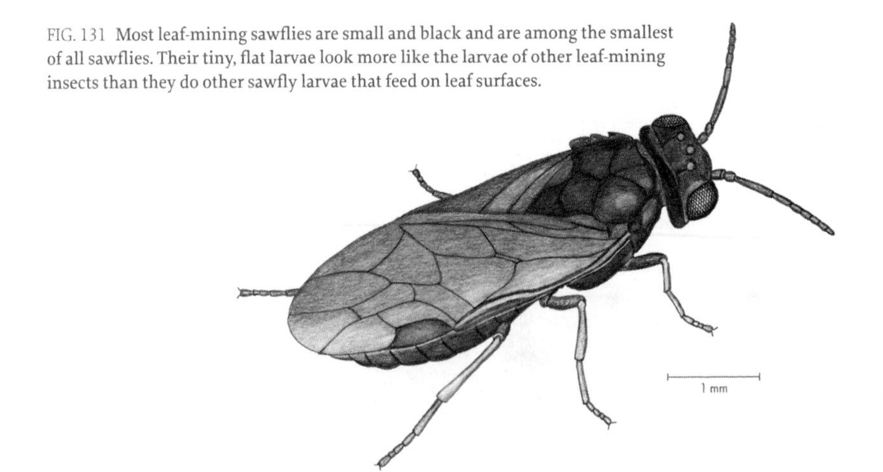

FIG. 131 Most leaf-mining sawflies are small and black and are among the smallest of all sawflies. Their tiny, flat larvae look more like the larvae of other leaf-mining insects than they do other sawfly larvae that feed on leaf surfaces.

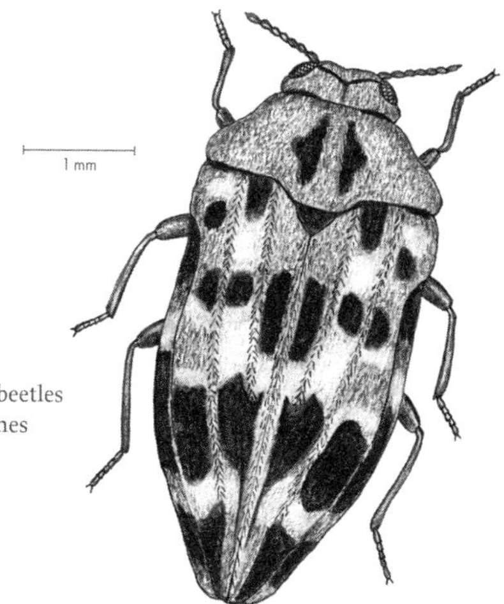

FIG. 132 Larvae of short, stout metallic wood-boring beetles in the genus *Brachys* (*brachy* = short) excavate leaf mines rather than wood tunnels.

the family mine tree leaves. The number of all these species of leaf-mining beetles in North America is around 200 (fig. 132).

Leaf beetles and leaf-rolling weevils are discussed earlier in this chapter, and wood-boring weevils are discussed further in chapter 4. A section on acorn weevils is included in chapter 5. Most leaf-mining weevils on trees are in the genus *Orchestes* (fig. 133), and one North American weevil in the genus *Prionomerus* (*priono* = a saw; *merus* = thigh, referring to the prominent spine on each front femur) mines leaves of sassafras and tulip trees. About 500 of the 40,000 described species of leaf beetles of the large family Chrysomelidae are leaf miners (fig. 134, 163) (Santiago-Blay 2005).

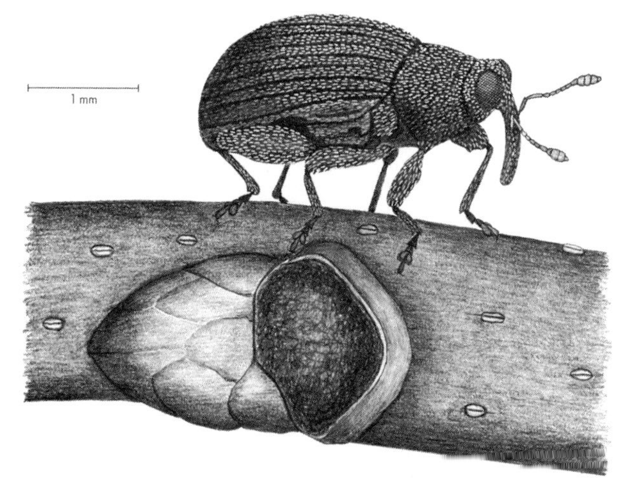

FIG. 133 An *Orchestes* (*orchest* = dancing) flea weevil perched above a bud on the oak tree whose leaves it mined as a larva.

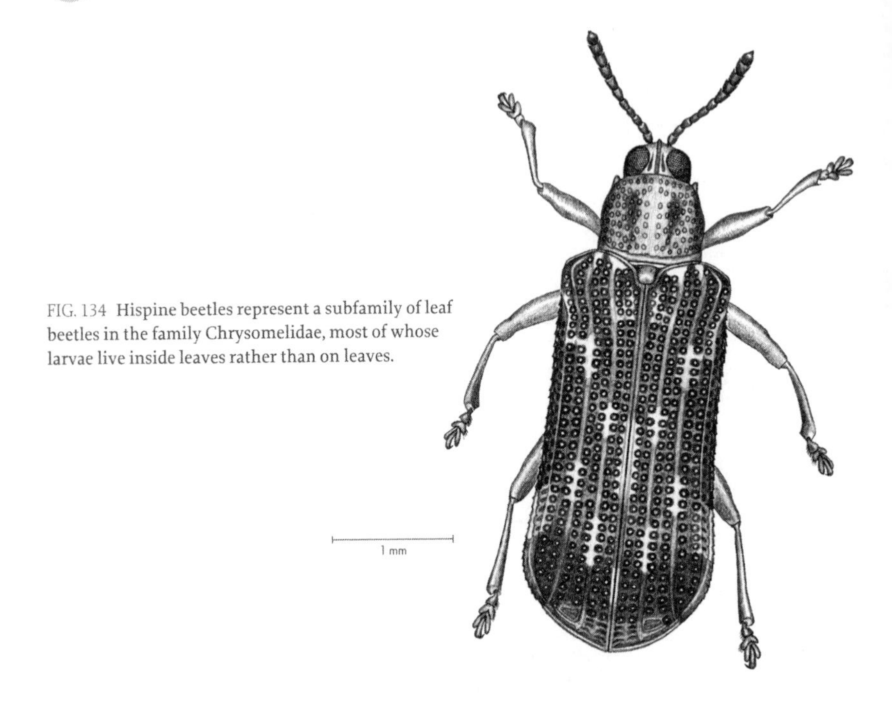

FIG. 134 Hispine beetles represent a subfamily of leaf beetles in the family Chrysomelidae, most of whose larvae live inside leaves rather than on leaves.

Leaf-Mining Housekeeping

Leaf miners keep the close quarters of their mines tidy by carefully segregating their droppings from their feeding quarters; some shift their droppings to the edge of the mine; others shift them to the vacant center of the mine. The winding mines of the gracillariid caterpillar *Phyllocnistis populiella* (*phyllo* = leaf; *cnistis* = insect living beneath; *populo* = poplar; *-ella* = little) create an ornate serpentine pattern on poplar leaves that is accentuated by the line of dark droppings arrayed exactly in the middle of the prominent white mine (fig. 135). Other miners maintain a dropping-free mine by passing their droppings through small holes in the leaf epidermis.

Not only the miners themselves are concerned about sanitary conditions in their mines; the wasps that parasitize them and eventually take possession of the mines have adopted fastidious housekeeping practices as well. All remains of the larval host are first consumed by the larval parasitoid before it pupates in the mine. Any decaying remnants would attract microbes that might harm the parasitoid before it emerges as an adult. Next, the parasitoid avoids defecating in the mine by retaining its waste in a pouch of the larval gut until it is time to pupate. Finally, the accumulated feces are expelled from the gut and fashioned into fecal pillars that extend from the floor of the mine to its roof, offering structural support and ample space for the parasitoid larva's safe passage, first to pupa and then from pupa to adult (fig. 136).

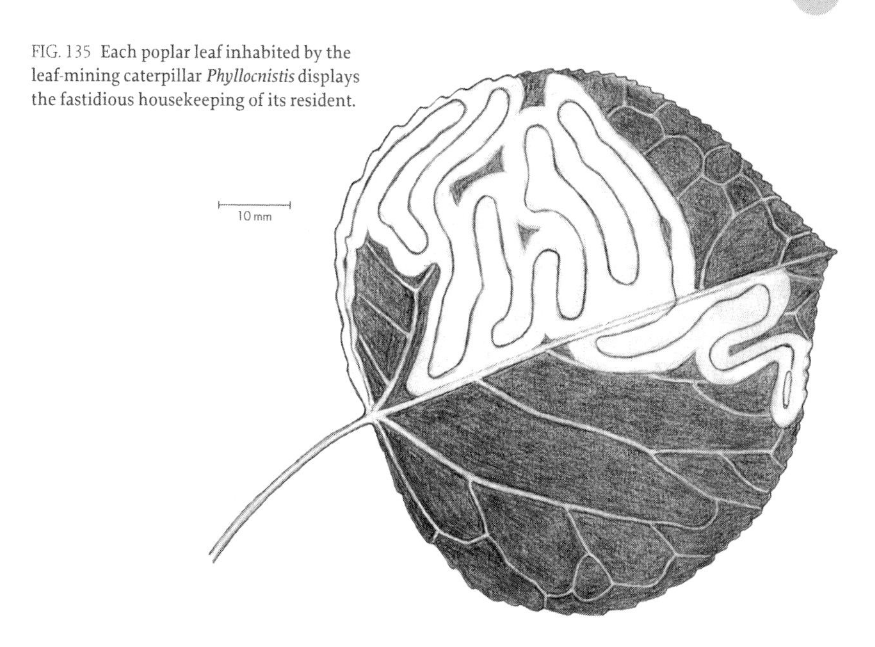

FIG. 135 Each poplar leaf inhabited by the leaf-mining caterpillar *Phyllocnistis* displays the fastidious housekeeping of its resident.

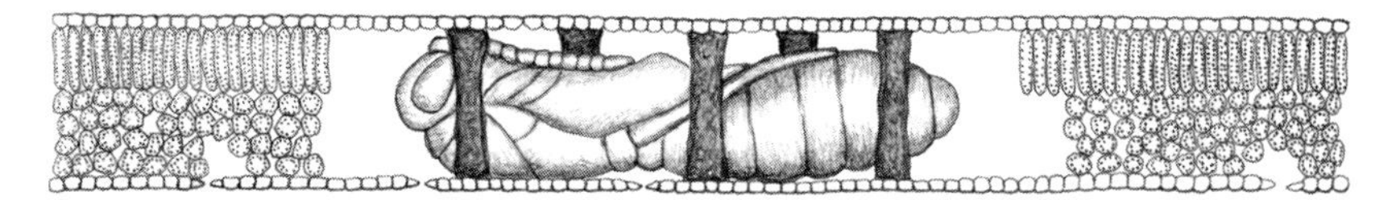

FIG. 136 As they prepare for pupation, the larvae of chalcidoid wasps that parasitize leaf miners use their own fecal waste to form pillars that reinforce the integrity of the mine until they emerge as adults.

Leaf Miners Prefer Certain Leaves and Manipulate Leaf Hormones

Leaf miners select areas of leaves that are both rich in nutrients and deficient in plant defenses. Moths selectively lay their eggs on leaves more centrally located on oaks rather than on leaves that are more peripheral. Peripheral leaves drop earlier than leaves closer to the center of a tree (Bultman and Faeth 1986).

Leaf miners confined to their narrow worlds between the waxy epidermal layers on the tops and bottoms of leaves compete with their host plants for the leaves' nutrients. At the end of summer, while the host plant is salvaging whatever nutrients remain in the old leaf to send into root storage, the leaf miner counters the tree's salvaging operation by promoting an environment in the leaf that discourages senescence and encourages photosynthesis. The miners and the symbiotic bacteria in their guts accomplish this reversal of leaf cell fate by continually transferring cytokinin hormones to the leaf tissues as they feed (Body et al. 2013). Experimental removal of symbiotic bacteria from the leaf miner accelerates senescence of its leaf. Not only miners that are confined to a single leaf but also other insects that are free to roam among leaves have adopted this strategy of manipulating tree hormones to achieve their self-serving goals. With addition of cytokinins produced by their salivary glands or with help from their microbial partners, these miners have managed to enhance the nutrients provided by aging leaves (Andreas et al. 2020; Zhang et al. 2017; Frago et al. 2012; Kaiser et al. 2010).

Leaf miners, gall makers, and leaf rollers clearly use the growth hormones of plants to their own advantage. By sharing the same hormonal language with trees, many herbivores and endophytes communicate their desires to their indulgent host trees. Trees tolerate the company of these companions whose demands on the trees' livelihoods are not too great or too threatening.

Trees, however, firmly object to excessive munching of their leaves or drilling of their branches. They emit volatile compounds that alert insect predators, insect parasites, and nearby trees that there are destructive agents in the neighborhood. In addition to recruiting insect allies for protection, trees release jasmonic acid and salicylic acid to initiate even more robust immune responses to these chewing and drilling intruders (fig. 8). As noted in the next chapter, however, trees face special challenges in responding to sap-sucking insects and mites that tap into the trees' circulatory systems and recruit microbes to help them subdue the trees' immune defenses.

TAPPING A TREE'S CIRCULATORY SYSTEM

3

SAP CELLS, BIRD BEAKS, AND INSECT BEAKS

A tree's circulatory or vascular system distributes water and nutrients from root tips to treetop. The two-way traffic lanes of the tree's vascular system arise from a thin layer of stem cells called the cambium that extends the entire length of the tree from its roots to its leaves and that lies between the central heartwood of the trunk and the outer bark. Stem cells of the cambium continually divide to produce not only more unspecialized dividing cells like themselves but also specialized cells of the vascular system that no longer divide. One set of specialized cells—xylem—acts as a conduit for water and mineral nutrients from the soil. The other set of specialized cells directs sugars produced in the leaves to destinations throughout the tree. The name *cambium* translates as "change," and that is what cambial cells are constantly doing, changing their fates as some divide toward the center of the tree to differentiate as xylem cells and others divide toward the periphery of the tree to differentiate as phloem cells. The generation of new cells proceeds in such an orderly fashion that a discrete ring of xylem cells is added to the heartwood with each growing season, and a new ring of phloem cells is added toward the bark. While the old phloem cells are continually sloughed off from the bark, the oldest xylem cells remain as small distinct rings in the center of the trunk, and the newest additions of xylem cells occupy the largest, outermost rings (fig. 137). Each ring of xylem cells leaves a record of a year's growth for the tree. Thick, robust rings are laid down in the heartwood during good growing seasons with sufficient rainfall and pleasant days; thin rings are the legacy of years with drought.

In a section of tree trunk, xylem cells divide between the cambium and heartwood, differentiate, lose their nuclei, and undergo programmed cell death, leaving only their skeletons or cell walls behind. Along the length of these long, hollow vessels, water and nutrients travel between the leaves and roots. The smaller members of the xylem—the tracheids—retain perforated end walls, but the larger vessel cells lose their end walls to form a chain of hollow cells that functions like a pipeline. Between the cambium and bark, phloem cells undergo

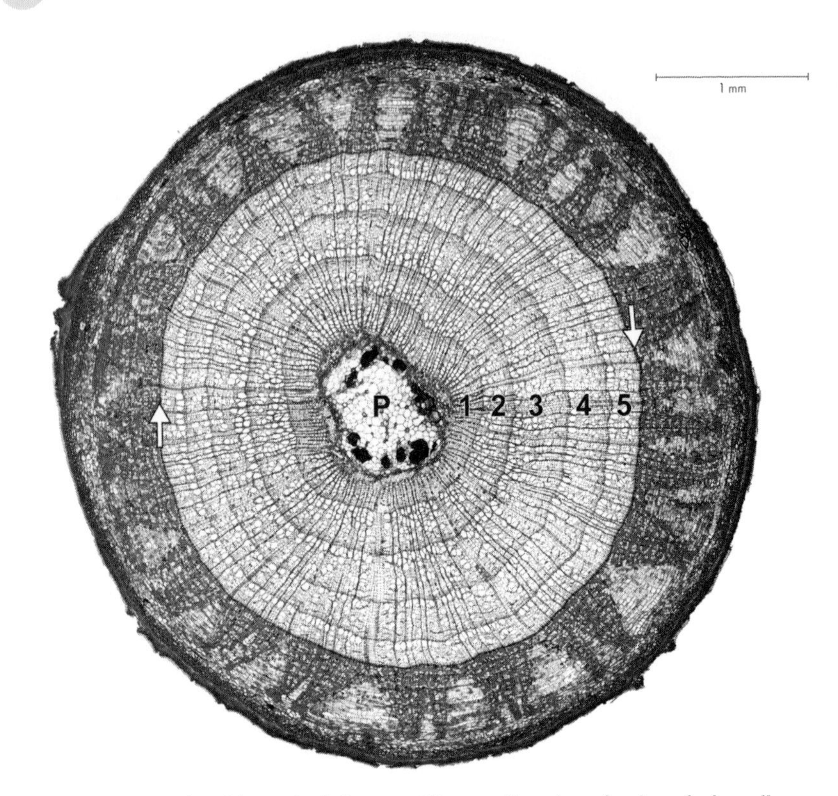

FIG. 137 A cross section of the trunk of a five-year-old basswood tree shows five rings of xylem cells, with a new ring being added each year to the heartwood between the central pith (P) and the cambium. Each year's addition is marked with a number, and the cambium ring is marked with two arrows.

division and differentiation to generate columns of sieve tube cells and companion cells. In contrast to xylem cells, these cells do not die; however, the larger sugar-conducting sieve tube cells of the phloem lose their nuclei. Their remaining cell contents are confined to the cell's periphery to create a channel in the center of the cell for the flow of sap. In the company of their smaller companion cells, the sieve tube cells transport sugars and water through their central channels and perforated end walls (figs. 138 and 139).

Sap flows between treetops and root tips. In the roots, sap picks up minerals and water. In the leaves, sap picks up the sugars produced by photosynthesis when carbon dioxide from the air is combined with water from the soil. After each part of a tree has taken its share of nutrients from the sap, there is still enough left over for the birds, mammals, and insects known as sap suckers.

A single apple tree, maple tree, or tulip tree may have several hundred tiny holes arranged in neat horizontal rows around its trunk and branches. This is the unmistakable handiwork of a woodpecker called the yellow-bellied sapsucker (fig. 140). This bird can spend hours tapping out neat rows of holes on trunks of trees with particularly sweet sap. Sap oozes out of these holes, and insects soon

FIG. 138 A longitudinal section (*left*) through the trunk of a chinkapin oak seedling (*right*) shows the symmetrical arrangement of phloem and xylem cells around the dividing cambium cells. A close-up of cells in the trunk section delimited by the square in the left figure is shown in figure 139 with differentiating and differentiated phloem and xylem cells surrounding the cambium cells of the trunk section.

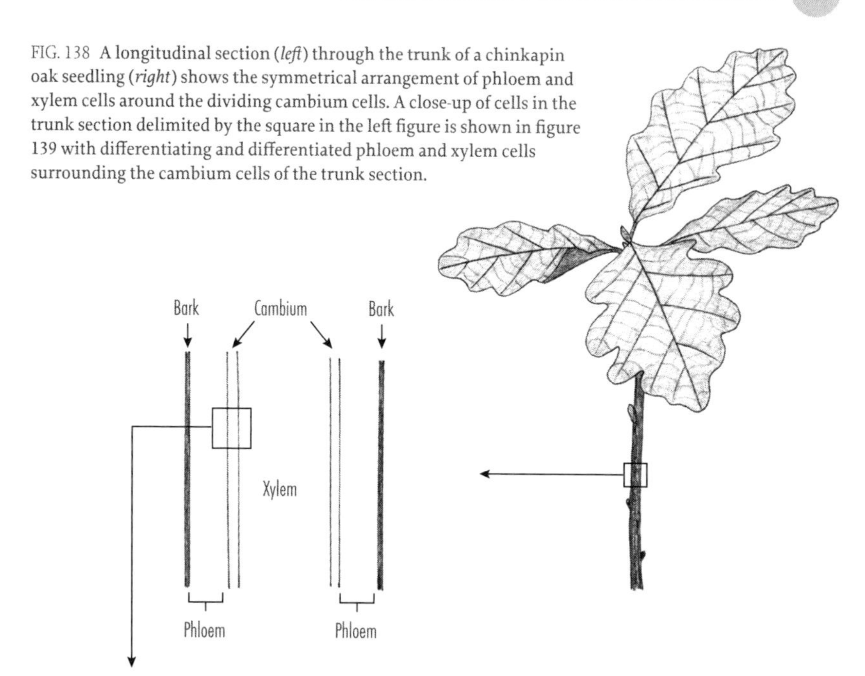

FIG. 139 All the phloem and xylem cells of a tree's circulatory system are generated from lateral divisions of the single ring of cambium stem cells.

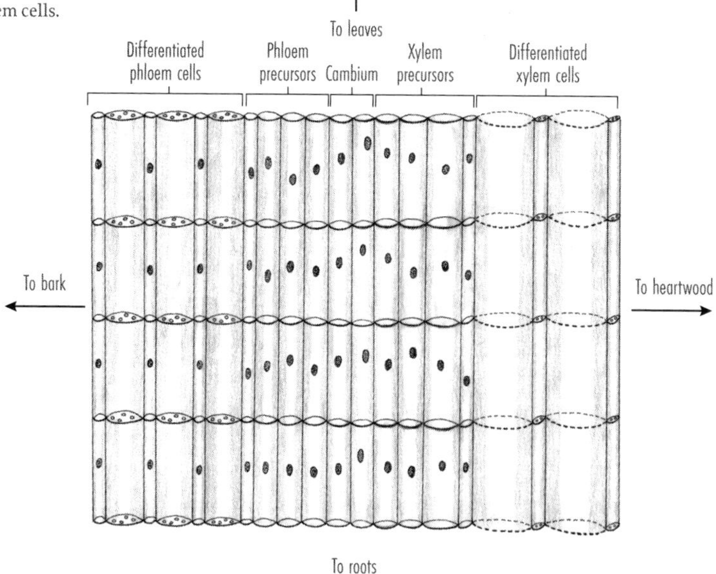

FIG. 140 As sweet sap rises from a tulip tree's roots in early spring, a yellow-bellied sapsucker uses its beak to tap its flow in the tree's circulatory system.

discover it. Sapsuckers are quick to snatch up many of these crunchy insects along with the sweet sap.

Squirrels "lap" sap from sapsucker holes; they also tap young maple trunks and branches with their teeth and return hours later to enjoy the sap whose sugar has now been enriched by evaporation of most of its water (Heinrich 1992).

An obvious feature that bird and insect sap suckers have in common is that they use beaks to tap trees for their sap. Sap suckers of all ilks can be considered botanical vampires. Their sharp beaks neatly pierce the outer layers of bark, twigs, or leaves to reach sap flowing beneath the surface. Insect beaks, however, are not like bird beaks. The mouthparts of sap-sucking insects fit together to form beaks with central channels. The piercing, sucking insect beak looks and acts like a straw while muscles in the head act like a pump, drawing fluid from the tree's vascular system. Other insects such as caterpillars, crickets, beetles, bees, and wasps have jaws rather than beaks, which they use for chewing rather than sucking. If you flip over an aphid, stink bug, or other sap-sucking insect, you will see a long beak lying against its underside. Watch one of these insects as it feeds from a tree, and you will see its beak extending from beneath its head into the tissues of the tree (fig. 141).

Almost all these sap-sucking insects belong to the order Hemiptera, the true bugs.

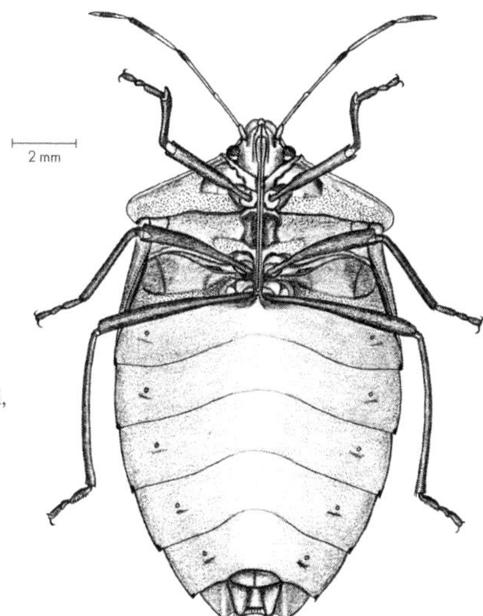

FIG. 141 The underside of a green stink bug. The long sap-sucking beak of a true bug lies along the midline of its belly, the line that separates left legs from right legs. The beak penetrates plant tissues and, like a straw, draws sap through its central channel.

Members of two of the three suborders of true bugs—Sternorrhyncha and Auchenorrhyncha—are strictly sap suckers. The few bugs that are predators suck blood and fluids from other animals. These predators are members of the suborder Heteroptera (chapter 1; fig. 28; chapter 2; fig. 95); however, some members of the Heteroptera are also sap suckers and are discussed at the end of this chapter (figs. 141, 163–165).

- TRUE BUGS, ORDER HEMIPTERA (*hemi* = half; *ptera* = wings) (107,000 species; 10,500 species NA; 1–150 mm).
- APHIDS, SCALE INSECTS, ADELGIDS, PSYLLIDS, SUBORDER STERNORRHYNCHA (*sterno* = sternum, breastbone; *rhyncho* = snout, referring to the beak appearing to arise from the sternum of the thorax) (18,700 species; 6,400 species NA; 1–12 mm).
- LEAFHOPPERS, TREEHOPPERS, PLANTHOPPERS, CICADAS, SUBORDER AUCHENORRHYNCHA (*auchen* = neck; *rhyncho* = snout, referring to the beak arising from the base of the head/neck) (42,000 species; 4,000 species NA; 1–50 mm).

As you watch an insect feeding on tree sap, you may also notice that some sap taken up by an insect's beak reappears at that insect's other end as droppings of honeydew. If there are enough sap-feeding insects on a single tree (and there are often millions), the leaves of the tree and the ground beneath the tree will be splattered with their sweet, sticky honeydew. Sometimes you can feel the fine droplets of honeydew falling on your neck and arms as you walk beneath the tree. Many molds grow well on the sugars and amino acids that make up honeydew, providing patches of microscopic pasture on which many chewing insects graze. Tree sap in the form of honeydew continues to nourish many different creatures such as beetles, ants, lacewings, and wasps even after it has been ingested, digested, and defecated by the many sap-sucking insects of a tree.

SAP SUCKERS AND MICROBIAL SYMBIONTS

Naturally a tree will mount defenses to protect itself from the sap-draining activities of insects with piercing beaks. Sap-feeding aphids, leafhoppers, and psyllids carry bacterial pathogens known as phytoplasmas from tree to tree. These pathogens have developed their own strategies to counter the defenses of their tree hosts. They have discovered a way to use the defenses of trees to their own advantage. The conspicuous witches' brooms that form on tree branches are the telltale handiwork of the microscopic phytoplasmas that dwell in the phloem cells of the tree's vascular system (fig. 142). In addition to altering the forms of branches and twigs, phytoplasma pathogens produce proteins that suppress the production of the tree's defenses initiated by its release of jasmonic acid, salicylic acid, and other hormones (Dermastia 2019; Sugio et al. 2011).

Insect sap suckers have developed a sinister partnership with bacteria and viruses that are plant pathogens. Leafhoppers that are vectors for the transmission of bacterial pathogens from tree to tree not only survive longer on their host plants but also produce more offspring than fellow leafhoppers that have not established alliances with phytoplasmas. These tiny bacterial pathogens are even known to alter the chemicals emitted by their host apple trees so that the apple host attracts, rather than repels, infected insects that carry the pathogen

FIG. 142 Bacterial phytoplasmas carried from tree to tree by insect sap suckers often leave a trail of witches' brooms among tree branches.

from tree to tree (Mayer et al. 2008). They capitalize on using the tree's secondary metabolites to promote their own agendas, undermining the tree's use of these metabolites as insect repellents. Metabolites of the tree that repel many insects can act as alluring attractants to other insects that are vectors of microbial pathogens. Microbes—bacteria, viruses, and fungi—recruit insect accomplices to transport and disperse them by inducing trees to simply increase their production of secondary metabolites that attract their insect allies. Rather than relying on the vagaries of wind and weather to carry them to their tree hosts, microbes can count on the wings and legs of their sap-sucking insect vectors to deliver them to their desired destinations.

Even viruses that are smaller than phytoplasmas and transmitted by aphids rather than leafhoppers can increase the attractiveness of host plants by elevating levels of plant volatile compounds. The aroma induced by these viruses entices additional aphids to the plants to feed, pick up the virus, and then transmit that virus to new host plants. Dispersal of aphids to new host trees ensures the survival and spread of these viruses that depend on insect vectors for their transport (Mauck et al. 2010).

A diet of phloem sap alone is rich in sugar but deficient in amino acids and vitamins. To make up for deficiencies, these sap-feeding insects share their bodies with bacteria that supply the missing amino acids and vitamins. In addition to enhancing the diet of sap, some bacterial endophytes provide protection from fungal pathogens (Lukasik et al. 2013). The relationship between bacteria and insects is so intimate that special insect cells called bacteriocytes have been set

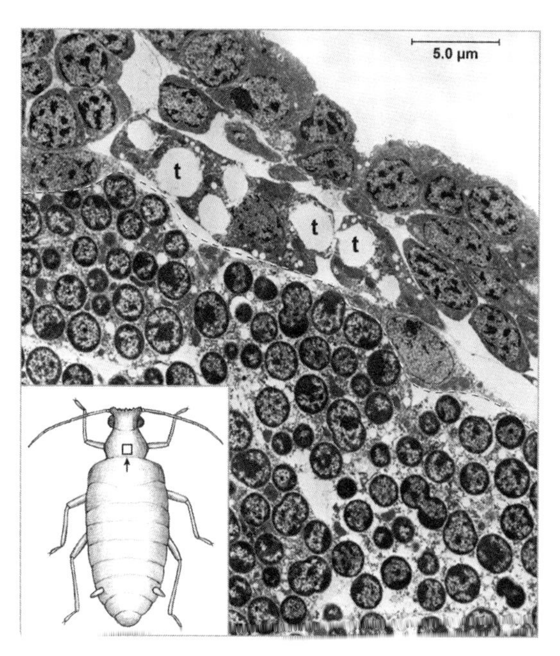

FIG. 143 Insects, such as this oak aphid (*lower left*), that survive on a diet of tree sap supplement their meals with nutrients supplied by bacterial partners that have taken up residence in special cells beneath the thoracic cuticle (*arrow*). The large aphid cells at top right surround the smaller, round cells of the endophytic bacteria that they host. The dashed line represents the interface between aphid cells and endophyte cells. Air tube cells (tracheoles) of the aphid are indicated with (*t*).

aside to provide a safe, secure haven for the bacterial cells (fig. 143). Within the bacteriocytes, the bacteria are removed from the insect's immune system, lose their outer coats, give up their days of free living, and are no longer recognized as foreign by their host. They are free to go about their business of producing amino acids and vitamins as they bathe in the welcoming environment of the host on whom they are totally dependent for their well-being.

— APHIDS, FAMILY APHIDIDAE (*aphy* = suck) (5,000 species; 1,350 species NA; ~ 3 mm).

Aphids can be found on either leaves or roots (LeafScape, 38). For most of the spring and summer, only female aphids are found on trees, and without ever mating, they continually give birth to live young that are also all females. When the leaf or root on which they are feeding becomes too crowded, the females begin to sprout wings and fly off to new feeding grounds. Only as summer wanes do male aphids make a brief appearance to fertilize the last eggs of the year that will overwinter on the tree.

Ants have a special fondness for sweet things and often travel far from their underground colonies to harvest the sap that passes completely through aphids and appears as droppings of sweet, clear honeydew (fig. 144). Some ants simply stay underground and adopt root-feeding aphids as nest mates and sources of honeydew. Ants may collect aphid eggs in the autumn and carry them to their colony. Here they tend the eggs during the winter until the newly hatched aphids wander off to feed on tiny roots. In the spring, ants sometimes carry the overwintering eggs from the base of a tree to the upper branches as new leaves unfold. In return for the secondhand sap they receive from the aphids, ants are diligent about protecting aphids from their predators and will boldly challenge the ladybird beetles, ladybird and hover fly larvae, or aphid lions that attempt to eat the aphids in their care.

Aphids can be unusually numerous on leaves, and their predators and parasitoids are correspondingly abundant and diverse. Ladybird beetles feed on aphids throughout their lives (LeafScape, 26, 28). Aphid lions are the larvae of

FIG. 144 Ants and aphids can be steadfast allies.

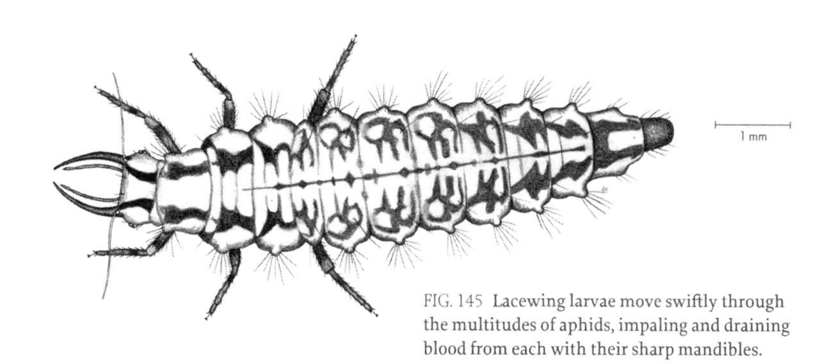

FIG. 145 Lacewing larvae move swiftly through the multitudes of aphids, impaling and draining blood from each with their sharp mandibles.

FIG. 146 The adult braconid wasp *Aphidius* leaves behind an aphid "mummy" when it exits its lifeless host.

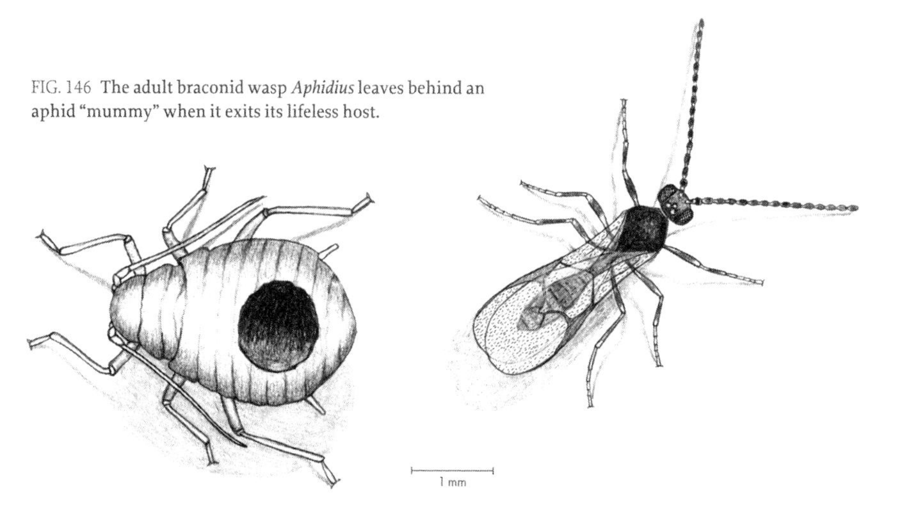

lacewings (figs. 31 and 32) and are particularly formidable-looking predators of aphids. With their hollow, sickle-shaped jaws, these fast-moving larvae impale and then drain blood from their aphid victims (fig. 145; LeafScape, 13; BarkScape, 27). Hover fly larvae use their sharp mouth hooks to rip into their victims. More than 400 species of tiny braconid wasps are known to parasitize aphids. A mother wasp lays one egg on each aphid, and the hatchling spends its entire larval life inside the living aphid. Once the larval wasp completes its development by consuming all internal tissues of the aphid, only a stiff mummy of the aphid remains (fig. 146). As the adult wasp emerges, it leaves a perfectly circular exit hole in the mummy's rear end (Starý 1988).

— APHID MUMMY WASPS, SUPERFAMILY ICHNEUMONOIDEA, FAMILY BRACONIDAE, SUBFAMILY APHIDIINAE (400 species; 125 species NA; ~ 2.5 mm).

One subfamily of aphids is noteworthy for being decked out in coats of white waxy fluff. The wax glands covering their bodies pump out long wax filaments to

FIG. 147 The slightest disturbance to a woolly aphid colony sets all of them twisting and turning in unison. Their swinging and swaying have earned these insects the additional name "boogie-woogie aphids." An individual woolly aphid is shown at lower right.

form fluffy white coats that make these tiny aphids stand out whether they are resting on twigs or gliding through the air (fig. 147). Their flamboyant fuzziness has earned them multiple names that convey the fluffy appearance of their finery: flying fuzz-balls, cotton fairies, fluffy gnats, poodle flies, and even flying mice.

— WOOLLY APHIDS, SUBFAMILY ERIOSOMATIDAE (*erio* = wool; *soma* = body) (310 species; 75 species NA; 2–4 mm).

The copious honeydew that a colony of woolly aphids drops on the lower leaves, twigs, and branches of a beech tree is a perfect substrate for a species of harmless fungus (*Scorias spongiosa*; *scoria* = dung; *spongia* = sponge) that grows only in association with the honeydew of the beech woolly aphid. The sooty mold fungus engulfs the honeydew-drenched branches and twigs with its hyphae. During late summer and early fall, the initially thin film of caramel-brown anamorphic (asexual) fungus continues to grow until it develops a spongy texture. After leaves have fallen, sooty molds are easy to spot as black, sooty lumps and clumps representing the fungus's teleomorphic (sexual) form.

Aphids have no shortage of predators and parasitoids. Aphid honeydew attracts the attention of harvester butterflies (fig. 155), whose larvae are not only the only meat-eating butterfly caterpillars in North America but also the only caterpillars that prey on aphids. Even some larvae of fly parasitoids have found homes in aphid bodies. The fly family best known for its gall-making prowess, Cecidomyiidae (chapter 2), has some small members whose larvae develop as parasitoids of the relatively large sycamore aphid (about 4 mm). These eccentric cecidomyiid parasitoids, aptly named *Endaphis* (*endo* = inside; *aphis* = aphid), emerge from the aphid as full-grown larvae that drop to the ground and pupate in the soil (Askew 1971).

— LEAFHOPPERS, FAMILY CICADELLIDAE (*cicada* = cicada; *-ella* = little) (22,000 species; 3,000 NA; 2–30 mm).

While aphids tap a tree's sugar-rich phloem cells for their nutrition, leafhoppers, cicadas, and treehoppers extend their beaks a little farther into stems to reach the xylem channels. As water transpires from the treetop, the water and nutrients in the xylem vessels continually flow upward from the roots. The nutrients in xylem sap are very dilute, and to obtain sufficient nutrition, xylem feeders like leafhoppers must take in vast volumes of this sap for their size. One calculation claims that one leafhopper in one day imbibes the equivalent of a human drinking 400 gallons in one day. From this sap, leafhoppers extract minerals, amino acids, important organic acids such as citric, succinic, and malic, and a little sugar as well (fig. 148; LeafScape, 35).

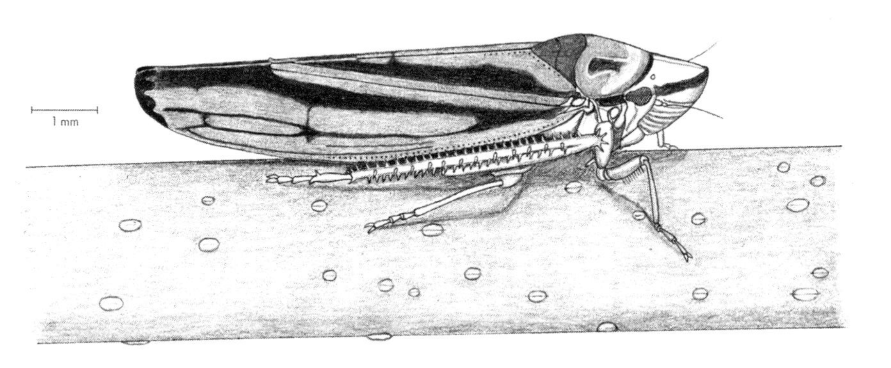

FIG. 148 Different leafhopper species have the same streamlined form but have different pigments arranged in multiple patterns.

Each spring maple sap rises in the xylem channels of maple trees, waiting to be tapped and concentrated for human use. Forty gallons of maple sap must be boiled down to generate a gallon of maple syrup. As it sucks up xylem sap from a maple tree, a leafhopper concentrates this sap by expelling and propelling a steady stream of water droplets from its rear end. Feeding leafhoppers firing off these honeydew droplets one by one, often accompanied by faint popping sounds, long ago earned these insects the name sharpshooters.

The honeydew trails that leafhoppers and fellow insect sap suckers leave on trees can spell their doom. Parasitoid wasps with a fondness for leafhoppers often follow these trails.

— PARASITIC PINCER WASPS, SUPERFAMILY CHRYSIDOIDEA, FAMILY DRYINIDAE (*drys* = oak) (1,900 species; 180 species NA; 1–11 mm).

These female wasps are wingless and look like ants (fig. 149); the males, however, have wings. Only the female has front legs with oversized tarsal claws, which she uses to grasp whatever ill-fated leafhopper she manages to track down. While she has the leafhopper in her firm grip, she thrusts her ovipositor into its

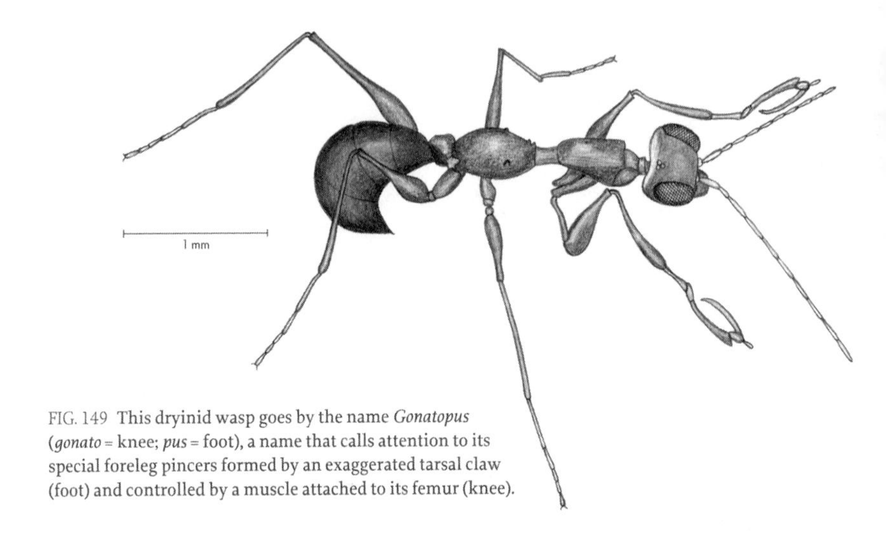

FIG. 149 This dryinid wasp goes by the name *Gonatopus* (*gonato* = knee; *pus* = foot), a name that calls attention to its special foreleg pincers formed by an exaggerated tarsal claw (foot) and controlled by a muscle attached to its femur (knee).

abdomen. As the parasitoid larva grows, it moves from the inside of the leafhopper to a chamber that it forms for itself from its molted skins. This chamber or sac protrudes from the side of the host, with only the mouthparts of the parasitoid larva still in contact with the body fluids of the host. Within its sanctum sac, it is protected from the immune response of the leafhopper.

— TREEHOPPERS, FAMILY MEMBRACIDAE (*membrax* = kind of cicada)
(3,500 species; 270 species NA; 2–20 mm).

There are treehoppers that, once they settle down and sink their beaks into a tree, look like just another spine or bump on the tree. They often appear outlandish, covered with spines and knobs, from the time they first hatch from their eggs and begin sucking sap. However, the spines and knobs on young treehoppers sometimes look very little like those of their parents. The nymph of a buffalo treehopper loses a row of spines along its back for two horns on its head when it transforms into an adult—one of many remarkable transformations of insect metamorphosis (figs. 150 and 151; LeafScape, 33).

— PLANTHOPPERS, SUPERFAMILY FULGOROIDEA (*fulgor* = lightning)
(12,500 species; 935 species NA; 1–90 mm), made up of 20 families
(Bartlett et al. 2011).

Whereas treehoppers and lace bugs have exaggerated and ornate expansions of their prothorax (first segment of the thorax), many planthoppers have exaggerated projections from their heads. One of the most famous—and one of the largest—of the planthoppers of the American tropics has an outlandish-looking head and a definite preference for the sap of quapinol trees in the pea family. The unmistakable resemblance of this tropical planthopper's head to a peanut and an alligator head has earned it the well-chosen names peanut-head bug and alligator bug (fig. 152).

A newly introduced and colorful planthopper from Asia, the spotted lanternfly, is one of those exceptional insects that prefer the sap of the unsavory *Ailanthus*

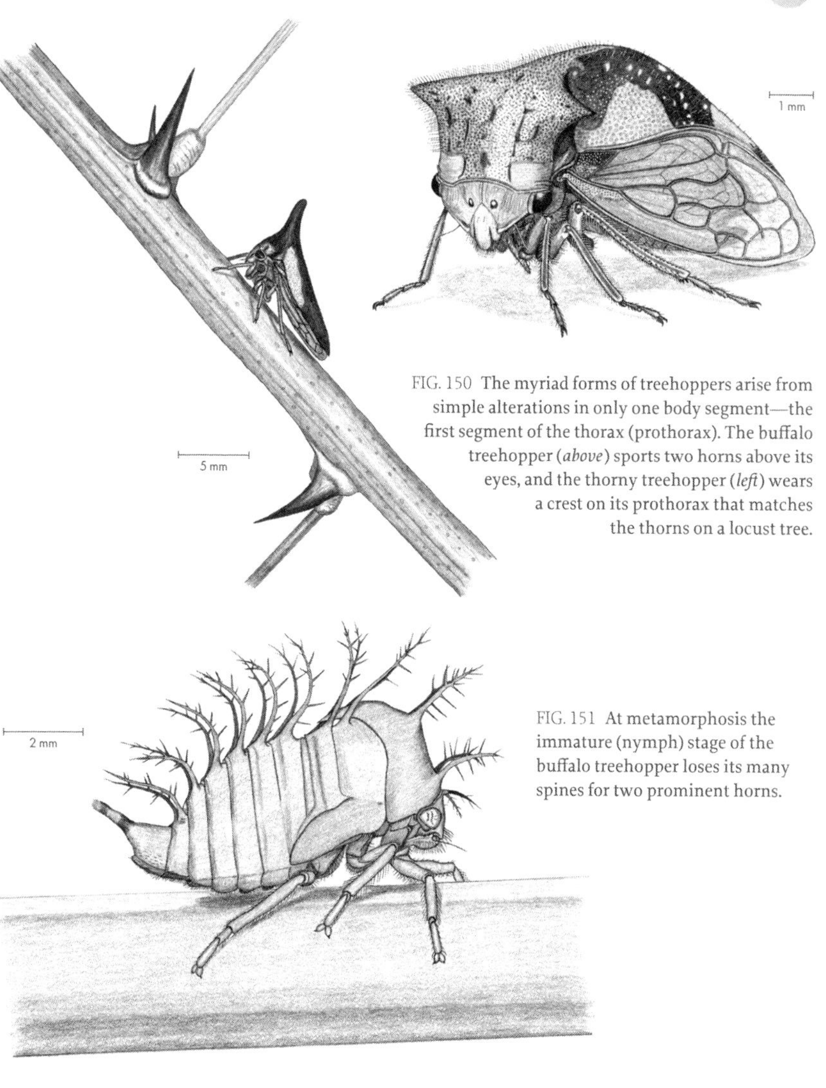

FIG. 150 The myriad forms of treehoppers arise from simple alterations in only one body segment—the first segment of the thorax (prothorax). The buffalo treehopper (*above*) sports two horns above its eyes, and the thorny treehopper (*left*) wears a crest on its prothorax that matches the thorns on a locust tree.

FIG. 151 At metamorphosis the immature (nymph) stage of the buffalo treehopper loses its many spines for two prominent horns.

FIG. 152 This tropical planthopper goes by several names including peanut-head bug, alligator bug, and lanternfly. An arrow marks the location of the eye on its unusual head.

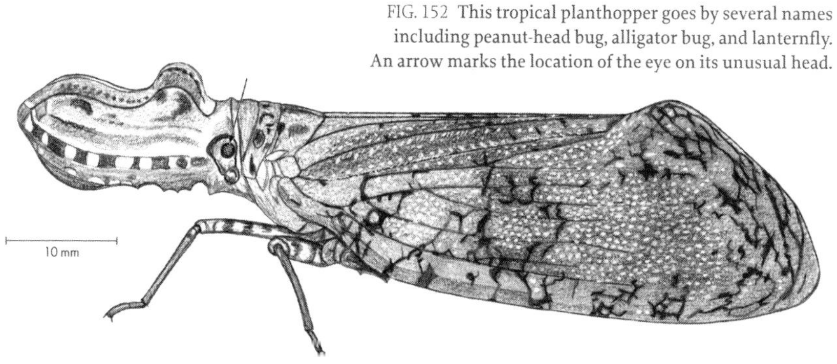

tree, another Old World introduction that arrived in America over 200 years ago. Unfortunately, this new insect arrival and rapidly spreading pest has discovered that the sap of not only the *Ailanthus* tree but also the saps of most of our native trees are acceptable, if not also tasty (Wakie et al. 2019).

Most planthoppers move with a slow gait but can hop for a quick escape. Most of the time they just sit peacefully sucking sap without ever being noticed. Not only do most planthoppers blend well with green leaves and twigs, but the venation of their prominent wings also blends seamlessly with the venation of surrounding leaves (fig. 153).

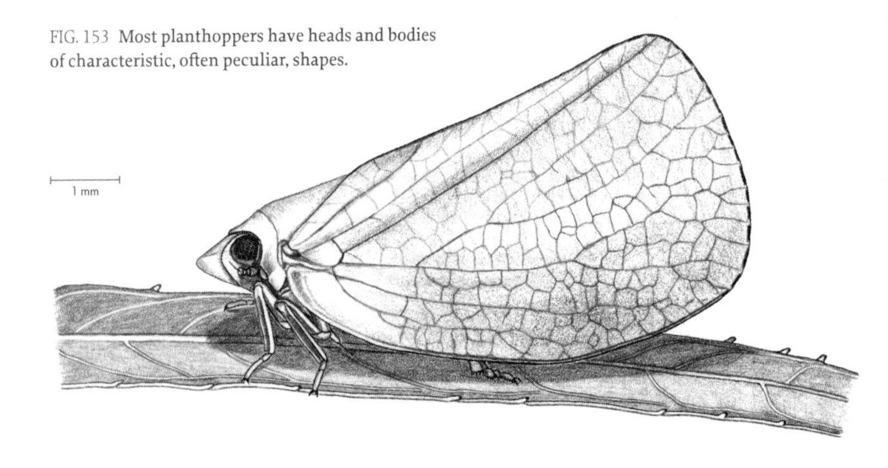

FIG. 153 Most planthoppers have heads and bodies of characteristic, often peculiar, shapes.

An exceptional connection exists between planthoppers and a most unusual family of moths—the only moths with parasitic caterpillars. For several weeks the tolerant planthopper donates its blood to the caterpillar. When the parasitic caterpillar finally parts company with its host, the planthopper ambles off without any obvious lasting harm.

— PLANTHOPPER PARASITE MOTHS, FAMILY EPIPYROPIDAE (*epi* = over; *pyr* = fire; *ops* = face) (40 species; 1 species NA; larvae 2–3 mm; wingspan 4–15 mm).

The caterpillars of this one small family of tiny moths have the unusual distinction of being the only caterpillar parasitoids of other insects (fig. 154). For some inexplicable reason, these caterpillars choose only planthoppers, and occasionally leafhoppers and cicadas (chapter 6), as hosts. Like some parasitic flies, beetles, and wasps, the mother epipyropid moth does not place her eggs on specific hosts but simply scatters several thousand eggs in planthopper habitat. Her very active larvae or planidia that hatch from these eggs must fend for themselves, either waiting for a host to pass by and then hopping aboard or actively tracking down a host. Once attached to its host, the larva settles down to a sluggish, sessile existence.

Each caterpillar latches on to the side or back of its host with silk lines, sharp claws, and sharp jaws. To lose its grip would mean certain death for the caterpillar,

FIG. 154 While the adult moths enjoy flower nectar like other moths, epipyropid caterpillars feed on planthoppers rather than on plants.

with no chance of finding a new host. The caterpillar hangs on for four to six weeks and survives on the blood that oozes from tiny punctures it makes with its jaws in the host's cuticle. After the caterpillar has reached full size, it relinquishes its tenacious hold on the planthopper to spin its cocoon on the closest branch.

While caterpillars of epipyropid moths may be the only parasitic caterpillars, the larvae of harvester butterflies in the family Lycaenidae happen to be the only truly carnivorous caterpillars that feed on plant-sucking insects. Harvester caterpillars have a fondness for aphids—especially woolly aphids found on alders and beech trees (fig. 147)—but they will also feed on scale insects and treehopper nymphs. The ants that often attend and protect honeydew-secreting aphids strangely accept these intruders among their aphid herds. Caterpillars emit an odor from their cuticles that mimics the odor released by aphid cuticles. The caterpillar's scent appeals to ant antennae and appeases whatever visual suspicions the ants might harbor. The ants probably not only accept the caterpillars but also actively protect them from other predators (Lohman et al. 2006).

Many other lycaenid larvae have adopted the tactics used by sedentary, vulnerable aphids and other sap suckers that attract and recruit ants as their defenders. Most lycaenid larvae are born with abdominal honey glands that produce a facsimile of the honeydew so esteemed by ants.

— HARVESTER BUTTERFLIES, FAMILY LYCAENIDAE, GOSSAMER-WINGED BUTTERFLIES, GENUS *FENISECA* (*feniseca* = harvester, mower) (1 species; 1 species NA; mature larvae 18 mm; wingspan 30 mm).

Adult harvester butterflies are also atypical in their tastes. Unlike almost other butterflies, the harvester does not contribute to flower pollination. Since its

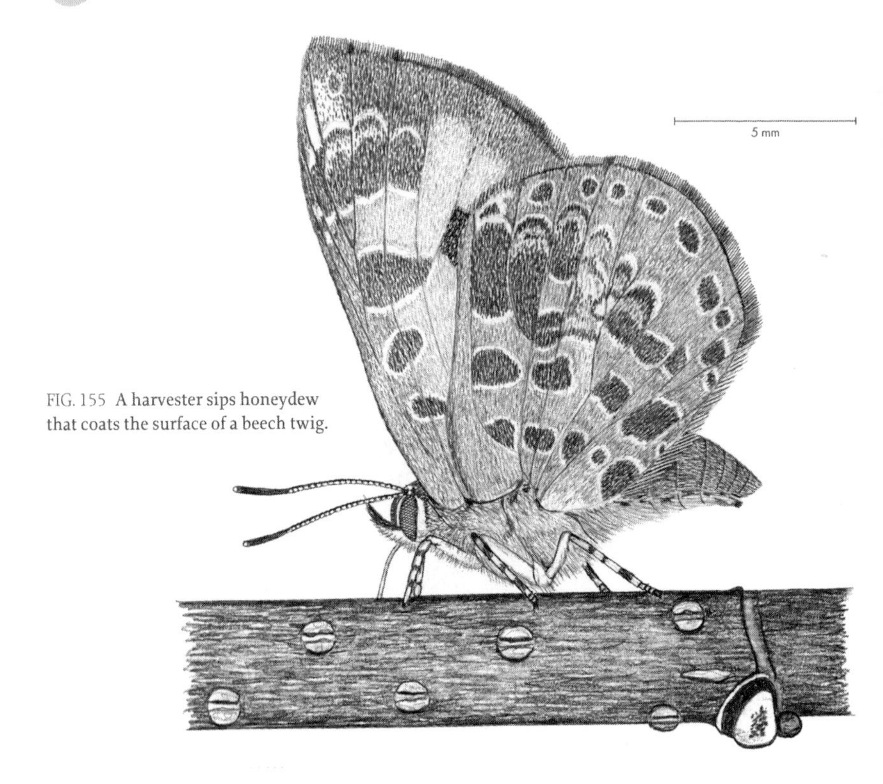

FIG. 155 A harvester sips honeydew that coats the surface of a beech twig.

proboscis is too short to reach flower nectar, it settles for sipping honeydew and rarely wanders far from its larval home, preferring to stay in the company of aphids and ants (fig. 155).

Members of another entire family of insects—this time a family of flies—are parasitoids of planthoppers, leafhoppers, and treehoppers. Only a few exceptional members of this family in the genus *Nephrocerus* (*nephro* = kidney; *cerus* = horn, antenna) prefer adult crane flies as hosts instead. While the parasitic larvae of epipyropid moths feed on the surface of these sap suckers as ectoparasitoids, the larvae of big-headed flies feed internally as endoparasitoids. In this way, the parasitoids partition their sap-sucking hosts to reduce conflict and competition. The adult flies have big heads that are unmistakable because almost their entire surface is covered by their bright red eyes (fig. 156). The body segments of the parasitic larvae blend and give them the appearance of slugs, with only a pigmented plate at the rear end marking the air entry for their respiratory system.

— BIG-HEADED FLIES, FAMILY PIPUNCULIDAE (*pipunculus* = little pumpkin, referring to their large heads) (1,400 species; 130 species NA; 2–10 mm).

— ADELGIDS, FAMILY ADELGIDAE (65 species; 22 species NA; 2–7 mm).

These sap suckers are found only on conifer trees, where they feed on twigs or needles. A few even form galls. Adelgids are best known as tiny insects concealed

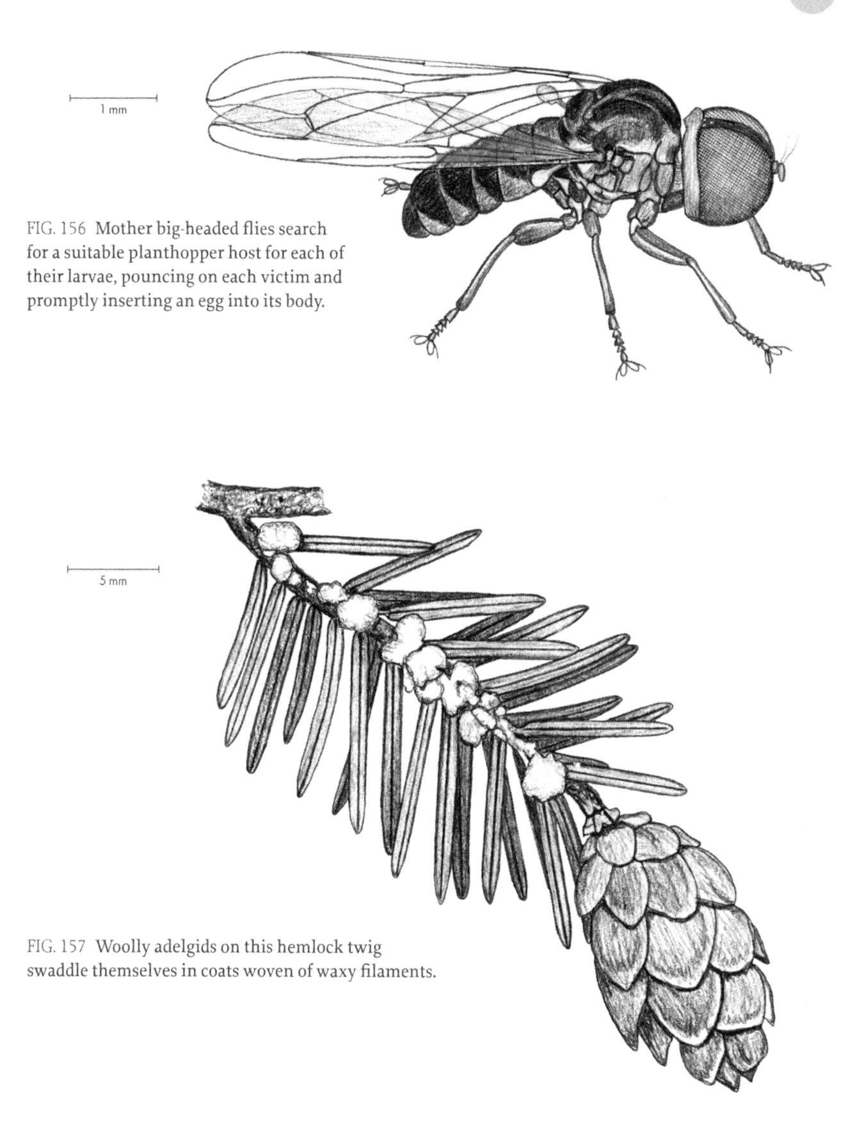

FIG. 156 Mother big-headed flies search for a suitable planthopper host for each of their larvae, pouncing on each victim and promptly inserting an egg into its body.

FIG. 157 Woolly adelgids on this hemlock twig swaddle themselves in coats woven of waxy filaments.

under woolly, waxy secretions on conifer twigs (fig. 157). They are rarely glimpsed unless one pulls aside their waxy blanket and peers beneath it. Without these waxy coats, adelgids look like psyllids (fig. 105) or tiny cicadas (fig. 343). These insects have life cycles as complicated as those of gall wasps (chapter 2). They can have sexual generations that alternate between one conifer host such as spruce and another conifer host such as pine, larch, or hemlock. But they can also have entirely asexual generations that remain on the same host from generation to generation. Unlike their close relatives the aphids, adelgids never give birth to live nymphs; they only lay eggs.

— SCALE INSECTS, SUPERFAMILY COCCOIDEA, with about 20 families (*cocco* = berry) (8,000 species; 1,900 species NA; 1–5 mm).

Among the strangest, if not the strangest, sap suckers are the scale insects. These peculiar creatures spend most of their lives on twigs, hidden beneath shells that the immature insects construct from a mixture of their shed skins and secretions from special wax glands (fig. 158). Late in the summer adult male insects emerge from beneath their shells. Although they appear to be normal insects with eyes, wings, and six legs, they have no mouthparts and are unable to feed. Their days are numbered, and they are fated to die soon after mating. Male scale insects are very ephemeral and rarely seen creatures. Female scale insects do have mouthparts, but they lack eyes, wings, and legs and never crawl out from beneath their shells—not even for mating or egg laying.

The life cycles of scale insects, like the life cycles of thrips discussed in the next section, depart in striking ways from the typical life cycles of other sap-sucking insect relatives for whom the immature nymphal stages resemble smaller, wingless versions of the adult stage. Metamorphosis with nymphal stages is referred to as incomplete metamorphosis and lacks the discrete stages of larva, pupa, and adult of those insects with complete metamorphosis such as beetles, butterflies, and bees. In scale insects, the conventional nymphal stages have been replaced by an active crawler stage that hatches from the egg and then molts to sedentary stages. Not only are there striking differences in form between female and male scale insects, but females also complete their rudimentary development with fewer molts and fewer nymphal stages (usually two or three). Development of the more refined and perfected forms of the males with their wings, eyes, and legs necessitates four molts and four nymphal instars (fig. 159), the last two of which resemble the pupae of those insects with complete metamorphosis and the "pupae" of the idiosyncratic thrips (fig. 161).

Bringing a twig with sessile female oyster scales indoors on a late winter day usually prompts the hatching of the many eggs that sheltered through the days of winter under their mothers' scales. After a few days in a warm room, the twig will be "crawling" with crawlers. The young scale insects, conceived beneath the shells of their mothers, set off from the shells to claim a new territory on their twig.

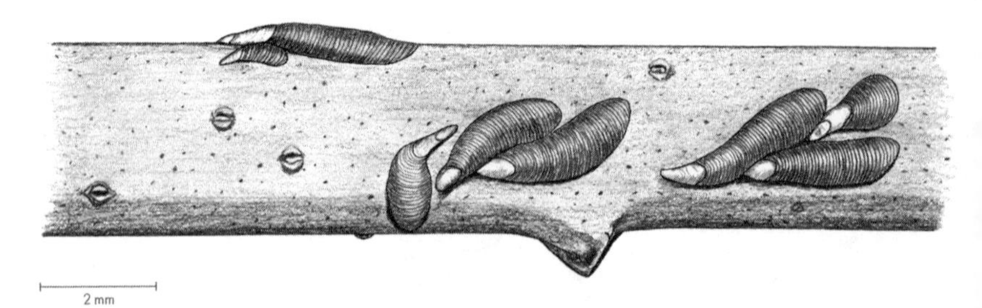

FIG. 158 Throughout the winter, the shells of female oyster shell scale insects shelter the eggs and the remains of the wingless females.

FIG. 159 Unlike the immobile adult female scale insect, the adult male scale insect (*left*) and newly hatched crawlers (*right*) move about on the tree.

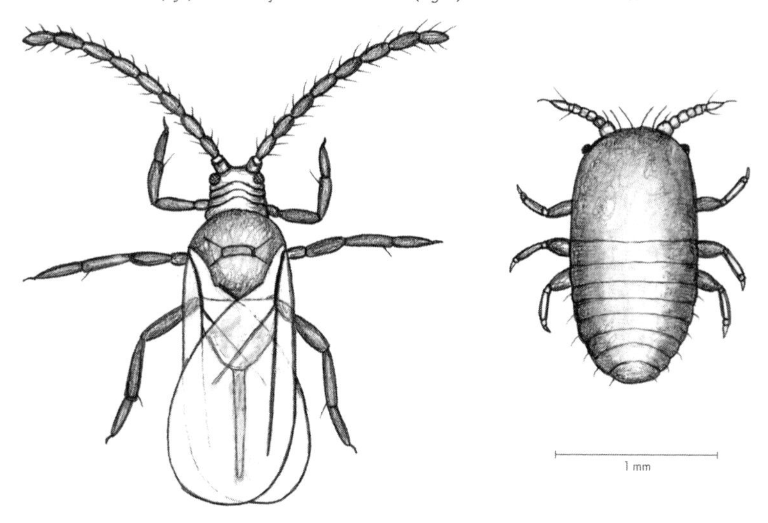

Many of these newly hatched and active nymphs or crawlers will settle down to build scales of their own and drill for sap until the end of summer (fig. 159).

Some of the scales will sport perfectly round holes on their surfaces (fig. 160). These are the exit holes of small parasitic chalcidoid wasps that have discovered these ill-fated scales. With scales on twigs often numbering in the millions per tree, these wasps encounter a vast untapped resource. What is so unusual about these wasps is how different the lifestyles of the male and female wasps are. The females of one family of chalcidoid wasps (Aphelinidae)—all of which are less

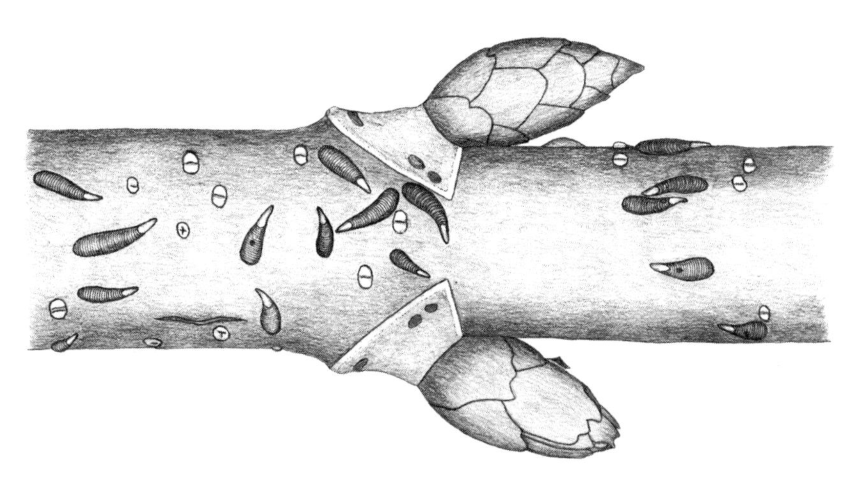

FIG. 160 Several members of this oyster shell scale colony have exit holes from which aphelinid or ceraphronid parasitoids have emerged.

than 2 mm long—specialize in parasitizing scale insects and aphids. The smaller male aphelinid wasps, by contrast, follow very different developmental pathways from those of the females, either as parasitoids of moth eggs or as hyperparasitoid cannibals of females of their own species or of related wasp species.

To add to the complexity of this already complex web of interactions, members of a very different family of parasitic wasps, Ceraphronidae, specialize as parasitoids that in turn feed strictly on these parasitoids and hyperparasitoids of scales and aphids. Other members of this diverse family are parasitoids of fly and lacewing larvae; many are wingless and stalk insects of the soil. These parasitic ceraphronid wasps are neither braconid nor chalcidoid wasps but belong to a separate superfamily of parasitic wasps. While all other parasitic wasps have their ovipositors attached anterior to the tip of their abdomens, parasitic wasps in the family Ceraphronidae have their ovipositors attached at the very tip of their abdomens. This feature is reflected in the name of the superfamily to which they belong: Proctotrupoidea (*procto* = rear end; *trup* = body).

— CERAPHRONID WASPS, FAMILY CERAPHRONIDAE (*cera* = horn; *phron* = mind, spirit) (~ 1,000 species estimated; 52 species NA; 2–3 mm).

— THRIPS, ORDER THYSANOPTERA (*thysanos* = fringed; *ptera* = wings) (6,300 species; 800 species NA; 0.5–14 mm).

Thrips make up an entire order of insects, the Thysanoptera, whose members are known for their many idiosyncratic ways. Immature thrips and many soil-dwelling thrips are wingless, but winged members of the order fly about on ornate fringed wings. The life cycle of thrips has features of complete metamorphosis with egg,

FIG. 161 In this view of the asymmetric face of a plant-feeding thrips, only the left piercing mouthparts function in drawing sap from a tree.

larval, pupal, and adult stages as well as features of incomplete metamorphosis with egg, nymphal, and adult stages. The jaws of thrips are likewise a combination of chewing mandibles and sucking maxillae. The right mandible of all thrips is the immobile and nonfunctioning mouthpart, while only the left mandible and left maxilla are articulated and capable of poking and penetrating tissues before suctioning out their contents. Why always left and never right on the lopsided head of a thrips? This invariable asymmetry of mouthparts remains one of several mysteries in the lives of thrips. Plant-feeding, pollen-feeding, and fungus-feeding thrips abound on trees, but predatory thrips control the populations of mites and other thrips that feed on plant tissues. Over tree surfaces, thrips scurry and glide about on inflatable toes that have earned them the name "bladder feet." The pads or arolia (= rolls of cloth) between their two tarsal claws can be inflated or deflated as their blood pressure waxes and wanes. These eccentric insects can be found on all parts of trees—on bark, under bark, in the soil, and on leaves, twigs, and flowers (BarkScape, 23). The winged plant-feeding thrips illustrated here shows the asymmetry of the insect's mouthparts, with only one functioning mandible present on the left side of its face (fig. 161). A predatory thrips that stalks other insects beneath tree bark is featured in chapter 4 (fig. 237).

— SPIDER MITES, CLASS ARACHNIDA, SUBCLASS ACARI, FAMILY TETRANYCHIDAE (*tetra* = four; *nyx* = puncture) (1,300 species; 250 species NA; 0.2–0.9 mm).

Mites have colonized all the continents—including frigid, uninviting Antarctica. They are at home on plants, on animals with backbones, on animals without backbones, in soils, lakes, rivers, oceans—even in hot springs and glacial meltwater. For this entire family of spider mites, the lower surfaces of tree leaves have turned out to be their most inviting habitat. These mites pierce the lower epidermis of leaves with their sharp mouthparts or chelicerae and then suck sap from the green mesophyll cells, leaving patches of yellow, dead cells in their wakes. As

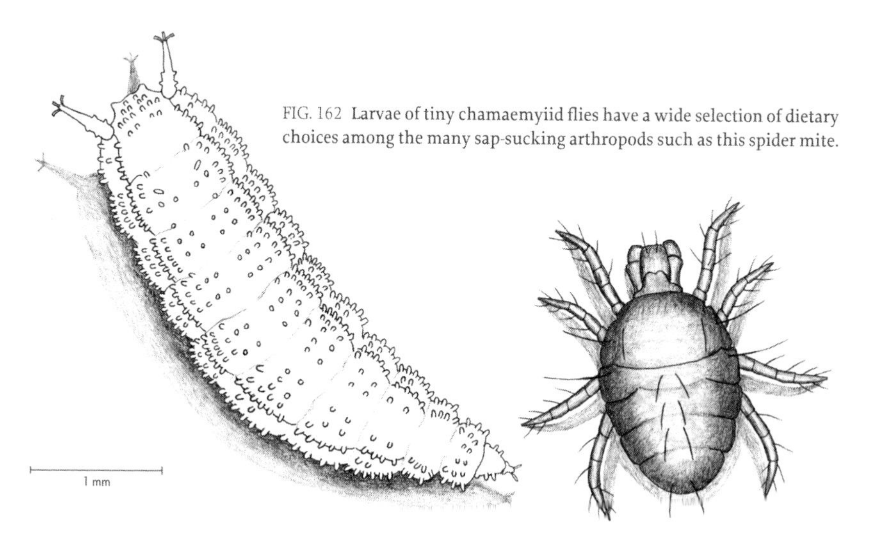

FIG. 162 Larvae of tiny chamaemyiid flies have a wide selection of dietary choices among the many sap-sucking arthropods such as this spider mite.

they move across the surface of the leaf, they leave a trail of silk that they spin from anterior appendages called pedipalps, rather than from posterior abdominal spinnerets that are the source of silk in their namesakes the spiders (fig. 162; LeafScape, 32).

— APHID FLIES, SILVER FLIES, FAMILY CHAMAEMYIIDAE (*chamae* = on the ground; *myia* = fly) (350 species; 150 species NA; 1–5 mm).

All larvae in this family are predators, and they have a definite preference for small sap-sucking insects. They belong to the ranks of aphid predators that also include larval and adult ladybird beetles (LeafScape, 26, 28), cantharid beetles (fig. 333), lacewings (figs. 31, 32, 145), and larvae of hover flies (chapter 4). The larvae of these small silvery gray flies (fig. 162) can easily overtake slow-moving mites and even slower-moving adelgids, aphids, and scale insects.

Growers of many fruit crops have benefited from the dietary needs of these tiny predators, employing them for effective control of mites, scale insects, and aphids in their orchards.

Spider mites and thrips reach the sap of trees by using their rasping mouthparts to scrape through leaf tissue. However, only the true bugs have hollow beaks that act like straws in drawing up tree sap or animal fluids. Most true bugs in the order Hemiptera feed on plant sap, and all members of the two suborders mentioned in the first half of this chapter—Sternorrhyncha and Auchenorrhyncha—are sap suckers.

True bugs of the third suborder, Heteroptera, are more diverse in their feeding habits. Predatory true bugs are discussed in chapters 1 and 2. Fungus-sucking bugs in the family Aradidae live under tree bark (chapter 4), and root-sucking true bugs in the family Cydnidae (chapter 6) inhabit the dark world of tree roots. The members of Heteroptera that are sap suckers aboveground are considered next.

All true bugs in the suborder Heteroptera, representing about 40 percent of all true bugs, have piercing mouthparts, most have stink glands on the dorsal surface of their abdomens, some have stink glands on the ventral surface of their thoraxes, and most have forewings described as hemelytra (*hemi* = half; *elytra* = cover). In contrast with beetle elytra or forewings, whose thick, sturdy cuticles provide protective covers for hind wings that do all the flying, each forewing of a true bug has a thick basal half and a thinner, membranous distal half. In their resting positions over the thorax and abdomen, beetle elytra always fold back and lie in parallel, never overlapping. However, as the forewings of a true bug in the Heteroptera fold and come to rest over its thorax and abdomen, the membranous halves overlap at the posterior midline of the bug.

— LEAF BUGS, PLANT BUGS, FAMILY MIRIDAE (*miri* = wonderful) (> 10,000 species; 1,950 species NA; 1.5–15 mm).

The family Miridae has more species than any other bug family, and its members come in a panoply of color combinations (fig. 163; LeafScape, 34). The number of species is at least 10,000, and new members are continually being added to the family. Like many true bugs, some adult leaf bugs retire to the litter on the forest floor as winter approaches, but many more spend their winters as eggs embedded

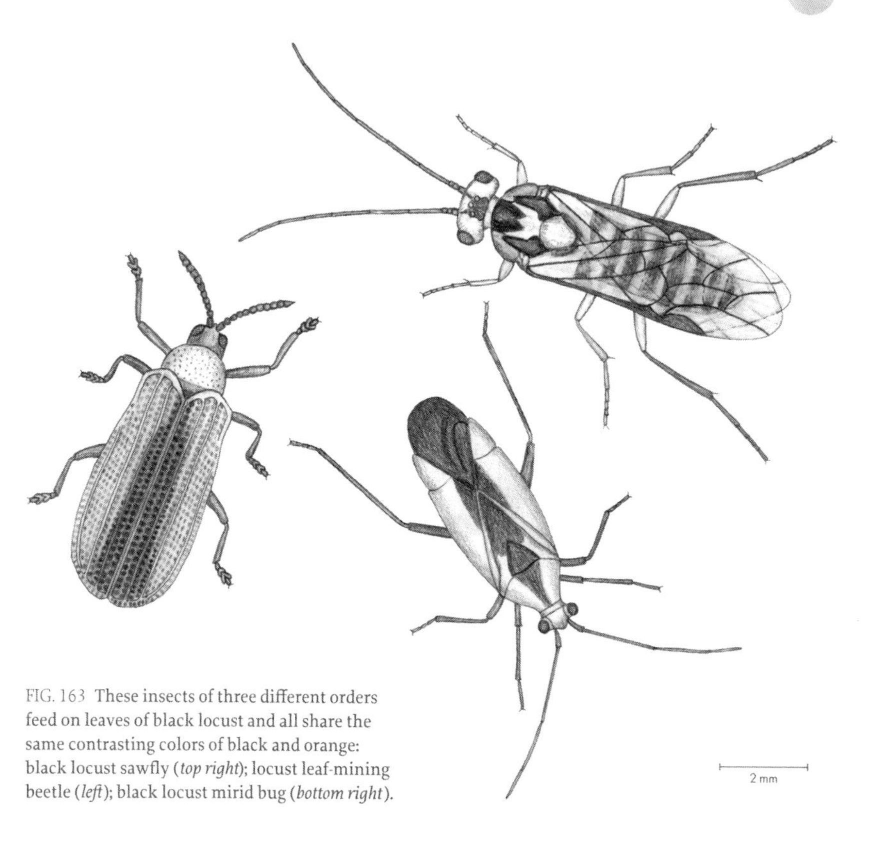

FIG. 163 These insects of three different orders feed on leaves of black locust and all share the same contrasting colors of black and orange: black locust sawfly (*top right*); locust leaf-mining beetle (*left*); black locust mirid bug (*bottom right*).

in plant tissue. Most leaf bugs, as their name implies, are leaf feeders; however, a few are predators of lace bugs (fig. 164), aphids, and mites. Even fewer, such as members of the genus *Cylapus*, appear to feed on the fungi and lichens of tree trunks and limbs. These bugs blend in beautifully with the bark and lichens over which they roam.

One mirid bug that catches the eye with its flashy orange and black colors is the black locust bug, *Lopidea robiniae*. Locust leaf-mining beetles that share locust leaves with *Lopidea* also happen to share the same body pattern of two longitudinal orange stripes flanking a central black stripe, black legs, and black antennae. The bright colors of this mirid bug announce to passersby that they are in the presence of an insect that can readily discharge a stinky mix of chemicals if one should dare to challenge it (Staples et al. 2002). The beetle has taken advantage of this bug's repellent reputation; by adopting the same intimidating appearance as this malodorous bug, the beetle also wards off unwanted threats. Even a sawfly species (*Nematus tibialis*) whose larvae feed on leaves of black locust has adopted this orange and black warning pattern. Even though they do not share the same ill-smelling odor as *Lopidea*, or even a similar one, sharing this bug's color pattern helps ensure safe passage for other insects that spend their days among locust trees.

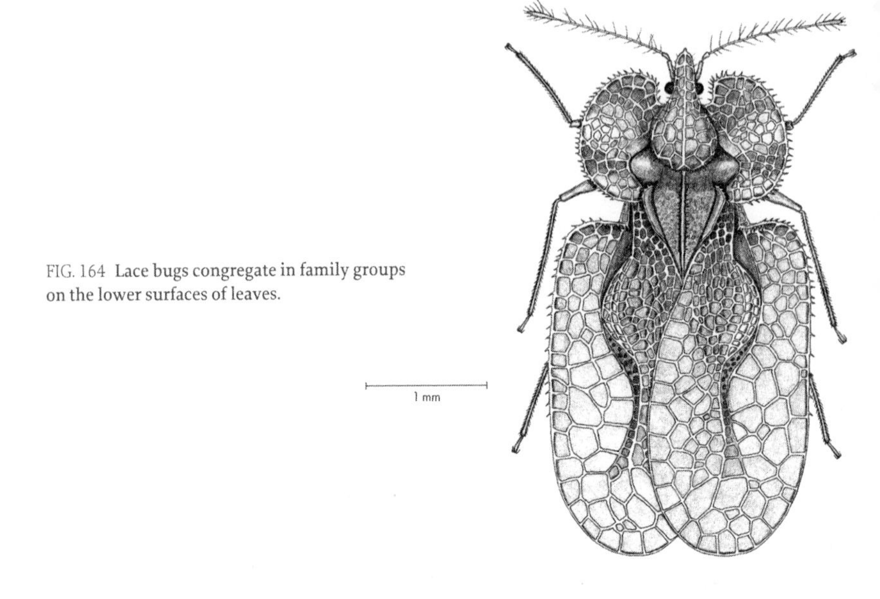

FIG. 164 Lace bugs congregate in family groups on the lower surfaces of leaves.

— LACE BUGS, FAMILY TINGIDAE (*tin* = stretched; *gan* = beauty) (2,350 species; 165 species NA; 2–8 mm).

Among the many species of sap-sucking true bugs, lace bugs are considered the most elegant looking; however, being only one-eighth of an inch long and dwelling on the undersurfaces of leaves, they almost always go unnoticed. Although inconspicuous, they are exceptionally abundant (fig. 164). A single sycamore tree can have more than a million lace bugs scattered throughout its leaves. They lay their eggs in domatia on the sycamore leaves' lower surfaces (fig. 94).

Some sap suckers have expanded their restricted diet of sap to include whole leaf cells. These are known as mesophyll feeders. Mesophyll cells occupy the spaces between leaf veins and between the leaf's two epidermal layers of cells (fig. 55), one on the dorsal surface of the leaf and the other on the ventral surface. Mesophyll cells are where green chloroplasts carry out the tree's photosynthesis and manufacture most of its sugars.

The salivary glands of mesophyll feeders contain a battery of enzymes that break down the pectin that holds leaf cells together and imparts strength and rigidity to these cells. These sap suckers and mesophyll feeders in the bug families Miridae (mirid bugs) and Tingidae (lace bugs) leave telltale signs of their dining on tree leaves as stippled and bleached patches of leaf tissue between the tough leaf veins that they leave untouched. A diet of tree sap supplemented with leaf cells apparently supplies well-balanced nutrition to lace bugs and mirid bugs. These bugs have not adopted bacterial symbionts like other sap-sucking insects to round out their diets (fig. 143) but apparently obtain all their essential nutrients from their own enzymes and digestion.

FIG. 165 Green stink bugs are well camouflaged on just about any green leaf, including this mulberry leaf.

FIG. 166 Sweet sap begins to rise from the roots of trees even before their first leaves appear.

— STINK BUGS, FAMILY PENTATOMIDAE (*penta* = five; *tomos* = sections, referring to their roughly pentagonal "shield-like" shapes and the five sections of their antennae) (4,700 species; 230 species NA; 5–18 mm).

Stink bugs distinguish themselves from members of other bug families by having roughly pentagon-shaped or shield-shaped bodies. The multiple colors of the pentatomid stink bugs blend in seamlessly with their surroundings in trees. The smooth green surfaces of green stink bugs match the leaves they move among (fig. 165), while the rough brown and gray surfaces of predatory *Brochymena* stink bugs (fig. 28; BarkScape, 26) are hard to distinguish from bits of bark on the trunks and branches they inhabit.

At the end of the summer these stink bugs and many of their true bug relatives, now fortified with nutrients from sap, dive into litter under trees where they can survive subzero temperatures and emerge unscathed in early spring, ready to tap into the sap that is once again rising in the trees from the roots to the branches and leaves (fig. 166).

THE WORLD BETWEEN BARK AND HEARTWOOD

4

READING THE LANDSCAPE OF BARK AND WOOD

Cells that occupy the world between bark and heartwood in the living tree generate wood cells (xylem) toward the heart of the tree and bark cells (phloem) toward the tree's periphery. These are stem cells referred to as cambium that continually divide as the tree grows in circumference. Each year these meristematic cells add a new ring of cells to the heartwood and a new ring to the bark. Most of the bark cells get compressed and eventually slough off; however, each year's addition of heartwood cells remains as a discrete ring that represents one growing season or one year in the life of the tree. During a good growing season with pleasant temperatures and abundant rainfall, trees add wide rings to their heartwood, but when rain is scarce and temperatures extreme, the tree's growth rings are thin and spare. However, defoliation by insects, drought, or fungi can result in a second sprouting of leaves and the addition of a second growth ring by the tree's cambium in one year. Since such occurrences are rare, establishing the age of a tree by counting its concentric rings of heartwood may not be exact but still represents a

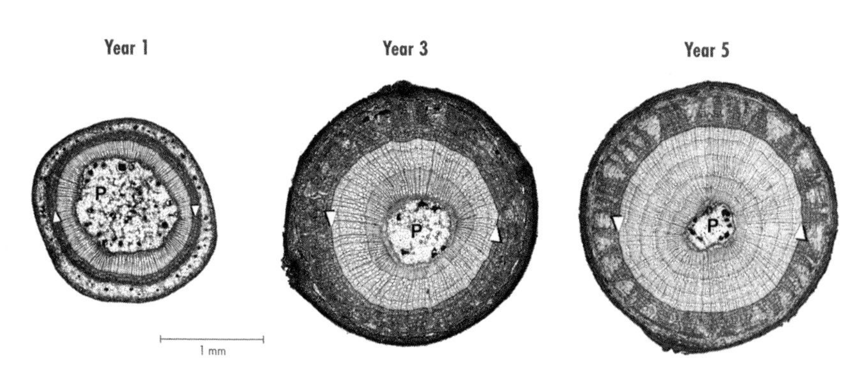

FIG. 167 In these cross sections of basswood seedling trunks, the xylem rings lie between the ring of cambium marked by the two arrowheads and the pith (P) in the center of the trunk. After a few years of growth, the pith ultimately shrinks to a point.

very good estimate of a tree's age. A tree's good years and bad years, its wet springs and dry summers are all recorded in the xylem rings of its heartwood (fig. 167).

The soft and tender cells of the cambium lack the thick, tough cell walls of the heartwood and bark. After the tree dies, these are some of the first cells to be colonized by microbes and arthropods. In the cramped space between bark and heartwood, these creatures soon carve out a habitat that becomes a refuge for insects and fungi adapted to live there (fig. 168). As the dead tree undergoes a natural succession of decay, the creatures that inhabit this space likewise come and go. Many of the insects that first colonize the cambium carve distinctive channels that eventually vanish with time. As fragments of bark fall off, the transient habitat between bark and heartwood fades away. Only the most resistant portions of the heartwood remain as the tree transforms into a well-decayed log. Eventually enough species of fungi apply their different enzymes to the job of decomposing the log and manage to return it to the soil.

Trees—living and dead—offer diverse habitats for immature insects, including many common species and some rare species. For so many larvae and pupae, life beneath the bark of an old tree seems to be a secure and sheltered habitat from which to set forth after metamorphosis to the variety of habitats frequented by adults. Insects that begin life under bark are diverse in their forms and habits—often even within a single family and within a single life cycle (fig. 168). They provide year-round dining for woodpeckers (fig. 169). Knowledge of the habits of members of many insect families is still rudimentary, and decisions about the families to which particular beetles, moths, flies, and wasps should be assigned often

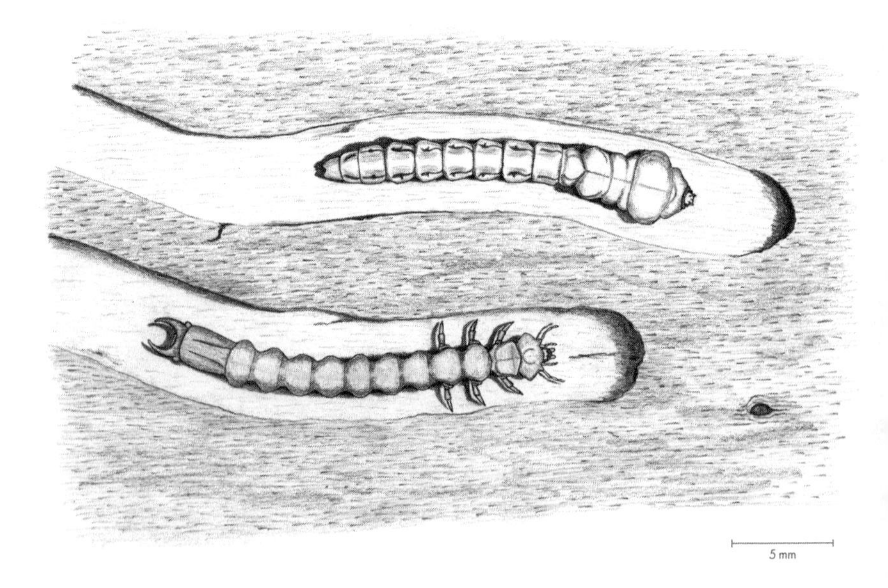

FIG. 168 Larvae of some species of metallic wood-boring beetles (*top*) and fire-colored beetles (*bottom*) mine the cambium, and other species drill into the heartwood.

FIG. 169 A downy woodpecker has excavated a hole in a willow tree where beetle larvae have already made tunnels under the bark.

change. However, so far, we have discovered so many fascinating features of their lives that we can be confident that even more enticing finds remain to be uncovered.

After its leaves are gone and its sap has dried, a dead tree remains an inviting place to creatures with special talents and symbiotic relationships. The actively dividing stem cells of cambium are rich with nutrients and have easily digested cells compared to the sturdy differentiated cells of the heartwood and bark, whose cell walls have been reinforced with extra polymers of cellulose, hemicellulose, and rigid, rot-resistant lignin. Dining on cells of bark and heartwood offers a diet rich in carbohydrates and carbon but poor in protein and nitrogen.

However, creatures inhabiting dead trees require a critical amount of certain nutrients to survive and thrive, nutrient requirements that a dead tree alone cannot fulfill. Healthy nutrition demands that creatures obtain a diet with a ratio of carbohydrates to proteins, or of carbon to nitrogen, of less than 15:1. To survive and grow on a spartan diet that is rich in carbohydrates and poor in proteins with a carbon to nitrogen ratio far greater than 15:1 requires an input of extra nitrogen and protein from some source. Insects that survive on this nutrient-poor diet have digestive tracts inhabited by bacteria and fungi that provide supplementary nutrients and enzymes. Even with help from their gut microbes, however, larvae

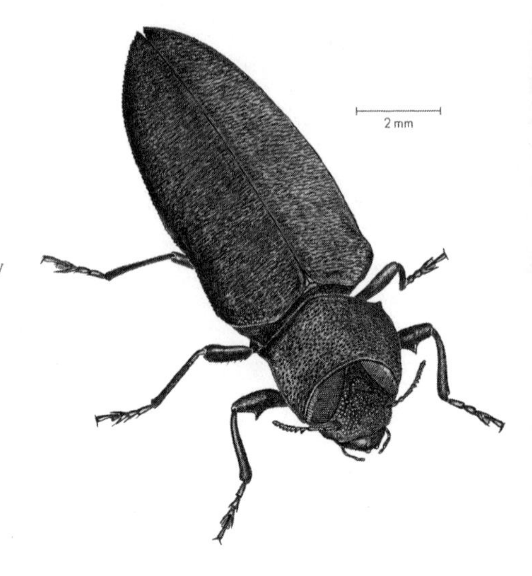

FIG. 170 After sometimes spending two or three years as larvae under bark, where they contribute to recycling dead trunks and branches, buprestid beetles fly to nearby flowers to contribute as pollinators.

that feed on this nutrient-poor diet often take several years to assimilate enough nutrients in this harsh environment to complete their metamorphoses.

After eating and navigating through dead wood, sometimes for several years, larvae halt their chewing and begin their metamorphoses. As their wings develop and then unfurl, insects leave their dark tunnels for life above the bark (fig. 170). With them they carry spores and filaments of specific fungi within special pouches on or near their body surfaces. Different species of fungi are associated with different species of wood-boring insects. These wood-digesting fungi are inoculated into the wood whenever and wherever the mother's eggs are laid. The fungi begin growing into the wood and preparing well-stocked nurseries for the larvae that will hatch in a few days.

Fungi benefit by gaining access to host trees, and fungi contribute to the well-being of their hosts by detoxifying the defensive chemicals of the trees and providing certain nutrients to their insect hosts such as amino acids and sterols. These fungi and insects also have bacterial associates that emit volatile compounds. These compounds selectively stimulate the formation of spores by some fungi and inhibit spore formation by other fungi (Adams et al. 2009; Scott et al. 2008).

WOOD WASPS AND THEIR PARASITIC WASPS

— XIPHYDRIID WOOD WASPS, FAMILY XIPHYDRIIDAE (*xiphy* = sword; *dry* = tree) (140 species; 10 species NA; 12–20 mm).

— HORNTAIL WOOD WASPS, FAMILY SIRICIDAE (*sirex* = kind of wasp) (150 species; 28 species NA; 10–40 mm).

Larvae of wood wasps are the few wood-boring members among the sawflies, almost all of whose other relatives feed on tree leaves (chapter 2). Mother horntails and wood wasps deposit a fungal inoculant along with each of their eggs. Special abdominal pouches on each female wasp harbor the fungus that supports the growing wood wasp larvae (figs. 171–173). The fungus acts as an ally for the developing larva as it tenderizes and enriches the wood of the stressed, weakened

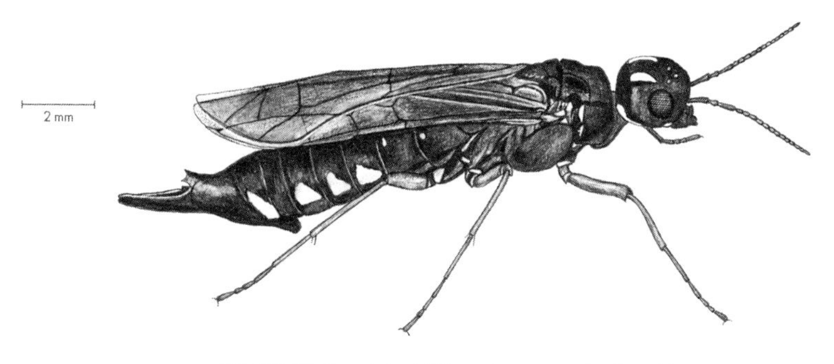

FIG. 171 Mother xiphydriid wood wasps prefer depositing their eggs and fungal partners in dead or dying maple trees.

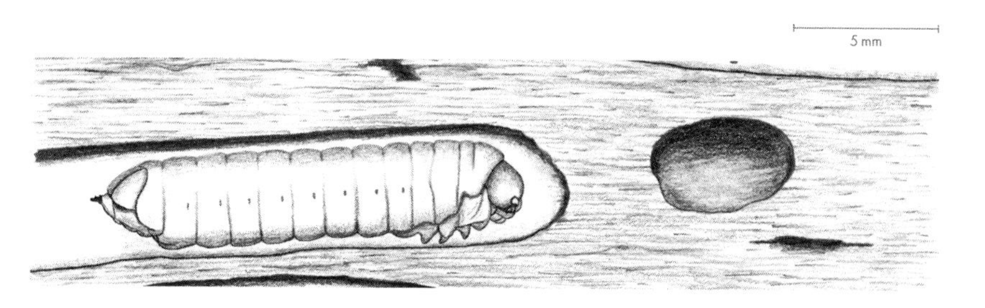

FIG. 172 Enzymes of neighboring fungi soften the heartwood and assist the excavating efforts of horntail wood wasp larvae.

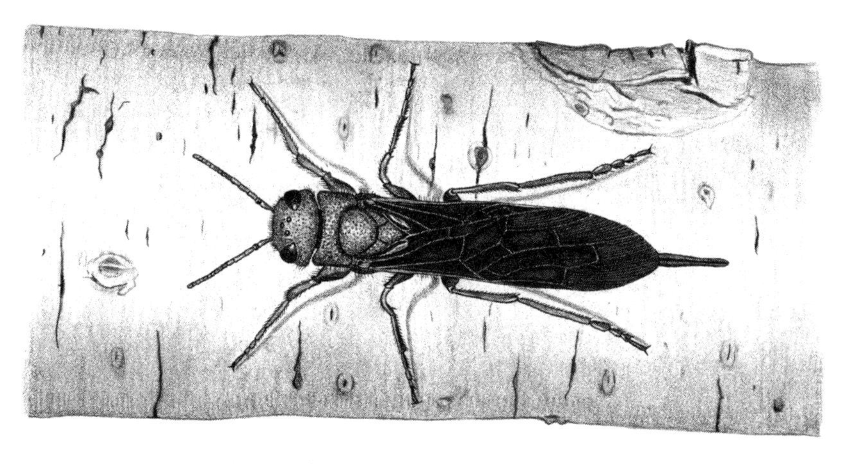

FIG. 173 As they lay their eggs, mother horntails deposit fungi that coinhabit the tunnels of their larvae.

trees that horntail mothers prefer, while the xiphydriid wood wasps usually choose dead wood. These larvae and fungi inadvertently prepare the wood to return to the soil, providing it as nourishment and habitat for generations of soil creatures that eventually liberate its mineral nutrients and leave behind humus for future generations of trees.

Nematodes as well as fungi play important roles in the lives of wood wasps. Both organisms depend on wood wasps for their distribution from tree to tree. Just as a species of fungus can adopt two very different forms—a fruiting sexual form (teleomorph) that decomposes wood, and a spore-bearing asexual form (anamorph) that feeds as a carnivore on nematodes (fig. 50)—nematode partners of wood wasps can assume two different forms—a fungus-feeding form and a form parasitic on wasp larvae. Some unknown factor, perhaps more than one, emitted by the fungus or the wood wasp larva triggers the nematode to choose either the fungivore path or the parasitic path. As the adult male and female wasps develop, the parasitic nematodes move to their developing eggs and testes, effectively sterilizing the wasps. However, the adult wasps nevertheless survive, mate, and lay sterile eggs, while spreading the nematodes throughout the forest's trees and its population of wood wasps (Morris et al. 2013).

Odors emitted by the wood wasp's symbiotic fungi are picked up by the antennae of parasitic wasps in the genus *Megarhyssa* and give away the whereabouts of their hidden wood wasp partners (fig. 174). The life cycle of one fungal symbiont of wood wasps has been well studied. After leaving the confines of its home as an

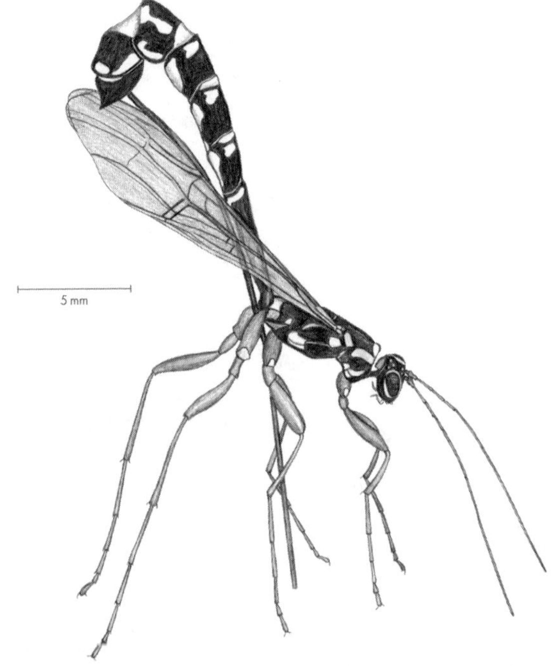

FIG. 174 With its long, serrated ovipositor, *Megarhyssa* drills deep into a tree's heartwood to reach the lair of a larval wood wasp.

anamorph in wood wasp pouches, the fungus assumes its teleomorph form as a light gray bracket mushroom named the mossy maze polypore (*Cerrena unicolor*), which sprouts from trunks of hardwood trees. Ovipositors of parasitic *Megarhyssa* wasps are long enough and sturdy enough to penetrate deep within the wood overlying the hidden larvae of the wood wasps, whose locations are inadvertently betrayed by the distinctive smell of this symbiotic fungus (Madden 1968).

— ICHNEUMON WASPS, FAMILY ICHNEUMONIDAE (*ichneumon* = tracker) (25,000 described species, estimated 60,000–100,000 species; 5,000 described species NA; 3–40 mm), SUBFAMILY RHYSSINAE (*rhyss* = wrinkled, referring to grooves on the dorsal surface of the thorax) (260 species; 15 species NA; ovipositors up to 200 mm; wingspan 12–60 mm).

Members of this subfamily are the largest members of the large parasitic wasp family Ichneumonidae. Some of the largest of these large parasitic wasps, appropriately named *Megarhyssa* (*mega* = great; *rhyss* = wrinkled), specialize as parasites of wood-boring horntail wasps. Even though these horntail larvae lie hidden in dark tunnels deep within the heartwood, they can still fall victim to these parasitic wasps whose far-reaching, highly penetrating ovipositors can bore as deep as 5 inches or 14 cm. With their very sensitive antennae, parasitic wasps can pick up odors and even slight movements of larvae below the dead tree's surface. Odors emanating from the wood-borer's fungal partner help the mother parasitic wasp locate its horntail host; in this inadvertent act of betrayal, the fungal partner divulges the whereabouts of its wood wasp partner.

Once a female parasitic wasp picks up the vibrations and scents of a host for her larva, she raises the ovipositor at the tip of her abdomen high in the air as she poises on outstretched hind legs. The backward-pointing ovipositor on the ventral surface of the abdomen begins rotating 90 degrees until it points downward and comes to lie within the inner surface of a compartment formed by stretching of the membrane separating the adjacent, more anterior segments of the abdomen. The ovipositors of many parasitic wasps are often much longer than their bodies, and their highly penetrating tips armed with small but sharp teeth are reinforced by the addition of zinc, manganese, and/or copper. Strong abdominal muscles contribute a downward force on each ovipositor. Two shafts interlock to form each ovipositor and surround the channel through which eggs pass one at a time. The secretion from the tips of the shafts locally softens wood fibers as the ovipositor slowly and methodically dives deeper into the heartwood. The teeth of one shaft catch the wood while the other shaft inches forward. The two shafts alternate in their forward advance. Lubricating secretions also coat the ovipositor as it begins boring into the wood. With a combination of physical and enzymatic penetration of wood fibers, the mother parasitic wasp uses the metal-reinforced tip of her ovipositor to position a single egg on her hidden host (Polidori et al. 2013; LeLannic and Nénon 1999).

Symbiotic fungi and bacteria of wood-boring insects provide a range of services to their hosts: they contribute compounds that detoxify defensive compounds of trees; they provide essential nutrients to their insect hosts; and they even influence which microbes should share the habitat with them and their hosts. All the benefits of hosting fungal and bacterial partners, however, may have a drawback. With so many species of wood borers associated with specific fungal and bacterial communities, the tracking of odors derived from their fungal and bacterial symbionts offers a widespread and effective stratagem that parasitic insects can use to locate their hidden wood-boring hosts. While fungi, bacteria, and wood-boring insects all benefit from their endophytic partnership, the parasitic wasps of wood borers likewise benefit by exploiting the escape of telltale odors to track down evasive invaders of a tree's heartwood. These parasitic insects act as tree allies, defending them from the ravages of wood-boring wasps and beetles.

The parasitic wasps in the three following related families also have an uncanny ability to locate hosts concealed under bark or inside twigs. They all have two features that set them apart from all other wasps. The last segment of their thoraxes and their entire abdomens are perched above the dorsal surfaces of their first two thoracic segments rather than attached, as in other wasps, posterior to these first two segments. As conveyed by the family name Gasteruptiidae, the bellies of these wasps appear to be bent backward. These wasps also have conspicuous "necks" that help distinguish them from the similar-appearing but usually more common ichneumonid wasps that have no necks whatsoever. These are the only three families belonging to the parasitic wasp superfamily Evanioidea.

— AULACID WASPS, FAMILY AULACIDAE (*aulac* = groove, probably referring to distinctive grooves on the basal trochanter and coxal segments of legs and/or between eye and antenna) (200 species; 30 species NA; 2–20 mm).

Adult aulacid wasps have mottled wings whose disruptive coloration helps them blend seamlessly with the mottled patterns of bark (fig. 175). The mother wasps locate the shafts drilled by ovipositors of their host wasps and beetles. Although the host larva and parasitoid larva start out at about the same stage and size, the parasitic aulacid larva delays and prolongs its development as its host grows to full size. Not until its host is well fed and fully grown does the parasitic larva quickly devour it. The aulacid larva, now itself fully grown, leaves the empty cuticular shell of its host, spins a cocoon, and pupates.

— CARROT WASPS, FAMILY GASTERUPTIIDAE (*gaster* = belly; *upti* = bend backward) (2,000 species; 15 species NA; 13–40 mm).

Adult carrot wasps appear to have a special fondness for the umbel flowers in the carrot family such as Queen Anne's lace, parsnips, and parsley (fig. 176). For their hosts, carrot wasp larvae prefer bees and wasps that nest in twigs of trees. (figs. 242–245).

— ENSIGN WASPS, HATCHET WASPS, FAMILY EVANIIDAE (*evani* = disappearing, perhaps referring to the disappearance of wing veins in various members of the family) (400 species; 11 species NA; 3–7 mm).

As they search leaf litter and tree trunks for the egg cases of their host cockroaches, these wasps pump their posterior ends rhythmically up and down. Their movement

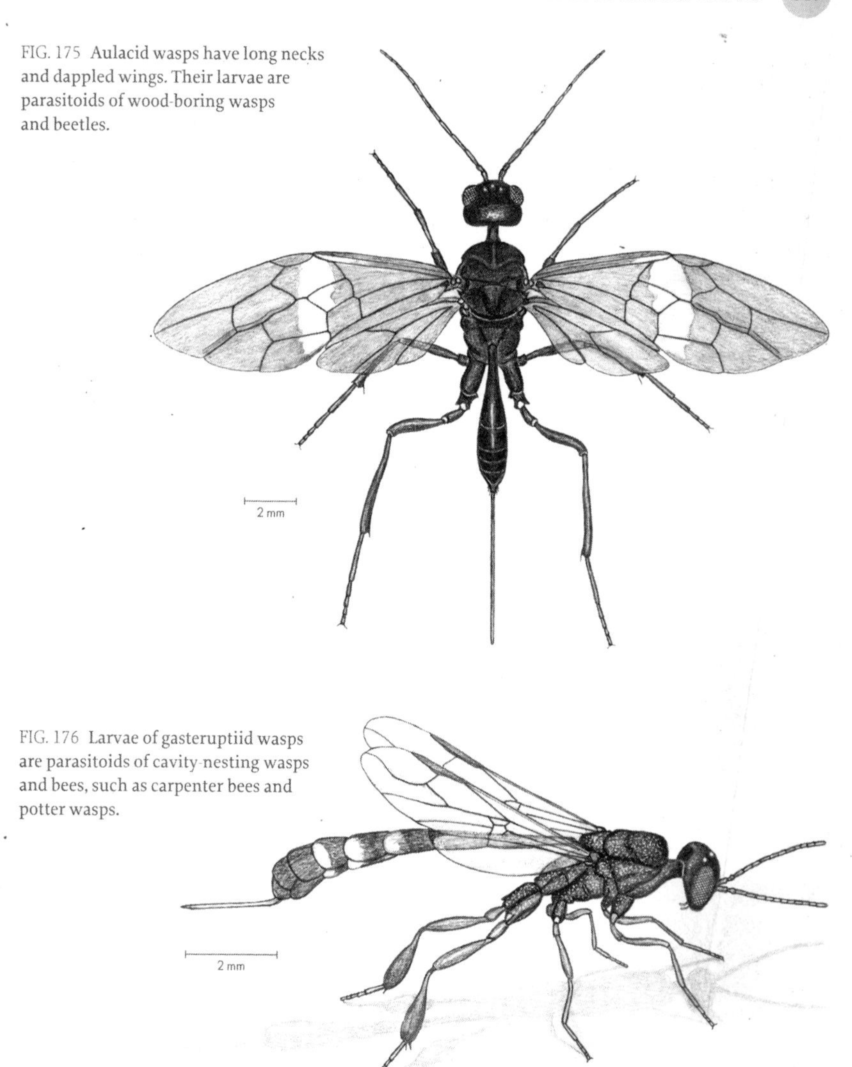

FIG. 175 Aulacid wasps have long necks and dappled wings. Their larvae are parasitoids of wood-boring wasps and beetles.

FIG. 176 Larvae of gasteruptiid wasps are parasitoids of cavity-nesting wasps and bees, such as carpenter bees and potter wasps.

suggests the chopping motion of a hatchet or the waving of an ensign flag. These wasps have very short and not particularly stout ovipositors, which they use to place each of their eggs in a cockroach egg case (fig. 177). Anticipating the challenges of penetrating a tough egg case with her short ovipositor, a mother wasp searches for egg cases that have been newly or recently laid and have not yet developed their hard, impervious coats. Once the mother wasp deposits her egg in the cockroach's egg case, the larva has access to abundant provisions since each case contains around 50 eggs. Consuming only a few of these eggs provides ample nutrition for a single wasp larva

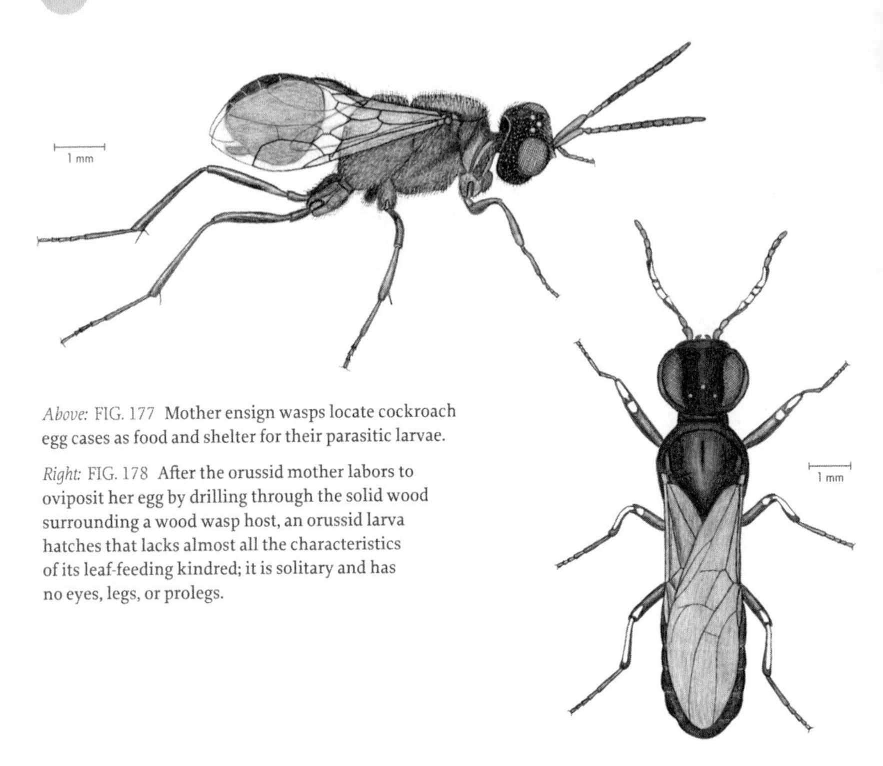

Above: FIG. 177 Mother ensign wasps locate cockroach egg cases as food and shelter for their parasitic larvae.

Right: FIG. 178 After the orussid mother labors to oviposit her egg by drilling through the solid wood surrounding a wood wasp host, an orussid larva hatches that lacks almost all the characteristics of its leaf-feeding kindred; it is solitary and has no eyes, legs, or prolegs.

— PARASITIC WOOD WASPS OR ORUSSID SAWFLIES, FAMILY ORUSSIDAE
(*oruss* = burrowing) (85 species; 7 species NA; 2–15 mm).

Orussid sawflies stand alone as the only sawflies that are not plant feeders. In their search for larvae of wood-boring beetles and wood wasps, these parasitic sawflies busily tap tree bark with unusual, kinked antennae that lie exceptionally close to the bark, arising just below the sawfly's large eyes and just above the jaws. Once the mother orussid sawfly locates a suitable hidden host, she begins the laborious drilling that will ensure her offspring a good start in life. Unlike the long, highly visible ovipositors of other parasitic wasps such as that of *Megarhyssa*, the ovipositor of the mother orussid sawfly lies hidden from view. Her concealed ovipositor, however, is longer than her body and is extruded only a short stretch at a time as it extends deeper into the wood. Accommodating this imposing ovipositor in such a small sawfly body is a feat worthy of a master illusionist. The ovipositor lies coiled within the abdomen and thorax, within a long, deep, cuticle-lined inpocketing at the end of the female's abdomen that extends almost the entire length of her body.

The antennae and legs of orussid sawflies are dappled with bright patches of white (fig. 178). These patches stand out and disrupt the otherwise distinct dark pattern of the sawflies. The optical illusion of this disruptive coloration helps these sawflies blend in with the rough patterns on tree trunks.

Parasitic wasps have mastered the art of locating wood-boring hosts hidden deep under tree bark. In addition to the horntail wood wasps and xiphydriid wood wasps that they so deftly track down using their sensitive antennae and sharp, penetrating ovipositors, they also come across many beetle larvae that tunnel through a tree's heartwood. Many of these slow-moving, plump beetle larvae serve as suitable—even preferable—hosts for the larvae of many parasitic wasps. These wood-boring beetles certainly present a plethora of choices.

BEETLES OF THE HEARTWOOD

— BEETLES, ORDER COLEOPTERA (*coleo* = sheath; *ptera* = wings) (390,000 described species; 28,000 species NA; 0.5–120 mm).

Relationships between beetles and trees go back a long way. In terms of the number of species involved in such relationships, no group of animals or plants that has been described by biologists can match the beetles. About 40 percent of all insect species are beetles, and about 25 percent of all animal species are beetles. The classification of beetles, like the classification of most plants, animals, and other creatures, is constantly being revised. As new information on the genetic material of beetles has emerged, new beetle groups have been proposed that include beetles once assigned to other groups. As a result of new classifications, some new beetle families have been created, some old families have grown in number of species, some have shrunk in species numbers, and some have remained as before. Currently there are around 180 families of beetles worldwide and around 130 families in North America.

Judging from the names that have been assigned to beetle families, tree bark and tree fungi feature prominently in the lives of many beetles. The beetle families whose common names include "bark," "timber," or "log" number 14. The common names of at least 11 beetle families include the word "fungus." Since many members of beetle families with such common names as spider beetles, darkling beetles, checkered beetles, click beetles, and false click beetles also lead lives under tree bark and among fungi, counting only the beetles assigned to the families whose common names include the words "fungus," "bark," "log," or "timber" provides a great underestimate of the beetles that live in, on, or under bark or that obtain most of their nourishment from fungi. Names of so many beetles reflect how the lives of trees are intertwined with the lives of beetles—the most diverse group of animals on our planet.

— POWDER-POST BEETLES AND BRANCH- AND TWIG-BORING BEETLES, FAMILY BOSTRICHIDAE (*bostrich* = type of insect) (700 species; 70 species NA; 2–24 mm, one exception = 52 mm).

— DEATH-WATCH AND SPIDER BEETLES, FAMILY PTINIDAE (*ptin* = feathered) (2,200 species; 420 species NA; 1–9 mm). This family was formed by the union of two older families—the Ptinidae and the Anobiidae.

The larvae and adults of these closely related beetles drill into wood, obtaining most of their nourishment from whatever starch they can extract from the surrounding cellulose. With their heads and jaws facing forward, the powder-post beetles (fig. 179) can be distinguished from the death-watch beetles (genus *Anobium*; *ano* = without; *bio* = life) as well as the branch- and twig-boring beetles, whose heads and mandibles face downward (fig. 180).

The powder-post beetles and the branch- and twig-boring beetles are elongate, cylindrical beetles that drill channels through wood, perforating it with a multitude of tiny holes and leaving behind wood reduced to a powder. When death-watch beetles channel through dead wood, their heads bob up and down, tapping against the wood surface with a regular rhythm of tick, tick, tick. This sound emanating from the wooden walls of their homes must have been perceived as ominous enough to earn them the name death-watch beetles. Some species of death-watch beetles have ventured into all sorts of dead plant matter that has been modified for human purposes—lumber, cigarettes, and dried herbs and spices. The success of these beetles at colonizing and surviving in such a range of habitats and food

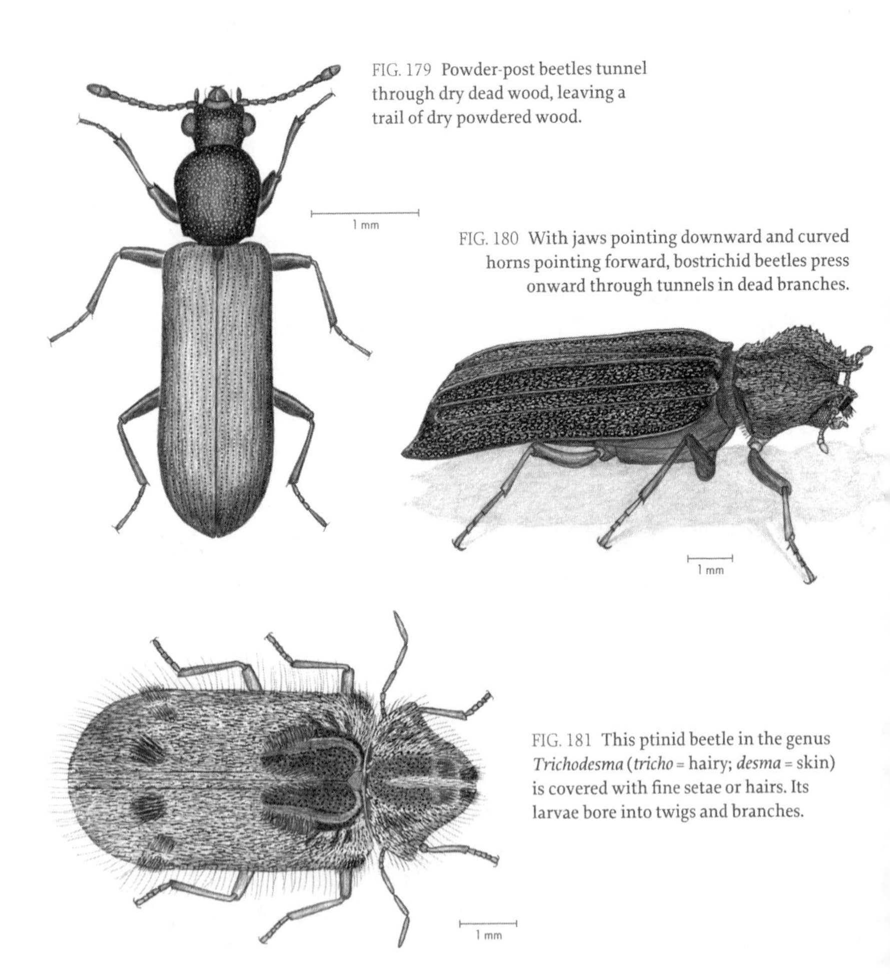

FIG. 179 Powder-post beetles tunnel through dry dead wood, leaving a trail of dry powdered wood.

FIG. 180 With jaws pointing downward and curved horns pointing forward, bostrichid beetles press onward through tunnels in dead branches.

FIG. 181 This ptinid beetle in the genus *Trichodesma* (*tricho* = hairy; *desma* = skin) is covered with fine setae or hairs. Its larvae bore into twigs and branches.

sources as unpalatable and as poor in nutrients as dry wood can be attributed to the bacteria and fungi that have colonized their digestive tracts (fig. 181). These microbes contribute their potent enzymes to consuming foods that are recalcitrant to digestion by many beetle enzymes. With the help of their microbial partners, these beetles accomplish such surprising digestive feats as consuming hot red pepper and nicotine-laden cigars.

— LONG-HORNED BEETLES, FAMILY CERAMBYCIDAE (*cerambyc* = type of beetle) (30,000 species; 1,000 species NA; 2–200 mm, typically 5–50 mm).

Practically all larval members of this large beetle family are wood borers of both living and dead trees; however, most of the adults in this family visit flowers and are pollinators. As expected from their name, adults of long-horned beetles have long antennae (fig. 182; LeafScape, 7; BarkScape, 11). The males of each species, however, have longer antennae that are especially attuned to the pheromones of females. The larvae, known as round-headed borers, excavate circular passageways through tissues of both conifers and deciduous trees, from pith and twigs to the heartwood of the sturdiest hardwoods such as hickory, oak, and black locust. The largest North American beetles of the family can be 75 mm; their larvae specialize in boring within roots of both living and dead trees. All round-headed borers leave behind a trail of winding tunnels packed with layers of their droppings as they eat their way through wood (fig. 183).

FIG. 182 The first thing you notice about long-horned beetles is their long antennae.

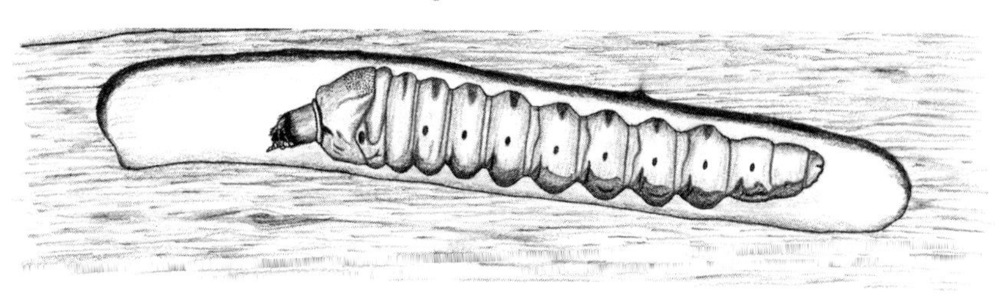

FIG. 183 The tunnels of round-headed borers penetrate the heartwood of trees.

The dilemma faced by insects that feed on wood—whether solid or decaying—is that wood is not only difficult to digest but is also low in essential nitrogen-containing nutrients. To circumvent this dilemma, beetle larvae have recruited intestinal microbes—bacteria and fungi—to assist with the digestion of tough wood fibers that contain both lignin and cellulose. Pouches protrude from the basal surface of each larva's intestine and accommodate millions of these microbial partners. In these pouches segregated from the food moving through the main lumen of the gut, microbes work and multiply unperturbed. Here the microbes contribute not only their enzymes to the digestive process but also vitamins and amino acids to supplement the nitrogen-poor diet of the beetle larvae (Mohammed et al. 2018).

For the twig girdlers among the long-horned beetles, girdling is the job of the mother beetle. She girdles each twig of hickory, apple, or elm at its base and then lays an egg at the end of the twig. The mother's girdling cuts through the peripheral bark and sugar-transporting phloem tissue of the twig, leaving the more central xylem tissue intact. Sugars continue to form in the leaves, but their transport beyond the girdle is halted even though inorganic nutrients and water transported by xylem from the soil continue to nourish the leaves at the end of the twig. Sugars and other organic nutrients normally transported from the leaves to stems and roots on the other side of the girdle are therefore trapped and concentrated for the use of the developing beetle larva—clearly a wise strategy for beetle success (Forcella 1982).

— ZOPHERID BEETLES, FAMILY ZOPHERIDAE (*zopherus* = darkness) (1,700 species, 110 species NA; 2–35 mm).

Until recently many members of this family were referred to as cylindrical bark beetles. Many have short, stubby legs and sleek, elongate forms for navigating narrow passageways. These beetles travel the dark, fungus-lined galleries of wood-boring beetles. Like so many hidden insects of trees, the larvae and adults of these beetles have features of their lives that remain undiscovered. In their dark and diminutive habitats, we get only brief glimpses of what transpires. These beetles may survive as predators on other wood-boring beetles, as fungivores, or as both carnivores and fungivores (fig. 184).

— SHIP-TIMBER BEETLES, FAMILY LYMEXYLIDAE (*lymo* = destruction; *xylo* = wood) (70 species; 3 species NA; 5–40 mm).

Larvae of ship-timber beetles have been found in both sapwood and heartwood of dead and even living trees. In the days of wooden ships, the larvae of these beetles were often found in the oak and chestnut timbers of ships. Fungal partners assist these larvae in their wood-working missions. The mother beetle stores the fungus *Endomyces* in a pouch next to her ovipositor (fig. 185). As eggs are laid, each becomes coated with fungal spores. In this way she ensures that each of her offspring receives a share of its fungal partner. To maintain conditions optimal for the growth of their fungal partners, the larvae keep their tunnels clear of debris with sweeps of their long, spiny tails (fig. 186).

One of the North American species of ship-timber beetles that was once common in the eastern deciduous forests is now rarely seen. The fungal chestnut blight introduced from the Old World at the start of the twentieth century decimated

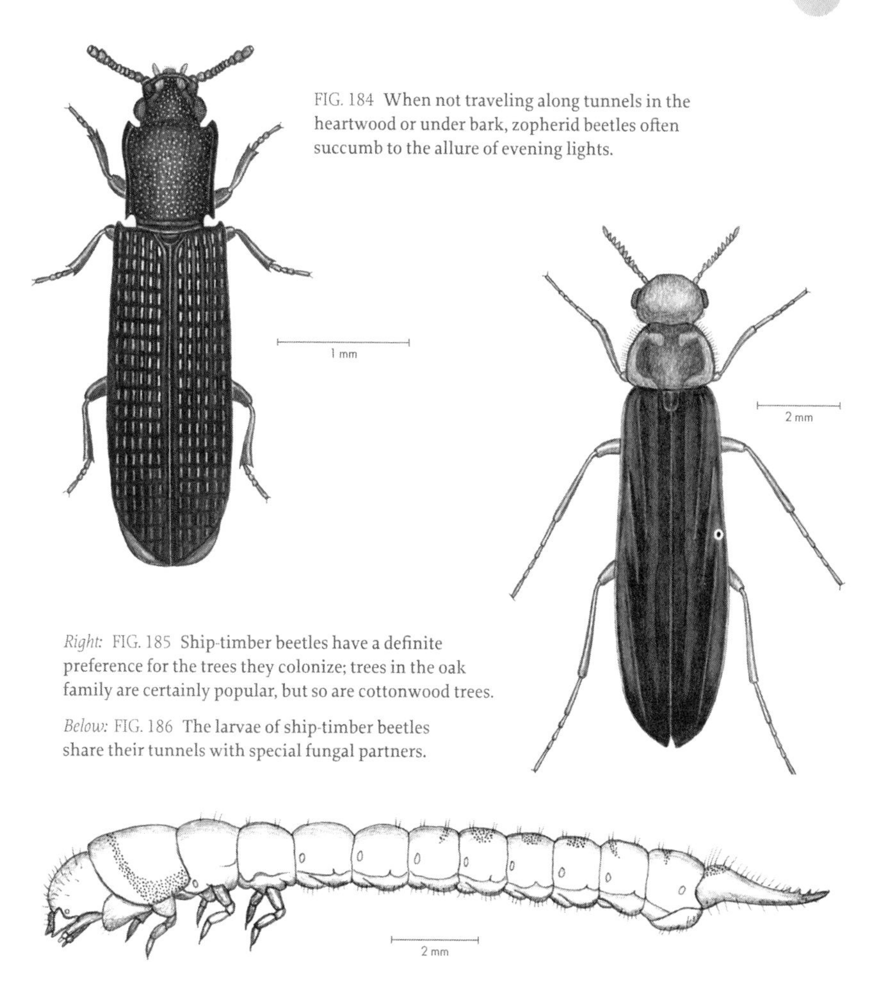

FIG. 184 When not traveling along tunnels in the heartwood or under bark, zopherid beetles often succumb to the allure of evening lights.

Right: FIG. 185 Ship-timber beetles have a definite preference for the trees they colonize; trees in the oak family are certainly popular, but so are cottonwood trees.

Below: FIG. 186 The larvae of ship-timber beetles share their tunnels with special fungal partners.

the populations of native chestnut trees. As its favorite chestnut trees disappeared, so did the lymexylid beetle known as the chestnut timberworm.

ANTS AND TERMITES OF THE HEARTWOOD

Carpenter ants and termites have little in common genealogically, and they have very different food preferences. However, both are found in the same habitats and are often mistaken for one another.

— ANTS, FAMILY FORMICIDAE (*formica* = ant) (estimated 22,000 species, 12,500 described; 700 species NA; 4–25 mm), CARPENTER ANTS, GENUS *CAMPONOTUS* (*campo* = bent; *notus* = back) (over 1,000 species; 50 species NA; 5–25 mm).

Carpenter ants are a diverse group (genus) of over a thousand species (fig. 187; BarkScape, 9). The conspicuous carpenter ants have the undeserved reputation of invading and destroying firm and sturdy wooden structures; however, that infamous reputation belongs to their fellow wood dwellers, the termites. Unlike termites, which excavate perfectly sound wood for consumption and the construction of colony corridors, carpenter ants chew into already decaying wood, not for dining but for constructing galleries and passageways for their colonies. Carpenter ants, unlike termites, cannot digest the cellulose of wood fibers. Carpenter ants are predators of live insects and scavengers of dead insects, often supplementing their diets with honeydew, nectar, and plant sap. Foraging workers communicate their discoveries by leaving pheromone trails to food sources. Colony members join them in the retrieving until the food source is depleted.

— TERMITES, ORDER ISOPTERA (*iso* = equal; *ptera* = wings) (3,100 species; 50 species NA; 4–15 mm).

Termites create habitat for carpenter ants by excavating tunnels in dead branches and tree trunks that the ants are eager to claim for themselves. After the termites have devoured and digested the wood of their tunnels, they prepare the way for carpenter ants to move in. The ants often extend the tunnels by only excavating and not eating their way into the wood. While many tropical termites survive without cellulose from trees, the one species of termite in northeastern North America—*Reticulitermes flavipes*—devours, digests, and thrives on the cellulose of damp or even dry wood (fig. 188). Plain, unadulterated wood cellulose is digestible for termites only because termite guts contain microbes whose enzymes help them digest this hard and fibrous food. The decomposing activities of termites greatly influence the properties of the soil, to which they add nutrients and undigested organic matter from the wood that passes through their guts. Nutrients that have been sequestered for years in the heartwood of a tree are now liberated to enrich the structure and chemistry of the nearby soil.

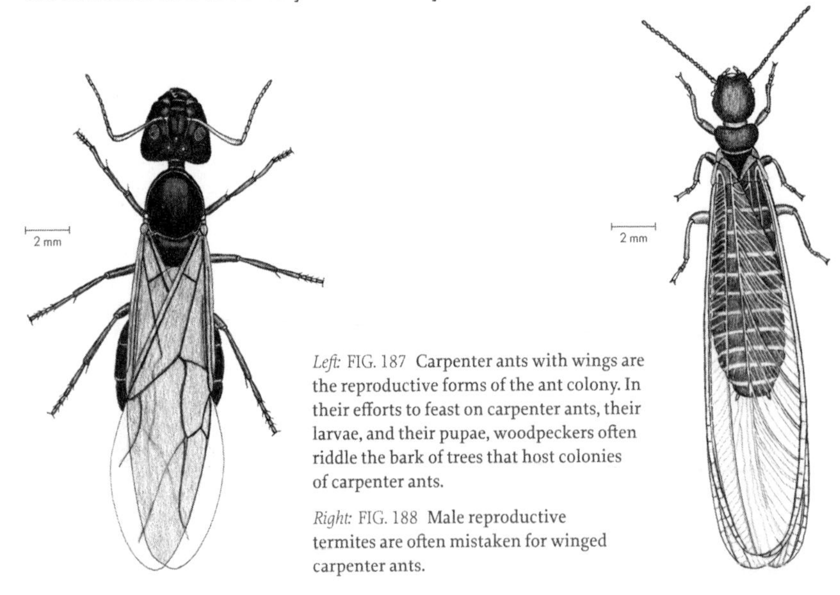

Left: FIG. 187 Carpenter ants with wings are the reproductive forms of the ant colony. In their efforts to feast on carpenter ants, their larvae, and their pupae, woodpeckers often riddle the bark of trees that host colonies of carpenter ants.

Right: FIG. 188 Male reproductive termites are often mistaken for winged carpenter ants.

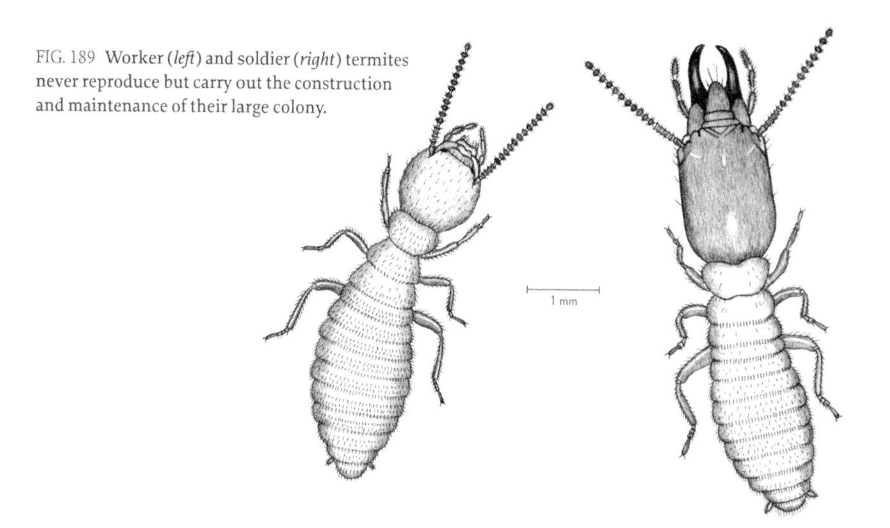

FIG. 189 Worker (*left*) and soldier (*right*) termites never reproduce but carry out the construction and maintenance of their large colony.

Workers and soldiers of a termite colony make up most of the population, which numbers several thousand individuals (fig. 189). In the spring or fall, however, the queen of the colony produces a brood of winged male and female termites that leave their hidden chambers to disperse, mate, and establish new colonies. A single couple will settle on a good location for a new colony, shed their wings, and begin laying eggs that hatch as the workers and soldiers who continually expand the territory of the colony into unexcavated wood and carry out housekeeping duties for the colony.

CATERPILLARS OF THE HEARTWOOD

Practically all caterpillars consume leaf after leaf—either from the outside as external feeders on leaves or from the inside as internal feeders on leaves—in preparation for their transformation, but a few caterpillars have chosen other paths to metamorphosis. These exceptional caterpillars have reputations for their unorthodox diets.

- CLEARWING MOTHS, FAMILY SESIIDAE (*sesia* = moth) (1,370 species; 135 species NA; larvae 12–65 mm; wingspan 8–48 mm).
- CARPENTERWORM MOTHS, FAMILY COSSIDAE (700 species; 50 species NA; larvae 20–75 mm; wingspan 25–85 mm).

The caterpillars of carpenterworm moths and clearwing moths are all borers, some in stems of herbs and berries but mostly in trunks and branches of trees. The low nutrient content of their woody diet—low in nitrogen, high in carbon—often prolongs the larval lives of these wood-chewing caterpillars from one to four years.

The elegant, handsome clearwing moths have a striking resemblance to wasps (fig. 190). Their slim bodies are not only shaped like those of wasps but also adopt poses like wasps, pretend to sting like wasps, and have clear wings like wasps.

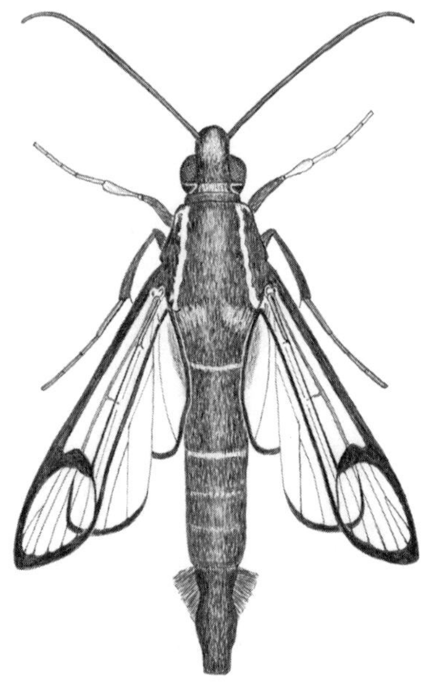

FIG. 190 These colorful clearwing moths, which are such good mimics of wasps, begin life as wood-boring caterpillars.

2 mm

Most of the scales so typical of moth wings have been lost from the wings of these moths. Remaining unmolested by predators is apparently a clear reason for these attractive moths to mimic stinging wasps.

One of the more striking moths in the carpenterworm moth family (Cossidae) is the leopard moth (*Zeuzera*; fig. 191), which closely resembles a moth in the distantly related family Erebidae, the giant leopard moth (*Hypercompe*). The attractive giant leopard moth and its spiny caterpillar are in the same subfamily (Arctiinae) as the fall webworm moth (fig. 61) and the lichen moth *Lycomorpha* (fig. 197). The leopard-like arrangement of spots on the wings of the leopard moth and the giant leopard moth is uncannily similar; the one subtle difference is that many spots on the wings of the giant leopard moth have white centers, while the wing spots of the leopard moth are solid black. However, the caterpillars of the two moths are as different in outward appearance as they are in their dining habits. The smooth, cream-colored caterpillar of the leopard moth feeds on wood; its giant leopard moth counterpart, most often referred to as the giant woolly bear, feeds on tree leaves. Each body segment of the giant leopard moth caterpillar is covered with black spines and is separated from adjacent segments by conspicuous red bands. The colorful features of the giant woolly bear caterpillar and the giant leopard moth, to which it transforms, serve as warnings to predators. The resemblance of the leopard or carpenterworm moth to this distasteful moth that it mimics so well has undoubtedly spared many tasty carpenterworm moths from becoming someone's meal. Both carpenterworm moths and their close relatives the clearwing moths have mastered the art of impersonating insects that predators prefer to avoid.

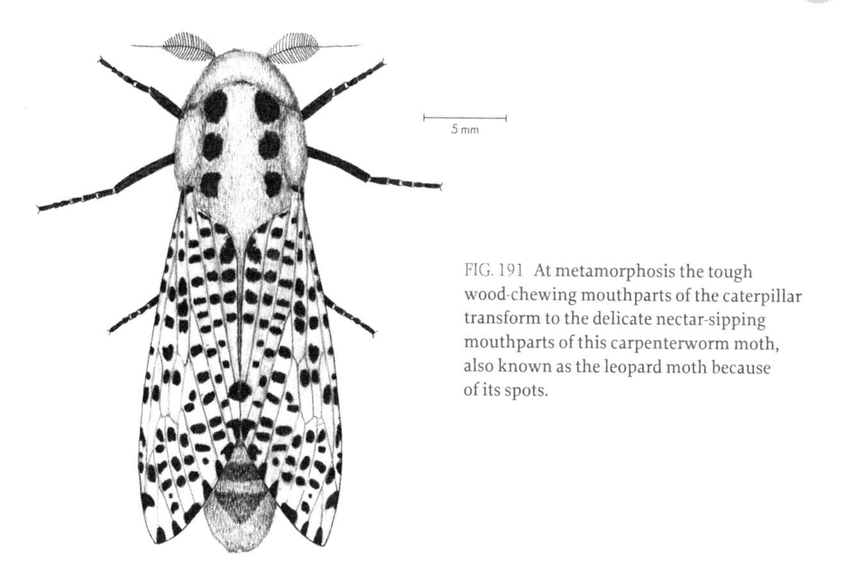

FIG. 191 At metamorphosis the tough wood-chewing mouthparts of the caterpillar transform to the delicate nectar-sipping mouthparts of this carpenterworm moth, also known as the leopard moth because of its spots.

Caterpillars of the leopard moth (*Zeuzera*) have been found in the wood of more than 125 tree species and are clearly open minded in their choice of heartwoods.

TREE BARK, EPIPHYTES, MOSSES, AND LICHENS

In the tropics, lush flowering plants, mosses, and lichens have colonized the trunks and branches of trees. These epiphytes (*epi* = upon; *phyte* = plant) find sufficient light, nutrients, and water in their aerial soils to grow independently of their tree hosts. In the temperate forests at higher latitudes, mosses and lichens have replaced the larger flowering plants as the dominant epiphytes on tree bark. Lichens and mosses establish forests of their own on trees of the forest.

The colorful lichens on tree bark are tiny organisms that are part fungi, and sometimes part algae or sometimes part bacteria (cyanobacteria). For many years these bacterial partners of lichens were referred to as blue-green algae. They not only carry out photosynthesis but also convert nitrogen in the air to ammonia, the form of nitrogen that trees and other plants can use. On some trees, lichens of different forms and colors cover almost the entire surface of the bark. The photosynthetic algal and bacterial members of a lichen partnership can live independently and are often found dwelling on their own; however, the fungal member depends on its photosynthetic partner for survival. Photosynthesis produces sugars and energy for the partnership, and the fungal partner provides mineral nutrients, extending its filaments into the dead outer bark of trees. Over months and years, lichens slowly decompose the dead bark and use whatever nutrients it can provide. These fragile-looking organisms can readily endure the extremes of heat and cold to which they are exposed on bark, but they are particularly sensitive to traces of air pollution.

As we zoom in with a magnifier or microscope for a closer look at lichens and mosses, we see that they form a forest of their own on branches and trunks of most trees (fig. 192; BarkScape, 15–17). Certain caterpillars and other insects live and grow in this forest. They feed on lichens and mosses and often live in portable cases that they construct with silk and fragments of lichens or mosses. Mites, fly and beetle larvae, moth caterpillars, adult beetles, and bark lice also move in the shadows of mosses and lichens. Even closer inspection with higher magnification reveals tardigrades, rotifers, nematodes, and protozoa feeding on the abundant fungi and bacteria that thrive and move about in the microscopic films of water on the bark (fig. 193).

The forests of lichens and mosses on branches and tree trunks thrive at the boundary between air and bark. The mosses and lichens intercept nutrients and water that flow down and around the trees; they gather and accumulate nutrients that are released slowly and would otherwise be rapidly washed and leached away

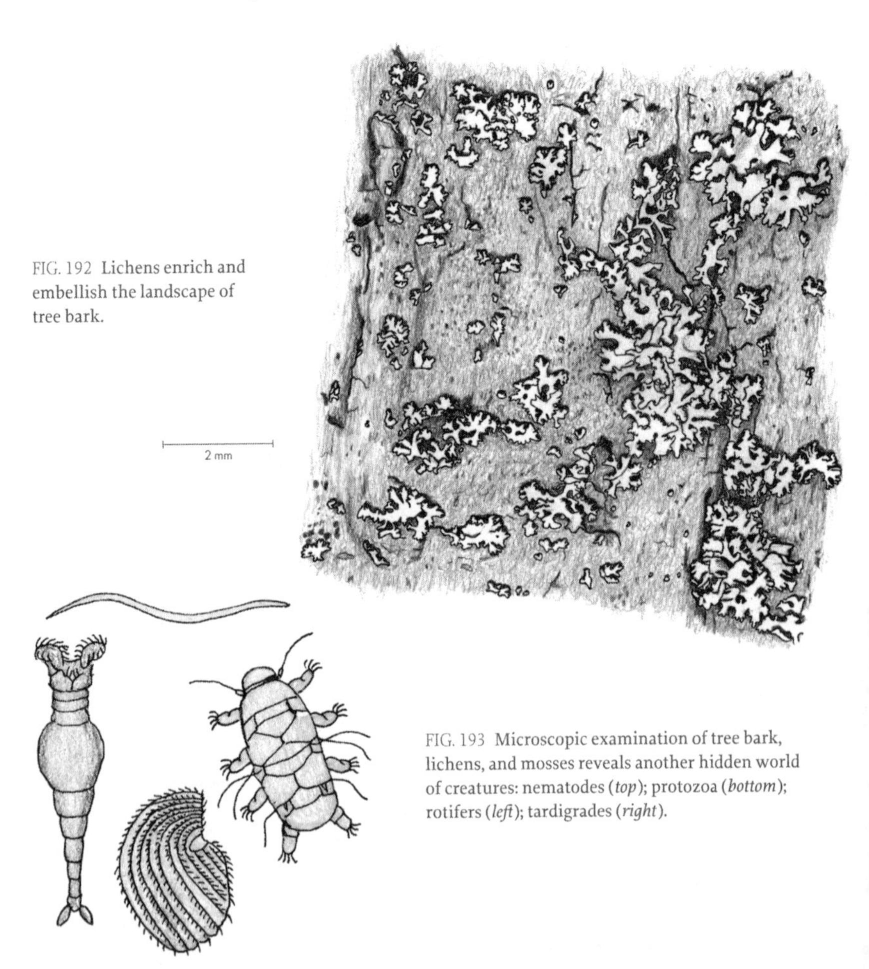

FIG. 192 Lichens enrich and embellish the landscape of tree bark.

FIG. 193 Microscopic examination of tree bark, lichens, and mosses reveals another hidden world of creatures: nematodes (*top*); protozoa (*bottom*); rotifers (*left*); tardigrades (*right*).

by the frequent heavy rains that pelt the bark. Over time these enduring plants repeatedly grow and decay, with their remains assuming the aboveground role that soil humus assumes belowground of tenaciously holding on to nutrients that would otherwise be leached from the soil. These tiny plants have never acquired vascular systems for conducting sap such as those of trees. Some species of mosses and lichens specialize in growing only on bark; the mosses' habit of draping over the surface of trunks and limbs has earned them the common generic name apron mosses. In the tropics and temperate rain forests these epiphytic nonvascular plants are especially abundant.

Through death, decay, and the action of decomposers, they transform to humus, retaining enough nutrients and dust from the air to create pockets of soil among tree branches. Soil that forms aboveground can support the lives of larger epiphytic vascular plants that take root in the canopies of trees.

Vireos, hummingbirds, pewees, and gnatcatchers often adorn their compact nests with lichens and mosses that they collect from tree bark (fig. 194). With a coating of lichens, each nest looks like just another knob on a branch. Other birds like brown creepers (BarkScape, 2) and nuthatches (fig. 9) scour the bark, poking their beaks into lichen and moss forests, for they have learned that every now and then they can pull out a tasty insect.

FIG. 194 Many birds such as hummingbirds add trims of lichens to their nests.

Lichens are a ubiquitous presence on tree trunks and branches. Any creature this abundant on trees is a resource to be used in as many ways as possible. Lichens are generously used as camouflage. The predatory larvae of lacewings (chapter 2) deck themselves out with lichens for disguise as they creep up on their prey (fig. 195). Some moths and caterpillars camouflage their bodies by mimicking lichens.

Moths tend to spend a lot of time on tree bark. Many of these moths are attired in blacks, grays, and whites, colors that blend splendidly with the colors and textures of bark. The wings of some moths are covered with patches of green, gray, and white, colors that match the patterns of the ubiquitous lichens growing on the bark of so many trees. Still other moths, such as noctuid moths, wear an all-purpose camouflage pattern of black, gray, green, yellow, and white splotches so they blend in with both unadorned bark and lichen-covered bark—and even bird droppings (chapter 2). Noctuid moths such as dagger moths are particularly talented deluders (figs. 12 and 196).

— DAGGER MOTHS, FAMILY NOCTUIDAE, SUBFAMILY ACRONICTINAE (*acro* = beginning of; *nicta* = night), GENUS *ACRONICTA* (150 species; 76 species NA; wingspan 40–62 mm).

Several of our most colorful moths that begin their lives feeding on lichens of tree trunks and branches are known as lichen moths. Lichens are known to have a variety of repellent chemicals, and they probably impart a distasteful flavor to all who feed on them. The bright colors of an arthropod are usually good indicators that predators should avoid these conspicuous insects. Lichen moths are a good

FIG. 195 From under their camouflage coats of lichen fragments, lacewing larvae move surreptitiously across the bark and through patches of lichens.

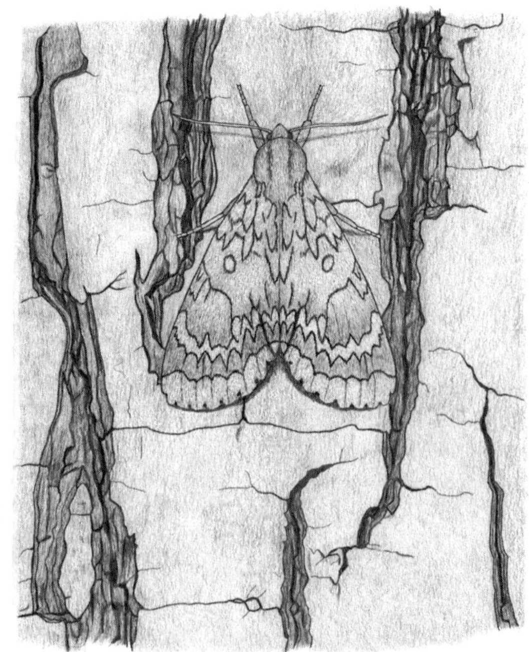

5 mm

FIG. 196 This dagger moth (*Acronicta*) is resting on the bark of a red oak tree.

example of brightly colored insects that are protected by virtue of their bad tastes and bright colors.

— LICHEN AND TIGER MOTHS, FAMILY EREBIDAE, SUBFAMILY ARCTIINAE (*arctos* = bear, referring to hairy caterpillars) (11,000 species; 265 species NA; larvae 10–35 mm; wingspan 28–40 mm).

This subfamily of hairy caterpillars is home to the well-known red and black woolly bear caterpillars of giant leopard moths (p. 156) that feed not only on many different trees but also on many weeds and wildflowers. A typical fuzzy tiger moth caterpillar is shown in LeafScape 19. We frequently encounter these fuzzy caterpillars in the fall as they scurry over tree trunks and the ground in search of sheltered spots where they can spend the winter as larvae or pupae.

About 3,000 members of this subfamily survive on meals of lichens, not only consuming the toxic secondary metabolites of the lichens but also sequestering these compounds for their protective value (Scott-Chialvo et al. 2018). Other members of this large family include fall webworms, tussock moths (chapter 2), litter moths (chapter 6), and the stunning underwing moths (this chapter).

Caterpillars of the black and orange lichen moth *Lycomorpha* (*lyco* = wolf; *morpha* = form) graze on lichens of tree trunks and branches (fig. 197). The blue-green algae of lichens are known to manufacture toxic microcystins that can be stored in the body of an insect and passed from caterpillar to moth (Kaasalainen et al. 2012).

Toxic lichen moths have such a bad reputation among insect-eating predators that toxic beetles (*Calopteron*) in the family Lycidae and toxic moths (*Pyromorpha*) in the family Zygaenidae have established mimicry complexes with these lichen moths. Members of this beetle family and this other moth family share the appearance, the forest environment, and the bad taste of black and orange lichen moths. Each of these mimics has its own special toxin. Each time a predator tastes any one of these three black and orange insects, their striking colors and unsavory flavors mutually reinforce each other's bad reputation. Members of other insect families—some toxic, some not—have discovered the benefits of resembling insects with toxic reputations.

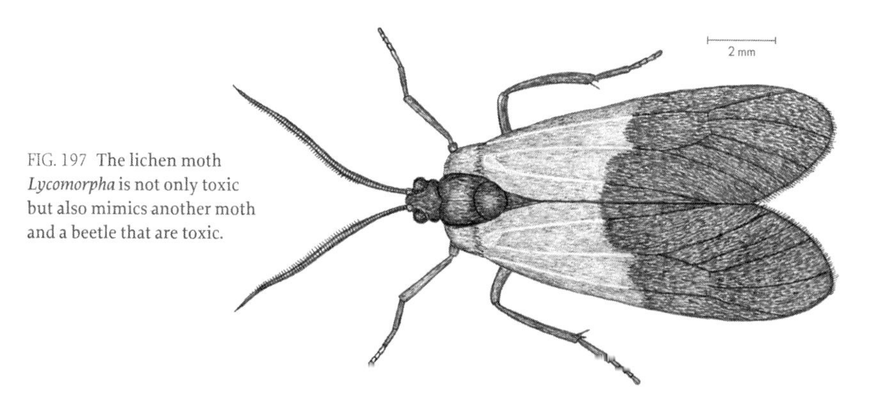

FIG. 197 The lichen moth *Lycomorpha* is not only toxic but also mimics another moth and a beetle that are toxic.

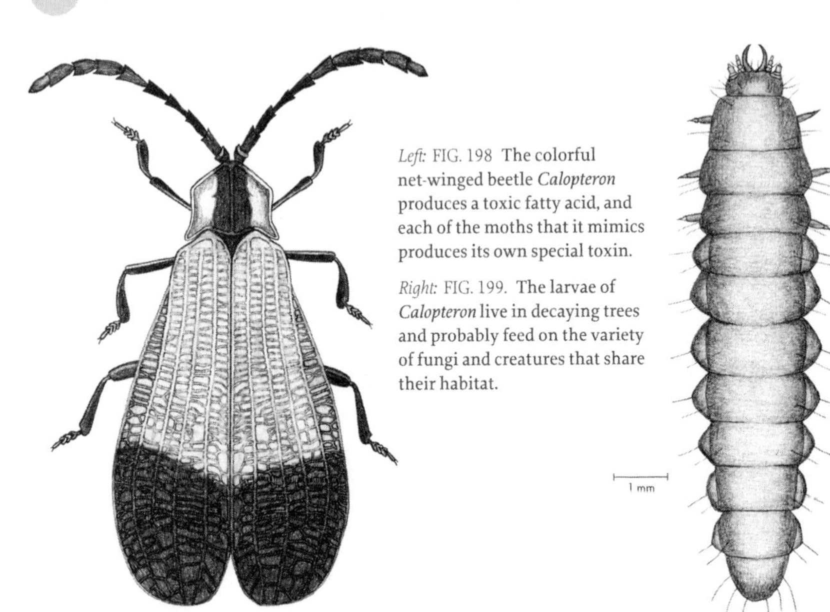

Left: FIG. 198 The colorful net-winged beetle *Calopteron* produces a toxic fatty acid, and each of the moths that it mimics produces its own special toxin.

Right: FIG. 199. The larvae of *Calopteron* live in decaying trees and probably feed on the variety of fungi and creatures that share their habitat.

— NET-WINGED BEETLES, FAMILY LYCIDAE (*lyco* = wolf) (4,600 species; 80 species NA; length 10–15 mm).

The colorful net-winged beetles have delicate, reticulated wings (fig. 198). These handsome wings have earned the beetles the genus name *Calopteron* (*calo* = beautiful; *pteron* = wings). These beetles are out during the day and can be found resting on leaves, flowers, and tree trunks in dense woods. They produce a toxic fatty acid called lycidic acid (Eisner et al. 2008).

Lycid beetles begin their lives as larvae in decaying tree trunks and branches. As is the case for so many insects that lead hidden lives, some uncertainty remains about how these beetles spend their larval days. They have been observed feeding on the fungi, slime molds, and mix of organisms that thrive in the liquid extracts of rotting wood. However, they have also been reported to use their distinctive sickle-shaped mandibles, each of which is inexplicably split lengthwise, to feed on other larvae that live under bark and in the passageways of rotten wood (fig. 199).

In America's eastern forests, net-winged beetles can easily be mistaken for not only black and orange lichen moths but also for the moths of another family. Both moths are of similar size, shape, and coloration; inhabit the same trees; and appear at the same times.

— LEAF SKELETONIZER MOTHS OR SMOKY MOTHS, FAMILY ZYGAENIDAE (*zygaena* = type of shark) (1,000 species; 25 species NA; larvae 10–25 mm; wingspan 16–28 mm).

Like sharks, these toxic caterpillars leave behind skeletons of their prey. The colorful, even flashy moths and equally flashy caterpillars that feed on oak leaves

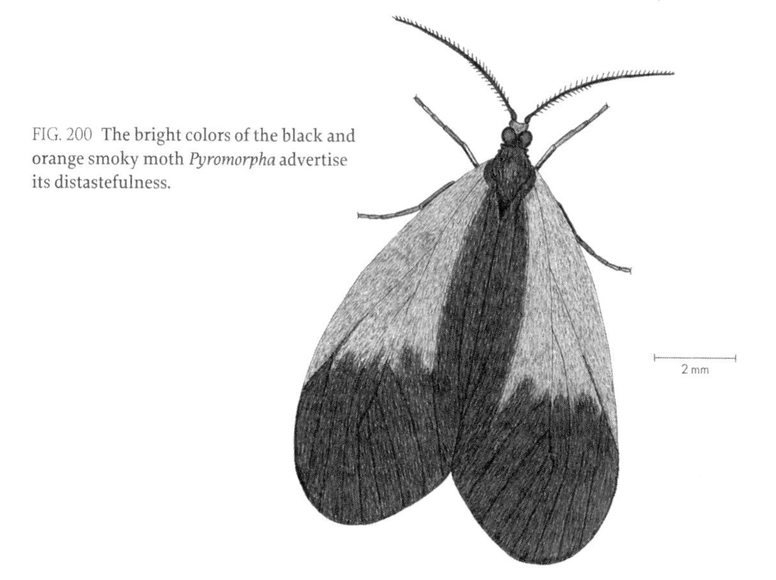

FIG. 200 The bright colors of the black and orange smoky moth *Pyromorpha* advertise its distastefulness.

contain hydrogen cyanide at all life stages. The toxic black and orange smoky moths in the genus *Pyromorpha* (*pyro* = fire; *morpha* = form) visually proclaim their distastefulness with their bright colors, and the generic name conveys the moth's striking appearance (fig. 200).

— FUNGUS MOTHS, FAMILY TINEIDAE (*tine* = a moth) (3,000 species, 200 species NA; larvae 6–50 mm; wingspan 7–36 mm).

Within this large family of small moths, most caterpillars feed on lichens, fungi, and dead leaves. Only a few species feed on living leaves. Some of the more infamous caterpillars of this family—the larvae of clothes moths—have the uncanny ability to survive on diets of wool, hair, horn, hooves, leather, feathers, or beaks. These different structures share one feature—their primary nutrient is the structural protein keratin. In this family and other moth families, there are many moths whose larval diets remain mysteries.

Many of the adult fungus moths masquerade as bird droppings—a ploy that has proven quite effective in protecting many other larger and smaller moths, as noted in chapters 1 and 2. With their wings folded closely against their bodies

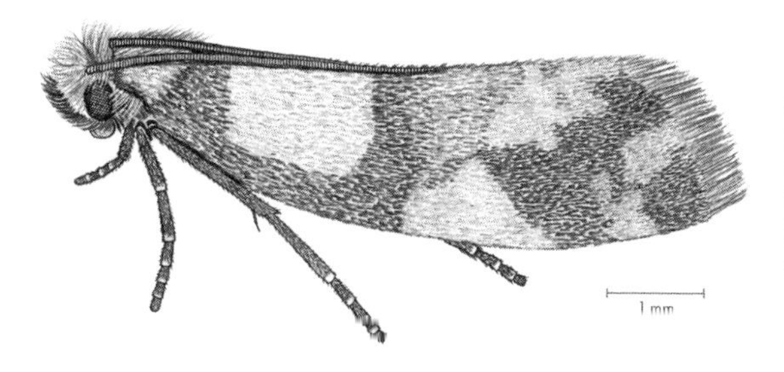

FIG. 201 Fungus moths are tiny, mottled, and sleek. Caterpillars of fungus moths have unorthodox diets of dead leaves, fungi, and even feathers and hair.

FIG. 202 A fungus moth whose wings have patches of black, brown, and white can easily be mistaken for a bird dropping as it poses on an elm leaf.

and with ruffled, shaggy scales atop their heads, the adults have the form and texture of bird dung (fig. 201). Blending of light scales with dark scales creates wing patterns that impart the expected diversity of bird droppings, easily fooling human eyes and deceiving the eyes of our vertebrate relatives and perhaps the eyes of many insect predators as well (fig. 202).

— UNDERWING MOTHS, FAMILY EREBIDAE, GENUS *CATOCALA* (*cato* = lower one; *cala* = beautiful) (Holarctic genus, 260 species; 125 species NA; larvae 35–75 mm; wingspan 20–98 mm).

When being camouflaged on lichens and bark fails to prevent detection, some insects have discovered that shock and awe can assist in rapid escapes from predators. Many well-camouflaged moths known as underwings suddenly flash their hidden bright colors to startle predators and aid their escapes (fig. 203; BarkScape, 10). A bird or mammal that ventures too close to an underwing moth is in for a startling experience. The resting moth folds its forewings over its strikingly colored hind wings and lies unnoticed against the bark of a tree. But if the moth is disturbed, it flies off, flashing the bright colors hidden under its camouflaged forewings, certainly startling its intruder long enough to ensure its escape (fig. 15).

Their caterpillars are just as adept at camouflage; on their dorsal surfaces, they blend as well with lichens as they do with bark (fig. 204). If they are abruptly disturbed and flipped on their backs, however, these caterpillars can flash brightly colored bellies. The contrast between belly and back certainly takes one by surprise. The art of combining camouflage with surprise is passed on from caterpillar to moth at metamorphosis.

FIG. 203 While resting on tree bark, underwing moths are camouflaged; only when perturbed do they lift their forewings to flash the bright colors hidden on their hind wings.

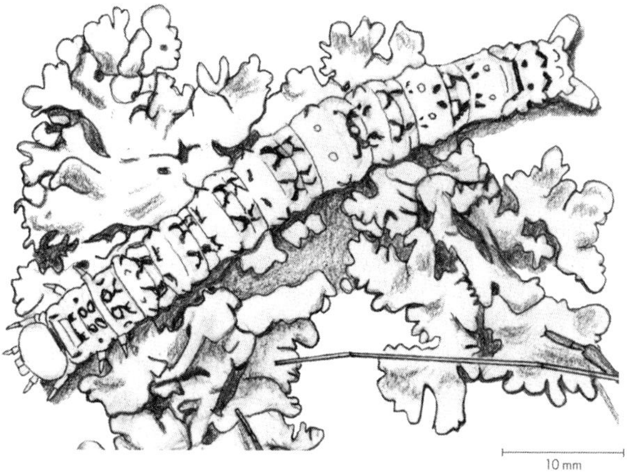

10 mm

FIG. 204 On its top side the caterpillar of the ilia underwing (*Catocala ilia*) blends beautifully with nearby lichens, but when it flips over, it startles with the flashy bright purple on its bottom side.

— BARK LICE, ORDER PSOCOPTERA (*psocus* = to grind or gnaw; *ptera* = wings) (5,500 species; 400 species NA; 1–10 mm).

Bark lice are related to thrips and true bugs, but each of the three groups is distinctive enough to be assigned to a different insect order (fig. 205; BarkScape, 24). Bark lice have strange mouthparts, unlike the piercing mouthparts of their relatives the thrips or true bugs. Mouthparts of bark lice are best described as grinders and scrapers. During feeding a bark louse gnaws with its mandibles while it braces its body with two stiff rods that extend from the other pair of mouthparts known as the maxillae.

They often aggregate in herds, nestled under diaphanous silk canopies that they spin from silk glands in their mouths. The male bark lice perform a courtship dance, accompanied by clicking sounds that arise from the coxae, the most basal segments of the hind legs. The sounds and sights of the courtship dance summon forth females from the lichen forest.

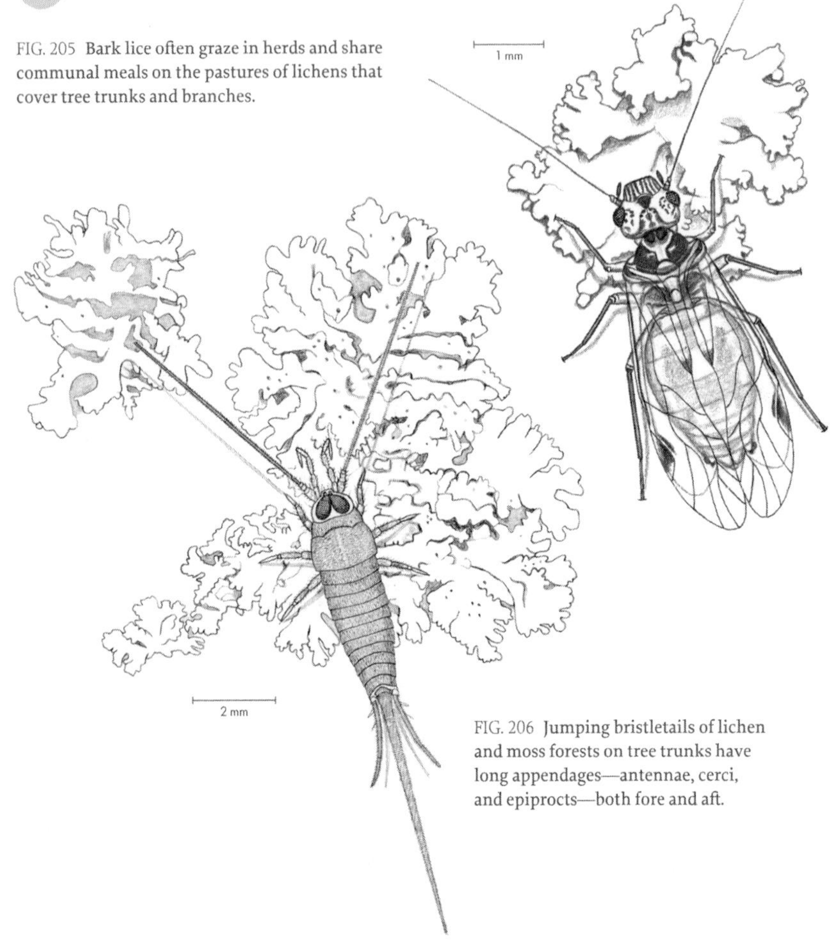

FIG. 205 Bark lice often graze in herds and share communal meals on the pastures of lichens that cover tree trunks and branches.

FIG. 206 Jumping bristletails of lichen and moss forests on tree trunks have long appendages—antennae, cerci, and epiprocts—both fore and aft.

— JUMPING BRISTLETAILS, ORDER MICROCORYPHIA (*micro* = tiny; *coryphe* = head) or ORDER ARCHAEOGNATHA (*archae* = ancient; *gnatha* = jaws) (350 species; 22 species NA; 8–20 mm).

For over a century, bristletails have had two alternative, commonly used names for their order.

Their two compound eyes are exceptionally large in proportion to their heads. These eyes meet at the top of a head that is small in comparison to the broad thorax and conspicuously arched back. The alternative name for this group of wingless insects refers to their jaws or mandibles, which differ from those of all other insects in having only one point at which they articulate with the head. In other insects, each jaw joins the head at two points (fig. 206; BarkScape, 25)

These most primitive of insects can be gregarious and are often found in groups grazing on lichens, fungi, mosses, and algae. You can encounter bristletails in moss and lichen forests on tree trunks, under bark, or under trees, wherever trees provide suitable habitat for these moisture-loving insects. Thin-walled vesicles on the ventral surfaces of their abdomens that can readily take up water protect the bristletails from dehydrating. The vesicles pop out of cuticular pouches to take up water and then retract into the ventral surface after sufficient water has been taken up. At the end of a bristletail's abdomen are three tails. The two lateral tails, known as cerci, are found on many other insects; however, the longer, middle tail, called the epiproct (*epi* = over; *procto* = anus), has endowed bristletails with their remarkable leaping ability. Suddenly slapping their tails against bark can launch them a foot into the air.

LIVING BETWEEN THE BARK AND THE HEARTWOOD

— FUNGUS GNATS, FAMILY MYCETOPHILIDAE (*myceto* = fungus; *philo* = love) (4,500 species; 620 species NA; 2–13 mm).

— DARK-WINGED FUNGUS GNATS, FAMILY SCIARIDAE (*sciaro* = dark) (1,700 described, estimated 20,000 species; 170 species NA; 1–7 mm).

Fungus gnats (families Mycetophilidae and Sciaridae) spend their larval days in great numbers under the bark of dead trees or in leaf litter where enough moisture persists to stimulate the lush growth of fungi that they constantly consume (fig. 207). These small larvae have black or dark brown heads that stand out against their translucent bodies. Their clear cuticles reveal the movement of their internal organs as these legless larvae wriggle between the bark and the heartwood. At metamorphosis they sprout legs and wings, emerging from beneath the bark to carry fungal spores to new substrates and to serve as minuscule pollinators who never receive the recognition for their contributions to pollination that is bestowed on the larger bees, beetles, and butterflies.

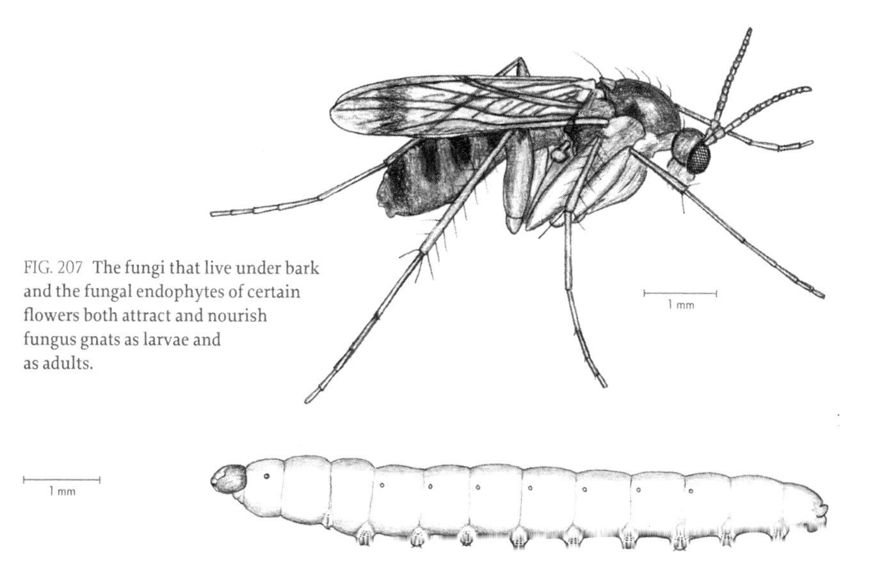

FIG. 207 The fungi that live under bark and the fungal endophytes of certain flowers both attract and nourish fungus gnats as larvae and as adults.

Midges and gnats spend their brief hours as adults in great swarms that we often observe hovering and levitating a few feet aboveground. The males emerge first and seem to prefer a sunlit location to establish a stage for their rollicking performances. After the males have assembled, females join the communal courtship dance, and mating occurs in midair among hundreds of flailing legs and flapping wings.

Rarely recognized for their contributions to pollination, these tiny flies are known to frequently visit the unusual spadix flowers of the forest understory—the flowers of jack-in-the-pulpit. The flies are enticed by the fragrance of fungi emitted by the flowers. Could the source of this fragrance be a fungal endophyte partner of this woodland flower? As the gnats flit around inside the male spadix, they become dusted with pollen. Once they find their way out of the male's floral chamber, they are laden with plenty of pollen grains. Now the gnats transport their load of pollen to the nearby female flowers of jack-in-the-pulpit. The gnats have an even harder time finding an escape route from the female flower. As they repeatedly collide with the flower's multiple stigmas, the gnats leave more than enough pollen to ensure fertilization of all the ovules. Each fertilized ovule is destined to form a single berry of the scarlet red fruit (Barriault et al. 2010).

- MIDGES, FAMILY CHIRONOMIDAE (*cheironomos* = one who moves the hands, referring to the habit of these midges of waving their front legs when at rest) (7,300 species, estimated 1,000,000 species; 1,050 species NA; 1–12 mm).
- BITING MIDGES, FAMILY CERATOPOGONIDAE (*ceratos* = horn; *pogon* = beard, referring to the fuzzy antennae of males) (6,000 species; 600 species NA; 1–3 mm).

Midge larvae are found in all sorts of habitats; they are common in just about every moist habitat—ponds, streams, decaying vegetation, soil, and under bark of dead trees (fig. 208). Wherever they are found, the larvae are present in vast numbers and—despite their tiny forms—contribute mightily to the breakdown and recycling of plant litter. A hint of their incalculable abundance is conveyed whenever the larvae leave their damp habitats and transform to adult midges (figs. 209 and 210). The males usually emerge from their larval homes before the females and form dense swarms that are often observed along roadsides, near bodies of water, in clearings in the woods, even in backyards. The swarm shifts up and down, sometimes from side to side as the individuals move in unison to avoid midair collisions that could result in entanglement of legs, wings, and antennae.

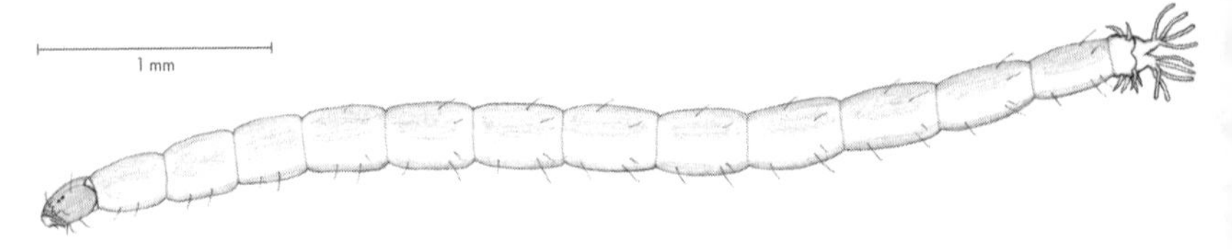

FIG. 208 Chironomid and ceratopogonid larvae are similar in appearance and share similar habitats under bark and in the soil.

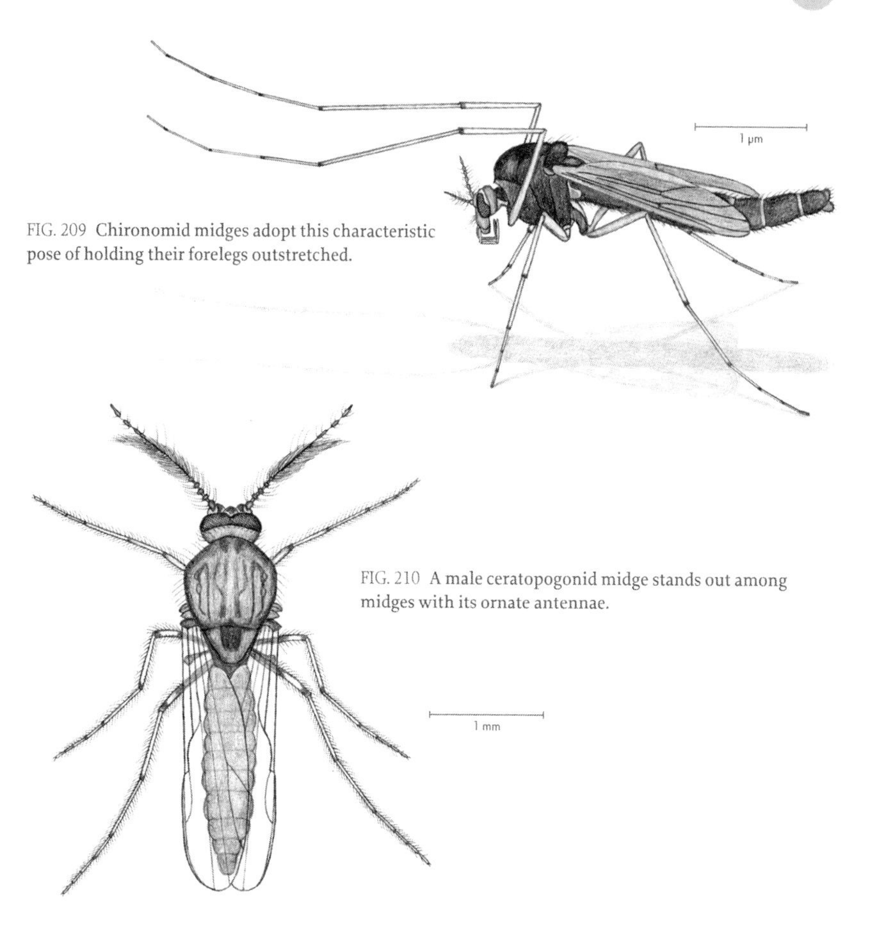

FIG. 209 Chironomid midges adopt this characteristic pose of holding their forelegs outstretched.

FIG. 210 A male ceratopogonid midge stands out among midges with its ornate antennae.

After the females join the swarm and matings have occurred, the swarm soon disperses. The immense number of midges and gnats that leave the soil where they labored at recycling as larvae soon expire to return to the ground and provide raw material for more recyclers.

- SOLDIER FLIES, FAMILY STRATIOMYIDAE (*stratio* = soldier; *myia* = fly) (2,700 species, 250 species NA; 3–30 mm adults).
- WOOD SOLDIER FLIES, FAMILY XYLOMYIDAE (*xylo* = wood; *myia* = fly) (140 species; 11 species NA; 5–14 mm adults).

Many adult soldier flies and wood soldier flies are such good impersonators of wasps that only a close inspection can establish their identity. Like wasps and bees, they frequent flowers, contributing to the pollination enterprise as they feed on nectar and pollen (fig. 211).

The diversity of the soldier fly family is reflected in its being partitioned into 12 different subfamilies. Two subfamilies begin life in shallow aquatic habitats—ponds, water holes, even tree holes—feeding on debris and algae. Larvae of

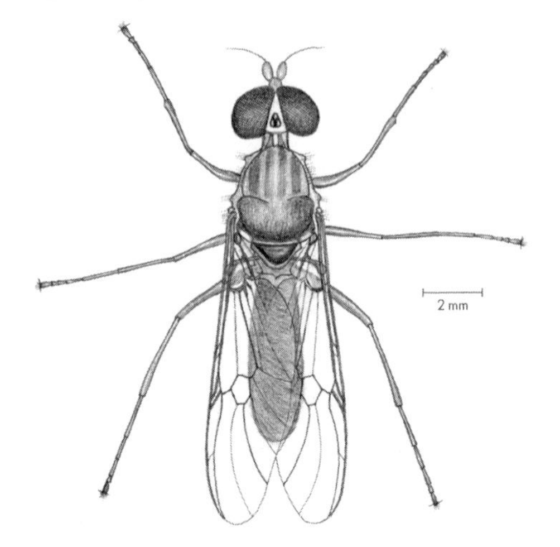

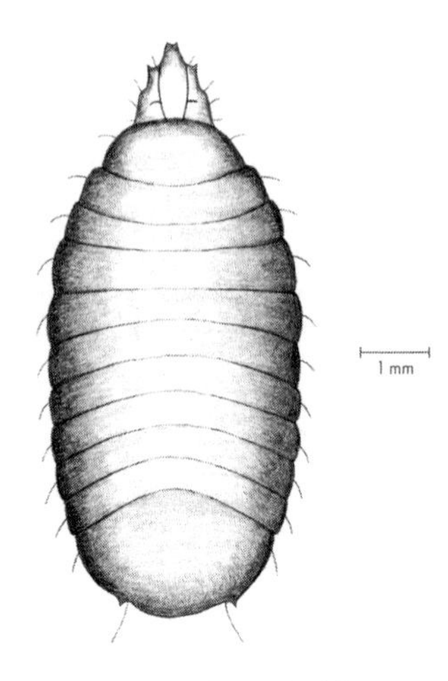

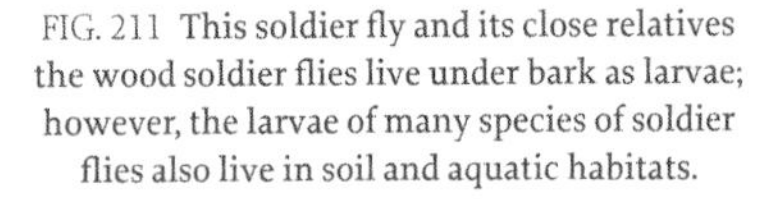
FIG. 211 This soldier fly and its close relatives the wood soldier flies live under bark as larvae; however, the larvae of many species of soldier flies also live in soil and aquatic habitats.

FIG. 212 This larva of a wood soldier fly was found under tree bark; as with the larvae of soldier flies, its skin is reinforced with mineral granules.

another subfamily dwell under bark of dead trees, where they prey on other insects that share their habitat. Many larvae of species in the other soldier fly subfamilies are important decomposers, feeding on whatever decaying organic matter they find (RootScape, 12).

Larvae of xylomyid and stratiomyid flies that live under bark are so similar in appearance that they are easily mistaken for each other (fig. 212). For a long time, wood soldier flies were considered another subfamily of the Stratiomyidae. They both have long, flattened bodies with especially tough, almost impenetrable skins or integuments. The many tiny plates of calcium carbonate embedded in their integuments contribute to their toughness and security. The one minor distinguishing feature between the larvae of the two related families is that these plates are missing on the dorsal surface of two thoracic segments in xylomyid larvae but are present on all segments of stratiomyid larvae. These larval differences have been considered significant enough to place wood soldier flies in their own family distinct from soldier flies.

— XYLOPHAGID FLIES, FAMILY XYLOPHAGIDAE (*xylo* = wood; *phago* = eat) (150 species; 28 species NA; larvae 15–20 mm; adults 2–25 mm).

Although the name Xylophagidae translates to "family of wood eaters," neither the larvae nor the adults ever feed on wood. One of the strangest-looking larvae I have ever encountered under bark is the larva with the strange name *Xylophagus*. Its cone-shaped head is hard to distinguish from its tapered tail. Hidden on the ventral side of its head, however, are the mouth hooks with which this larva stabs and ingests other insect larvae. This peculiar larva transforms into one of the handsomest flies of the woods; its slender shape and striking colors mimic the features of the parasitic wasps, sawflies, and clearwing moths that share its habitat (fig. 213).

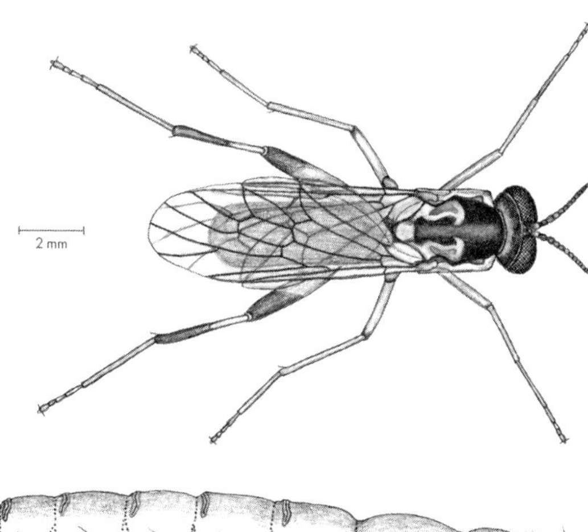

FIG. 213 Xylophagid flies are some of the loveliest insects of the forest, and their larvae are some of the most bizarre looking.

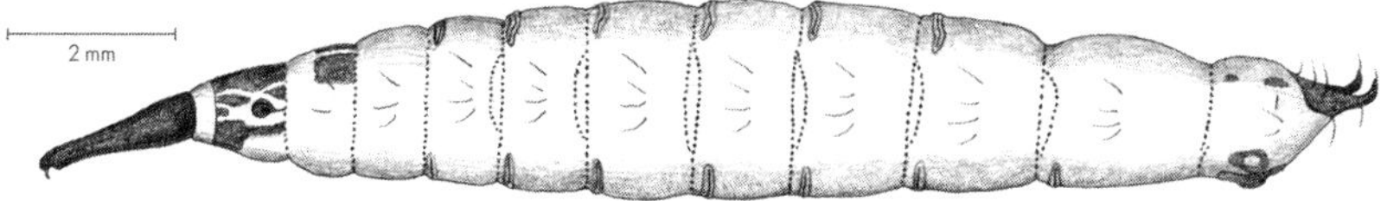

— LANCE FLIES, FAMILY LONCHAEIDAE (*loncha* = lance) (580 species; 125 species NA; 5–8 mm).

The wide wings of these bright, shiny flies are almost twice the length of their short abdomens. Each wing has an expanded lance-shaped window outlined by wing veins at the outer edge of its base (fig. 214). Most lance flies begin life under bark feeding on decaying cambium and wood, but some larvae that have been found in the galleries of bark beetles are suspected of preying on beetle larvae. A few

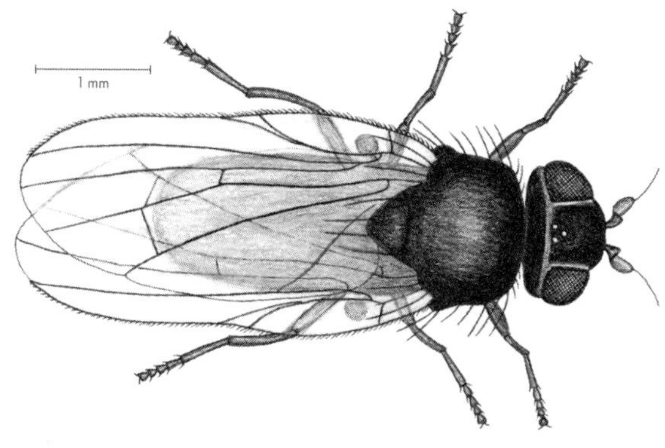

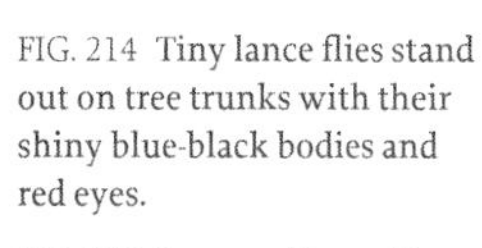

FIG. 214 Tiny lance flies stand out on tree trunks with their shiny blue-black bodies and red eyes.

FIG. 215 Larvae of lance flies are at home under tree bark and are apparently satisfied with the wide choice of food items they encounter there.

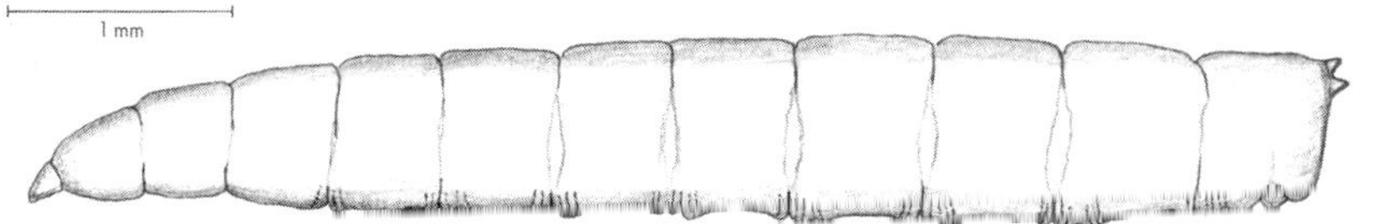

species, however, feed in fir and spruce cones, and one species matures on sweet fig fruits. To human eyes, the undistinguished larvae of these handsome flies have all the features of any other fly maggot; they have a smooth, tapered anterior end without a well-defined head or any lateral projections (fig. 215).

— JEWEL BEETLES, METALLIC WOOD-BORING BEETLES, FAMILY BUPRESTIDAE (*buprestis* = type of beetle) (15,500 species; 760 species NA; 3–80 mm).

Flat-headed borers chew winding galleries beneath bark, killing a tree as they devour the living phloem and cambium between the outer bark and inner heartwood (fig. 216). Prolonged development is a general characteristic of wood borers. The larvae take their time chewing on wood fibers, which are often rather poor in nutrients, often taking several years to complete their larval development. Not only are the heads of these borers flat, but their thoraxes are also flat and clearly wider than their abdomens, earning them their other name, hammerhead borers. Their elliptical galleries sometimes collide with the circular galleries of round-headed borers. Most beetle larvae have six legs, but round-headed borers and flat-headed borers have legs that are greatly reduced or completely lost. In their confined chambers under the bark where mobility is so restricted, six legs are superfluous.

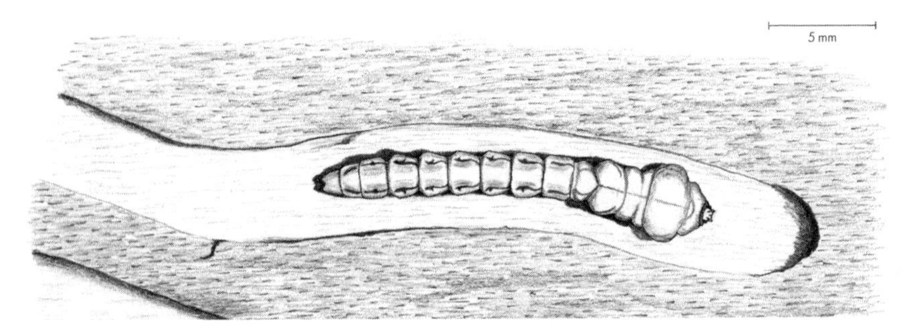

FIG. 216 Flat-headed borers, the larvae of buprestid beetles, carve their winding channels between bark and heartwood.

The most infamous buprestid beetle is the handsome metallic-green emerald ash borer that is decimating the ash trees of North America. Since its arrival from Asia in 2002, this beetle has destroyed millions of ash trees throughout the eastern United States and Canada. Our native woodpeckers, insect predators, and insect parasitoids are beginning to recognize this exotic beetle as desirable prey and as a suitable host; in addition, parasitic wasps introduced from the same forests in Asia inhabited by emerald ash borers have been recruited in attempts to stem ash borer spread and destruction (Bauer et al. 2015). The combined efforts of these different allies of trees have been very promising.

Among other members of this family, however, are short, stout beetles and their larvae that have forsaken lives under bark for lives between the two epidermal layers of leaves (fig. 132), where they chew galleries. On their more nutritious meals of leaf mesophyll cells, these leaf-mining buprestid beetle larvae develop far more rapidly than their wood-boring relatives (fig. 217) and complete their life cycles in a single year.

— FIRE-COLORED BEETLES, FAMILY PYROCHROIDAE (*pyro* = fire; *chroma* = color) (170 species; 50 species NA; 4–20 mm).

These adult beetles can be common visitors to evening lights and summer flowers. They stand out with their dashing red and orange colors usually tempered by contrasting black elytra. They sport long, branched antennae whose forms are often exaggerated to resemble antlers (fig. 218). The long, flat larvae are commonly encountered under tree bark, where they feed on decaying wood and fungi (fig. 219). These crescent-tailed larvae can spend several years attaining full growth on their meager diet. Representatives of all larval stages can be found under the bark of a single tree trunk or limb.

— HISTER BEETLES, CLOWN BEETLES, FAMILY HISTERIDAE (*histrio* = actor) (4,500 species; 440 species NA; 1–20 mm).

The forms of hister beetles say a great deal about their lifestyles and habitats. All hister beetles are shiny and slick, and with their short, stubby legs they easily maneuver in tight spaces. The more rotund of these beetles are drawn to the

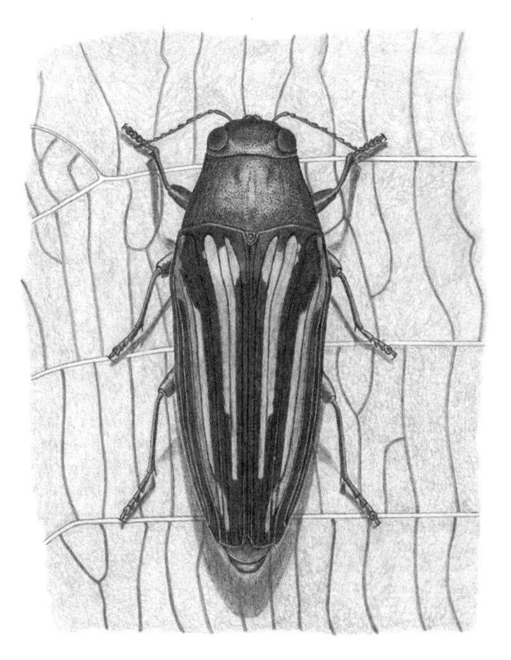

FIG. 217 The ornate cuticles of most buprestid beetles allow them to live up to the names jewel beetles and metallic wood borers.

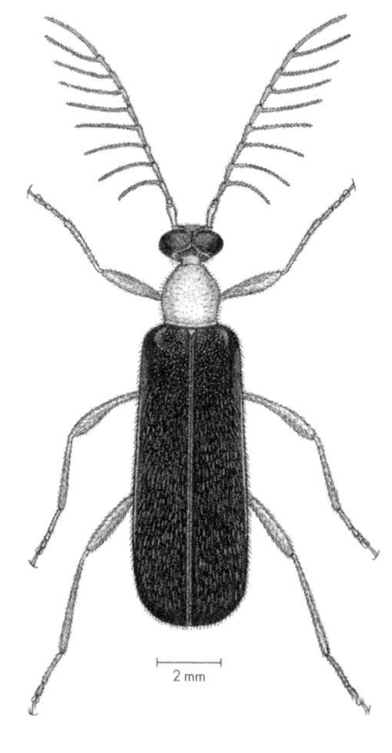

FIG. 218 Adult fire-colored beetles have feathery antennae and heads with distinct necks.

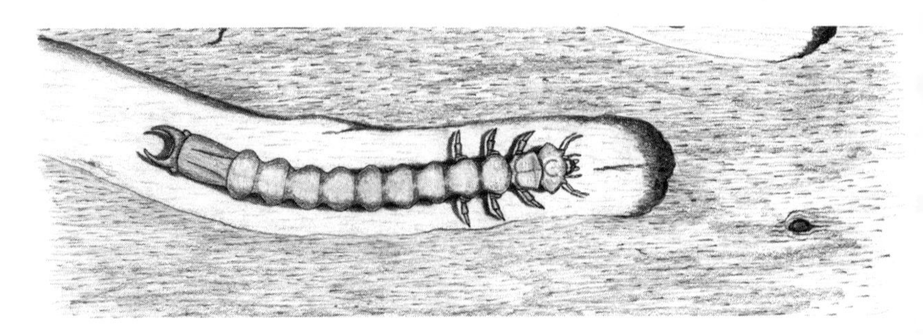

FIG. 219 Larvae of fire-colored beetles have characteristic two-pronged tails, or urogomphi, as larval beetle tails are known.

5 mm

odors of decaying carcasses, rotting vegetation, and dung, where they can find abundant eggs and larvae of flies and beetles on which to feed. Adult beetles and their larvae are biocontrol agents of mites as well as fly and flea larvae in nests of birds, mammals, and some reptiles. The adult hister beetles and their larvae that live on trees are also hunters. They can be either flattened to inhabit the flat spaces under bark or cylindrical to scurry down round tunnels through the heartwood (fig. 220).

Ant and termite nests are also the only habitats where some hister beetle species are found. Exactly what goes on between beetles and their hosts in the underground chambers of these social insect nests is still not well understood. They may survive as freeloaders as well as predators on their hosts. The odors these beetles produce mimic those produced by their hosts, and with these simple chemical passwords, the beetles fool their termite and ant hosts into accepting them as welcome guests.

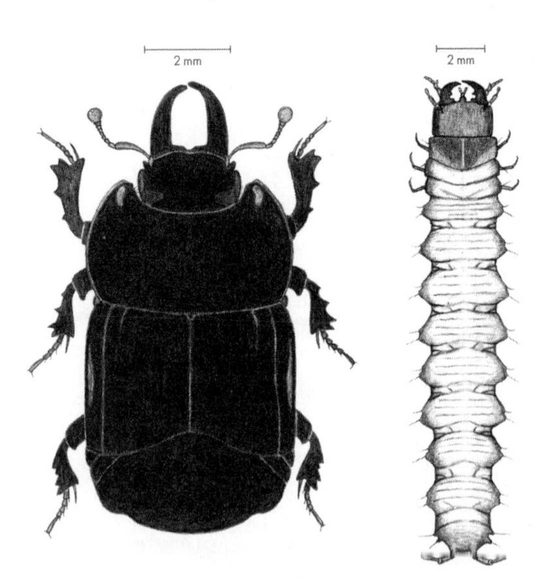

FIG. 220 This hister beetle adult and larva share many of the same features: flat bodies; short, stubby legs; and prominent jaws.

— CHECKERED BEETLES, FAMILY CLERIDAE (*cleri* = kind of insect) (3,600 species; 500 species NA; 3–24 mm).

These beetles with bristly hairs stalk passageways under bark for beetle, fly, wood wasp, and sawfly larvae (fig. 221). Successful predation of clerid beetles and their larvae on bark beetles and wood-boring beetles can be attributed to their antennae, which recognize the aggregation pheromones of their prey. As bark beetles emit aggregation pheromones to attract their fellow bark beetles, they unwittingly also attract the attention of these clerid beetle predators.

While larvae are primarily predators of tree-dwelling insects, some members of this family raid termite, bee, and wasp nests for eggs and larvae or feed on nests of grasshopper eggs. Larvae of many species of checkered beetles remain predators after metamorphosis, but some adult species prefer feeding on pollen or scavenging the remains of plants and animals (fig. 222).

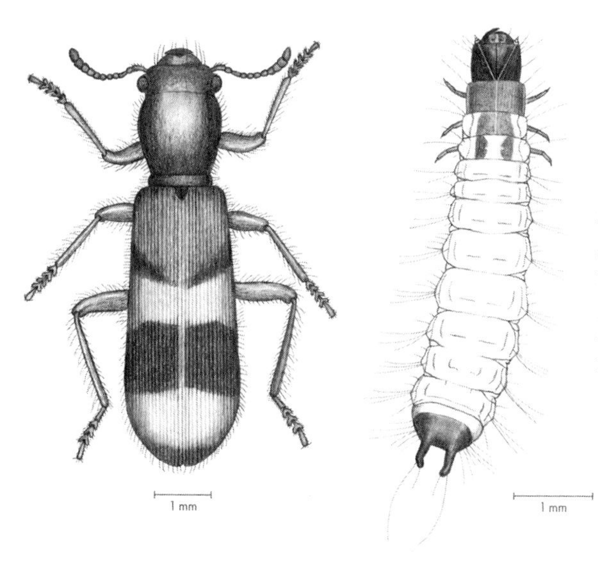

Far left: FIG. 221 Checkered beetles often sport pleasing checkered arrays of colors including reds and oranges.

Left: FIG. 222 Both larvae and adults of checkered beetles prey on other beetles.

— BARK-GNAWING BEETLES, FAMILY TROGOSSITIDAE (*trogo* = gnaw; *sit* = food) (600 species; 60 species NA; 5–20 mm).

The name of this beetle family is misleading. Rather than being gnawers of bark, these beetles are predators that prefer high-protein diets of insect meat. Trogossitid beetles share their habitats and prey found under tree bark with their relatives the checkered beetles (fig. 223). Most bark-gnawing beetles are dark colored and flattened, but a few are a striking iridescent green. Their jaw muscles are powerful enough to deliver a painful pinch to a poking finger in addition to dispatching their wood-boring neighbors. The cuticles of the head, thorax, and abdomen are thoroughly stippled with tiny pits. The last three segments of each antenna are conspicuously enlarged. The odors the antennae detect guide the beetles to the best hunting grounds in the tree's neighborhood.

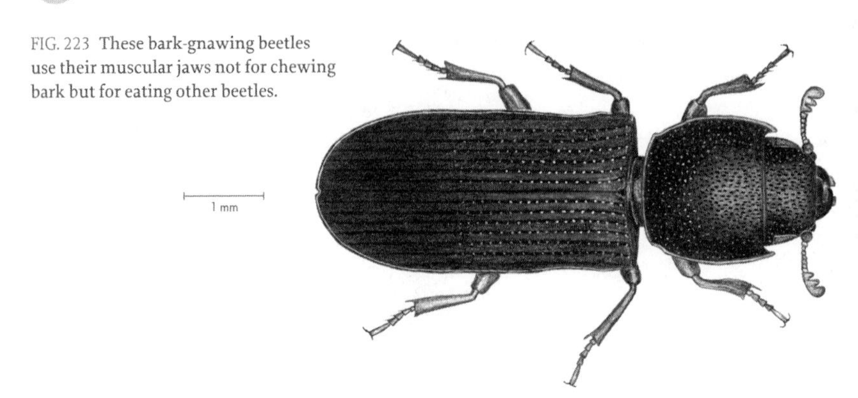

FIG. 223 These bark-gnawing beetles use their muscular jaws not for chewing bark but for eating other beetles.

— FLAT BARK BEETLES, FAMILY CUCUJIDAE (*cucuj* = kind of beetle) (60 species; 7 species NA; 6–25 mm).

The flattened forms of these adult beetles and their larvae are clearly suited for life under tree bark (figs. 224 and 225). Like beetles in many other families, the members of this family have a range of dietary preferences—many prey on other wood-dwelling beetles, some feed on fungi, and some eat dry plant matter. This one family, now reduced to only 60 species, at one time included several hundred species of other flat bark beetles that are now considered different enough to establish the following three new families.

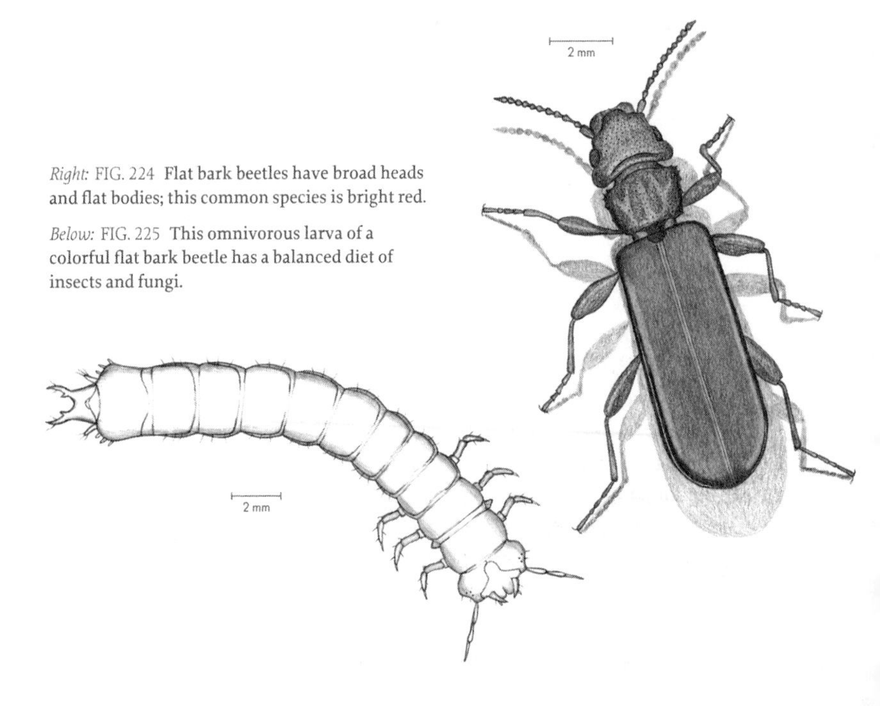

Right: FIG. 224 Flat bark beetles have broad heads and flat bodies; this common species is bright red.

Below: FIG. 225 This omnivorous larva of a colorful flat bark beetle has a balanced diet of insects and fungi.

— SILVAN FLAT BARK BEETLES, FAMILY SILVANIDAE (*silva* = woods, trees) (500 species; 32 species NA; 1.5–15 mm).

Many species of this family are fungivores that find homes under bark and in leaf litter, in habitats where fungi also thrive (fig. 226). In Australia, some of these beetles have taken up residence as freeloaders in ant colonies, while in the tropics of America, others rely on honeydew secreted by aphids and mealybugs. A minority of silvan flat bark beetles, however, are infamous as pests of stored grains, as hinted by their aptly chosen genus name *Oryzaephilus* (*oryza* = rice; *philus* = lover of).

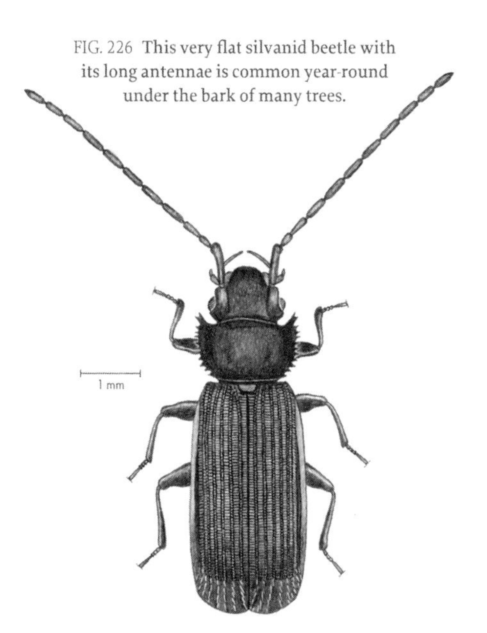

FIG. 226 This very flat silvanid beetle with its long antennae is common year-round under the bark of many trees.

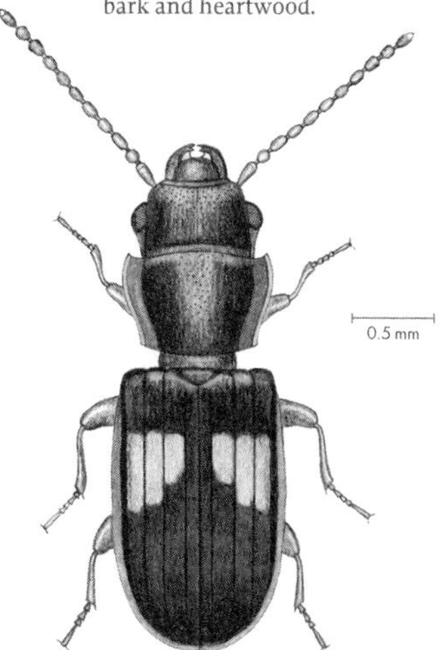

FIG. 227 Lined flat bark beetles are just one of several families of flat beetles, along with one family of flat bugs (Aradidae), that live in the narrow world between bark and heartwood.

— LINED FLAT BARK BEETLES, FAMILY LAEMOPHLOEIDAE (*laemo* = throat; *phloe* = bark) (450 species; 52 species NA; 1–5 mm).

These tiny, compact, flat companions of trees have many untold tales of their special talents and idiosyncrasies. The adult beetles stand out among the many bark beetles in having the first segments of the thorax bordered on the sides by fine grooves or lines (fig. 227). The larvae are well known for their novel production of silk from secretory cells on their thoraxes.

In the history of life, many creatures have gone down roads less taken. Spiders spin silk from abdominal spinnerets, caterpillars spin silken cocoons from a pair of mouthparts called labial glands, and lacewings construct cocoons with silk derived from modified cells of their internal excretory organ, the Malpighian tubules. However, the mature larvae of these lined flat bark beetles expel silk strands from large cells in their thoraxes to construct a cocoon as a sheltering refuge during their vulnerable transformation from larvae to adults. No other insects or arthropods are known to produce silk and build cocoons in this very unusual manner. Laemophloeid larvae stand out among all insects in having this novel silk-producing talent.

Larvae of at least two families of tree-dwelling beetles, the Bothrideridae and the Passandridae, specialize as parasites on the larvae and pupae of wood-boring beetles. After metamorphosis the sleek, elongate adult beetles stay around the galleries under tree bark where they spent their early days.

— PARASITIC FLAT BARK BEETLES, FAMILY PASSANDRIDAE (110 species; 3 species NA; 3–35 mm).

This group of beetles distinguishes itself by having parasitic larvae. Their insect hosts are the larvae and pupae of beetles and wood wasps that share their habitat under bark. Until recently these beetles, whose antennae and whose flatness are clearly pronounced, were considered members of the same family as other flat bark beetles in the family Cucujidae. Now they have attained new status by moving from a subfamily of the Cucujidae to a family of their own (fig. 228).

— DRY BARK BEETLES, BOTHRIDERIDAE (*bothri* = trench; *dero* = skin) (400 species; 18 species NA; 1.5–13 mm).

Dry bark beetles represent another family of beetles whose larvae are external parasites on larvae and pupae of wood-boring beetles and wood wasps (fig. 229). These larvae have been considered good candidates for controlling populations of invasive and native wood-boring insects that have developed reputations as

Below left: FIG. 228 Although they often leave their homes under tree bark to fly to lights on summer evenings, mystery still enshrouds the lives of these beautiful red-brown beetles.

Below right: FIG. 229 With their long, narrow bodies often clothed with long, recurved setae, dry bark beetles explore many tunnels under bark as they track down larvae of other beetles.

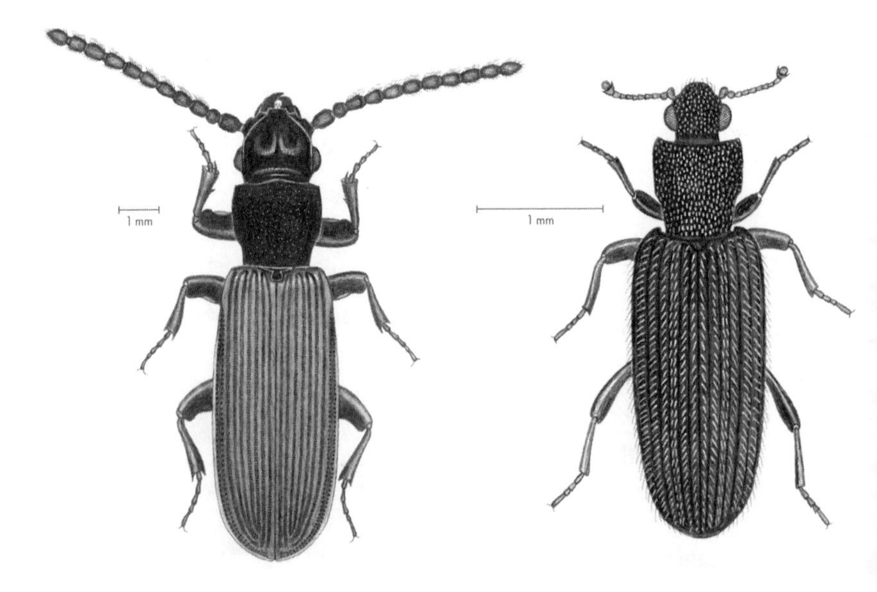

devastating pests of trees and forests. Until recently these parasitic beetles were considered members of the same family as many members of the newly created family Zopheridae (fig. 184).

— DEAD LOG BARK BEETLES, PYTHIDAE (*pytho* = decay) (23 species; 7 species NA; 7–22 mm).

— CONIFER BARK BEETLES, BORIDAE (*bori* = meat) (4 species; 2 species NA; 11–23 mm).

Dead log bark beetles prefer dead conifers such as pine, hemlock, spruce, and fir. Not only do they prefer conifer logs, but they also prefer that the bark on these logs be loose but still intact. Their large jaws, often iridescent colors, quick movements, and overall forms remind me of members of the earth-dwelling beetle family Carabidae (fig. 230 *left*). Another, even smaller family of even larger beetles, the Boridae, or conifer bark beetles, occupies these same habitats (fig. 230 *right*). The large, conspicuous jaws on these beetles and the larvae in both these families suggest that these insects are predators that roam beneath tree bark. Until recently members of both these two small families were included in the family Salpingidae, or narrow-waisted bark beetles. In the revision of the original Salpingidae, these two new families were created, and the remaining members of the original family now make up the latest and smaller version of the family.

— WRINKLED BARK BEETLES, FAMILY CARABIDAE, SUBFAMILY RHYSODINAE (*rhysso* = wrinkled; *-odes* = like) (350 species; 4 species NA; 5–8 mm).

In the ground beetle family Carabidae, known for its ground-dwelling, fast-moving predators, wrinkled bark beetles are the black sheep in appearance, behavior, and

FIG. 230 Dead log bark beetles (*left*) and conifer bark beetles (*right*) represent two small beetle families whose large, conspicuous members live under the bark of dead evergreen trees.

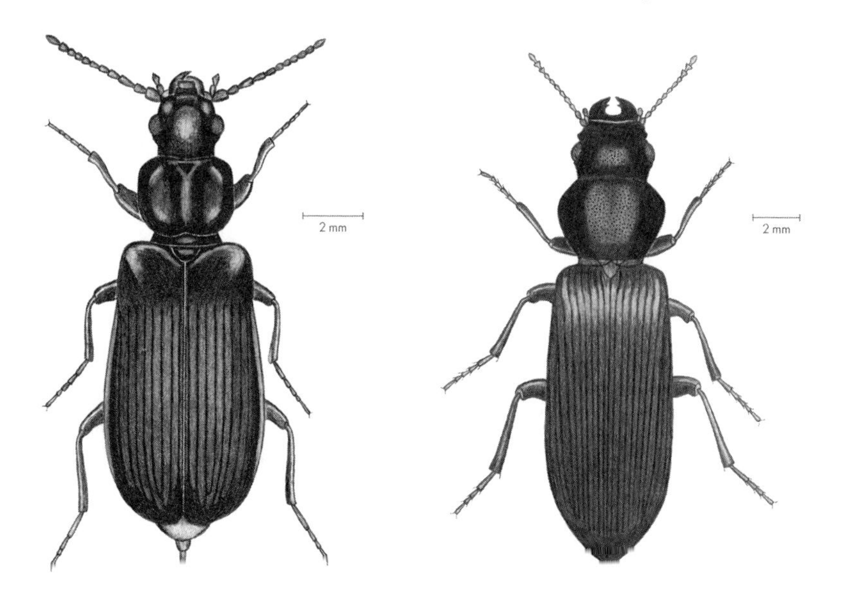

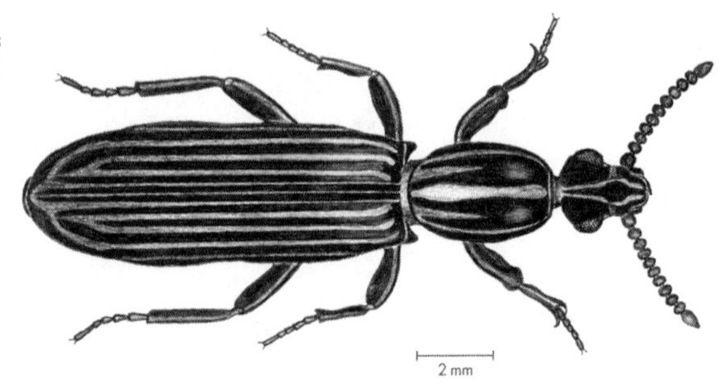

FIG. 231 Wrinkled bark beetles are among the few members of the large ground beetle family Carabidae that live on trees rather than under trees.

habits (fig. 231). During the winter these sluggish beetles often gather under the bark of moist, decaying trees and logs. Their close relatives the predatory ground beetles use strong jaws to subdue their prey. Rather than using jaws to eat, however, wrinkled bark beetles feed by swinging their heads to slice off fragments of decaying wood and fungi with the sharp edge of their "lower lip," or labium. The larvae of wrinkled bark beetles are slow moving, construct short galleries in the wood that they pack with wood flakes, and show little resemblance to their larval relatives among the other ground beetles. The pair of prominent projections at the posterior end of all other ground beetle larvae, called urogomphi (*uro* = tail; *gomphus* = bolt), are completely missing from the languid larvae of wrinkled bark beetles (figs. 29 and 328). However, the genetic material of wrinkled bark beetles says that they do not merit a family of their own but are specialized ground beetles; for this reason, they have recently been placed in the family Carabidae. For decades they were placed in the family Rhysodidae.

— WOOD-BORING WEEVILS, FAMILY BRENTIDAE (2,500 species; 151 species NA; 2–30 mm).

What distinguishes these weevils from true weevils is their straight antennae; true weevils have "elbows" in their antennae (figs. 133, 234, 277, 278). Some smaller members of the wood-boring weevil family feed on leaves and seeds of herbaceous plants, while some of the larger members excavate tunnels in living and dying trees.

The wood-boring weevils that live under the bark of dead trees also go by the names straight–snouted weevils, primitive weevils, and timberworms. With the small but stout mandibles at the tips of their slender, straight snouts, female oak timberworm beetles drill holes in wood and place one egg in each. The hole-drilling process is painstakingly slow and can take several hours. The jaws of the males are not designed for drilling (fig. 232). They do not look like the jaws of other weevils; they are relatively large, curved, and toothed. In his *Field Book of Insects*, first published in 1918, Frank Lutz describes how the male beetles patiently stand guard over the females during their prolonged labor of egg laying and defend them from strange males and other intruders. While the handsome adult beetles dwell under the bark of trees, their larvae bore into the heartwood, where their larval days are protracted for two to three years.

The New York weevil in the genus *Ithycerus* (*ithy* = straight; *cerus* = horn) was for a time placed in a family of its own, the Ithyceridae, but it is now considered a

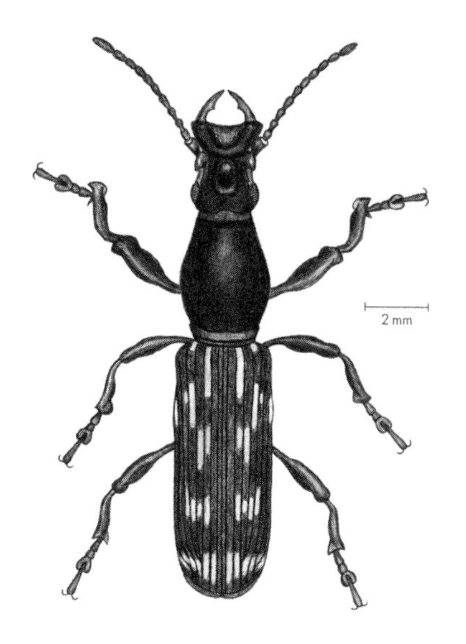

FIG. 232 The forms of male (*left*) and female (*right*) wood-boring weevils are easily distinguished.

member of the Brentidae. The adults are large gray weevils dappled with spots of white and black. While the larvae feed belowground on tree roots, the adults feed aboveground on buds and shoots.

— TRUE WEEVILS, FAMILY CURCULIONIDAE (*curculio* = weevil) (83,000 species; 2,500 species NA; 1–40 mm).

— BARK AND AMBROSIA BEETLES, SUBFAMILIES SCOLYTINAE (*scolyt* = shorten) and PLATYPODINAE (*platy* = broad; *pod* = foot) (Scolytinae 6,000 species; Platypodinae 1,400 species; 1–10 mm).

Weevils make up the second largest beetle family. The 20 subfamilies now embrace the former ambrosia beetle family Platypodidae and former bark beetle family Scolytidae. These small weevils share just about every molecular feature with most weevils, but not that many morphological features. Their most conspicuous omission is the iconic weevil snout. The easiest way to tell these two similar groups of beetles apart is to look at their eyes. The eyes of platypodine beetles are round and bulging, while the eyes of scolytine beetles are oval and vertically elongate (fig. 233).

FIG. 233 These compact, sturdy beetles carry fungi that help them surmount the defenses of their host trees.

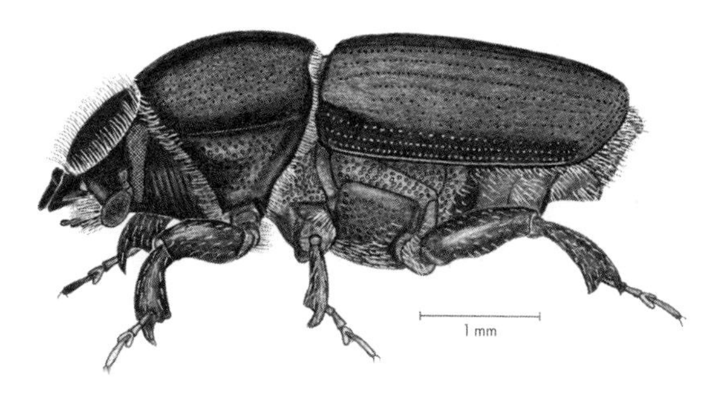

Several species and genera of pathogenic fungi have been identified as symbiotic associates of bark beetles in the genus *Dendroctonus* (*dendro* = tree; *ctono* = kill). Different fungal symbionts produce different mixes of volatile chemicals for their beetles and differ in their abilities to recruit beetles to infected trees. Beetle larvae feeding on different fungi show vastly different growth rates. The fungi induce host trees to produce defensive secondary metabolites. Fungi show differing tolerances to these defensive compounds, while the same defensive compounds show differing attractiveness to the beetles. The fungi penetrate the xylem heartwood, extract its nutrients, and concentrate them on the surfaces of the beetle galleries. These fungi have a remarkable influence on the intercourse between beetles and trees.

Fungi benefit by gaining access to host trees, and they contribute to the well-being of their beetle hosts by detoxifying the defensive chemicals of trees and providing certain nutrients such as amino acids and sterols. These fungi and beetles also have bacterial associates that emit volatile compounds. These compounds stimulate the formation of spores by some fungi and inhibit spore formation by other fungi. Furthermore, one of the primary defensive compounds of coniferous trees that host beetles, fungi, and bacteria also influences the interplay among all these associates (Lu et al. 2010; Adams et al. 2009).

Bark beetles in the genus *Hylurgopinus* (*hylurgo* = carpenter; *pinus* = pine tree) carry a fungal pathogen with the malevolent name *Ophiostoma* (*ophio* = snake; *stoma* = mouth) from elm tree to elm tree (fig. 234). The Dutch elm disease fungus owes much of its devastating success to its ability to not only trigger but also increase the release of secondary metabolites normally produced by elm trees. These chemicals prove alluring to bark beetles. Working with its beetle accomplices that transport and spread its spores and filaments, the fungus soon overwhelms the defenses of elm trees (McLeod et al. 2005).

Reading the tracks left by beetles beneath the bark can tell you a lot about these hidden lives (fig. 234, top). The tracks left by engraver beetles are particularly revealing. When the mother beetle finds a branch where she decides to lay her

FIG. 234 Engraver beetles in the genus *Hylurgopinus* (*below*) etch the record of their travels beneath the tree bark (*right*).

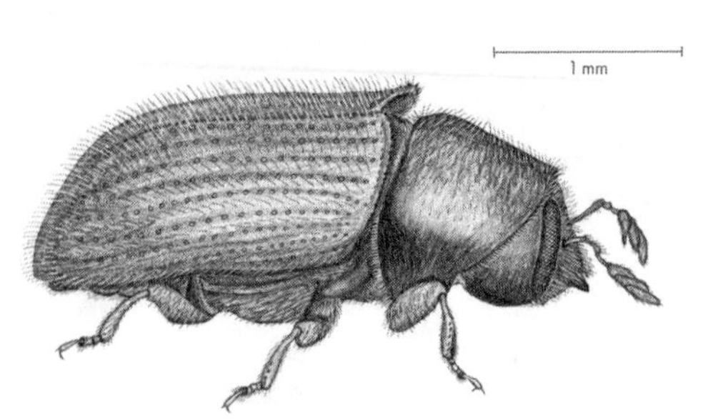

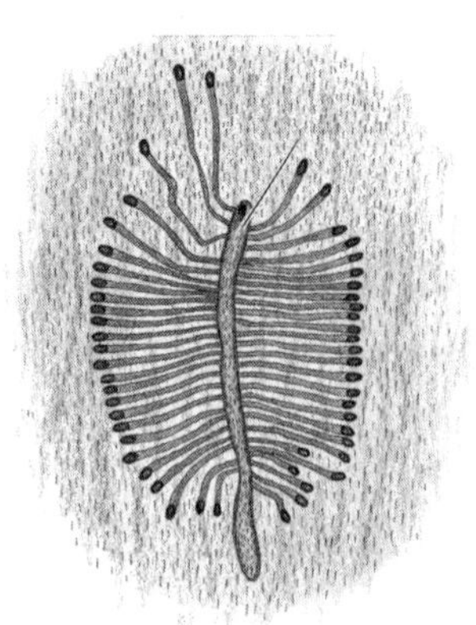

eggs, she first drills through the bark and then carves a trough called a brood gallery. Along the length of this gallery she distributes her eggs, and when the larval beetles hatch, each gnaws into the wood and leaves a trail that radiates from the central gallery. Somehow each larva steers a course between the trails being gnawed on each side by its brothers and sisters. Their trails neither cross nor collide. At the end of its trail, the larva pupates. When the adult beetle emerges, it drills its way to the surface of the bark and flies off. Each artistic engraving on the bark's lower surface is the work of an entire family of engraver beetles.

— FALSE BLISTER BEETLES, POLLEN-FEEDING BEETLES, FAMILY OEDOMERIDAE (*oedo* = swollen; *meri* = thigh, referring to the swollen hind femurs of certain members of this family) (1,500 species; 87 species NA; 5–22 mm).

While the larvae of blister beetles are parasites of soil-dwelling insects, larvae of false blister beetles inhabit decaying trunks, roots, and branches where they presumably feed on fungi rather than on other insects. As larvae, members of the two related families contribute very different gifts to the tree and its company.

Although different enough to be placed in separate families, adult false blister beetles and adult blister beetles have similar elongate bodies with soft cuticles and share a passion for pollen and flowers (fig. 235). Many adult species of both families are embellished with flashy metallic green, red, orange, blue, or copper colors that proclaims that they are best left undisturbed. If provoked, these beetles exude blood containing a blistering compound known as cantharidin. Wounds left by cantharidin can be painful and persistent. Although some adult blister beetles can ravage the foliage of certain herbaceous plants, adult false blister beetles are important contributors to tree pollination, and members of

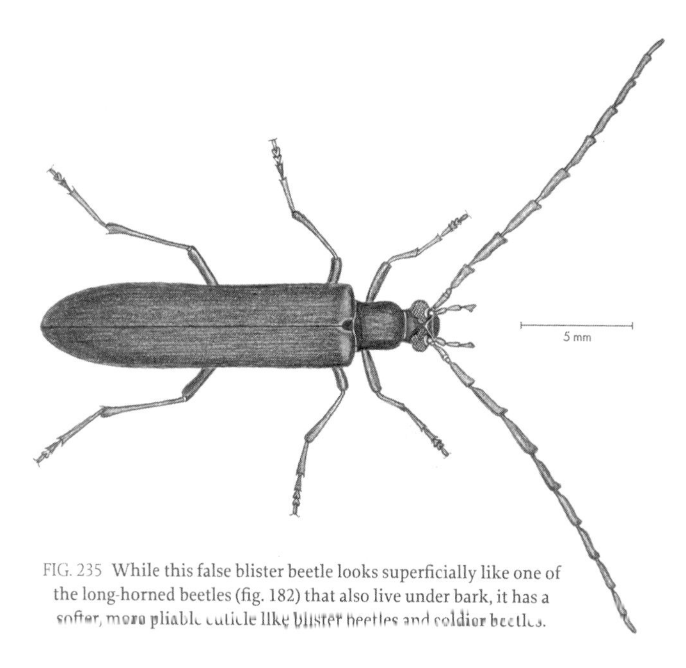

FIG. 235 While this false blister beetle looks superficially like one of the long-horned beetles (fig. 182) that also live under bark, it has a softer, more pliable cuticle like blister beetles and soldier beetles.

one common genus (*Asclera*) are frequent spring pollinators of willow flowers (fig. 270). An alternative common name for the Oedomeridae is pollen-feeding beetles, a name that emphasizes their positive contributions to the lives of trees rather than their negative reputations as blistering agents.

— FLAT BUGS, SUBORDER HETEROPTERA, FAMILY ARADIDAE (*aradus* = rattling, rumbling) (1,900 species; 130 species NA; 3–8 mm).

All true bugs feed with piercing, sucking mouthparts where typical insect jaws have been replaced with beaks or stylets. Almost all the true bugs that inhabit trees are either predators or sap suckers. Predatory bugs impale their prey with piercing beaks, expel potent salivary enzymes that begin the extra-oral digestion of the prey, and finally ingest the predigested body fluids of the prey. Sap suckers consume mostly sap but sometimes a few tender leaf cells as well. These all roam about on the surface of trees, on leaves, bark, or branches.

One very odd family of true bugs, however, lives under the bark of dead trees or in leaf litter, in the same habitats frequented by beetle larvae, fly larvae, and other, noninsect arthropods (fig. 236). Aradid bugs are configured so they have almost all dorsal and ventral surfaces, with hardly any room for lateral surfaces. In other insects, spiracles for air intake are located on lateral surfaces, but aradid bugs are so flat that some of their spiracles are positioned on their dorsal or ventral surfaces. Although these unusual bugs have always been assumed to feed on the fungi that are so abundant under the bark of old, dead trees, they also happen to be attracted to the same pheromones that attract bark beetles. This association with bark beetles has led to the supposition that these slow-moving bugs could also be predators. The fungivores among them are very particular about which fungi they associate with and are believed to seek out those trees that have been colonized by only particular species of fungi. Like many vertebrates that are opportunistic feeders, many invertebrates like these flat bugs are undoubtedly omnivores and probably diversify their diets to include other insects as well as fungi.

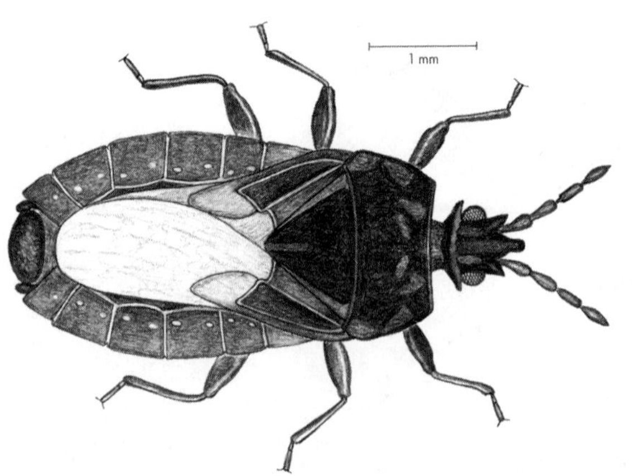

FIG. 236 Unlike other true bugs that are predators or feed on plant sap, flat bugs dwell under the bark of dead trees where the "sap" of fungi, not tree sap, is readily available.

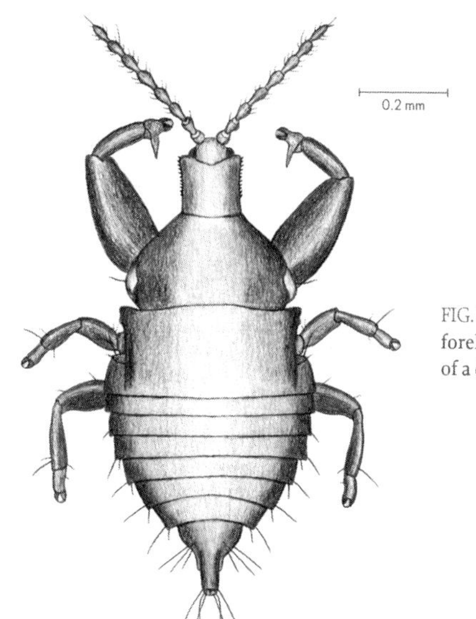

FIG. 237 The disproportionately massive forelegs of this thrips found under the bark of a dead beech tree tell of a rapacious life.

— THRIPS, ORDER THYSANOPTERA (*thysanos* = fringed; *ptera* = wings) (6,300 species; 800 species NA; 0.5–14 mm).

Thrips are known for their numerous idiosyncrasies, for their ubiquity on trees, and for the breadth of their food choices. Thrips come in winged and wingless forms (fig. 161; BarkScape, 23). Thrips with flat, broad forms scoot about in the narrow space between bark and heartwood. Such thrips are an exception among this exceptional group of insects (fig. 237). These thrips, like their predatory counterparts the praying mantids, are renowned for their massive raptorial (*raptor* = seize) forelegs. The tiniest residents of the landscape under tree bark, such as newly hatched larvae of beetles, flies, mites, springtails, and nematodes—those that are disregarded by larger predators—are the prime foci for these insect hunters.

LIVING IN TREE HOLES

— HOVER FLIES, FLOWER FLIES, FAMILY SYRPHIDAE (*syrpho* = gnat) (6,500 species; 820 NA species; 4–25 mm).

Members of this family of colorful flies are often mistaken for the bees and wasps that they so closely mimic in behavior and appearance. While the adult flies frequent flowers for their nectar and pollen, the larvae have followed many different career paths. Some free-loading maggots live in ant, termite, or bee colonies where they fool the workers into accepting them as members of the colony and providing protection and nourishment. Numerous predatory syrphid larvae can be found on leaf surfaces feeding on aphids and thrips, a few reside inside leaves feeding as leaf miners, and others live under bark or in tree holes feeding on detritus and microbes.

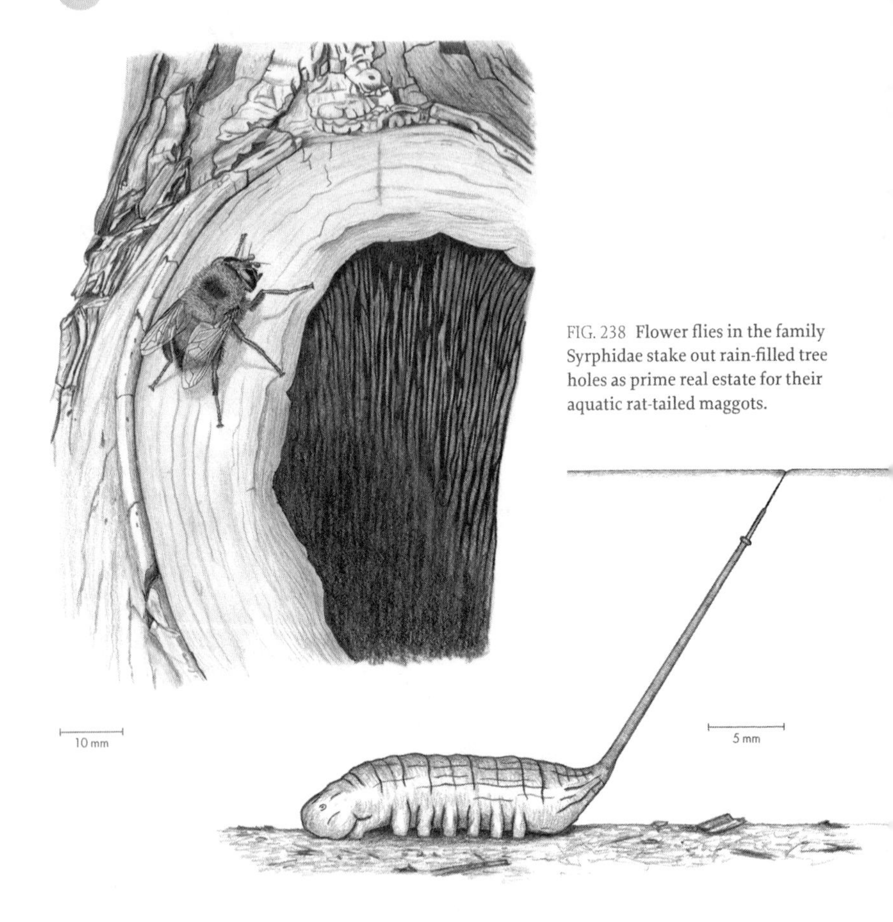

FIG. 238 Flower flies in the family Syrphidae stake out rain-filled tree holes as prime real estate for their aquatic rat-tailed maggots.

Tree holes serve as basins for rainwater and create tiny aquatic habitats in the treetops. Insects and mites quickly populate the tiny pools of water where they are safe from the jaws of fishes, frogs, and dragonfly nymphs. Creatures that live in these basins feed on either debris in the hole or creatures smaller than themselves. Many flies and mosquitoes return generation after generation to these pools to lay their eggs. In many forests, tree holes that fill with water are few and far between. Even though the water often becomes unsavory for human tastes, these pools are especially coveted by a number of interesting creatures. Flower flies are among those insects that search for these scarce, rain-filled holes. Once a male flower fly discovers a tree hole, he guards it diligently until a female joins him. By looking and sounding like bumblebees, these flower flies also fend off hungry birds and tree frogs. After the fly eggs hatch, their larvae inherit the tree hole from their parents. These rat-tailed maggots have a unique way of breathing as they feed and move about in the debris at the bottom of the pool. An air tube acts as a telescoping snorkel that can stretch several inches or shrink to less than an inch (fig. 238).

In their old age, trees become the dominion of borers and woodpeckers, owls and raccoons, and even some flies and their larvae. Even though their roles may change, old trees, like old people, can continue to be productive and useful members of their communities.

— TREE HOLE MOSQUITOES, FAMILY CULICIDAE (*culex* = gnat; 3,500 species; 175 species NA; 3–15 mm), GENUS *TOXORHYNCHITES* (*toxo* = bow; *rhyncho* = snout; *-ite* = part) (71 species; 2 species NA).

This genus includes the largest species of mosquitoes, with some tropical species having wingspans up to 24 mm. Even though these intimidating, gargantuan mosquitoes have long, curved proboscises like elephant trunks, they do not bite, and females do not need a blood meal to mature their eggs (fig. 239). Instead, both males and females survive and thrive on plant nectars and juices alone. Their large size and long trunks have earned them the common name elephant mosquitoes. Not only are these adult mosquitoes large and harmless but they are also adorned with iridescent blue, purple, green, and silvery white scales.

During the short lifetime of *Toxorhynchites*, it can assume two very different occupations in a tree community—as adult pollinator and as larval predator. *Toxorhynchites* larvae share the same aquatic habitats with larvae of their smaller relatives that are destined to become the annoying, biting mosquitoes. While these smaller larvae feed and grow on algae, protozoa, and other microbes in the water of the tree hole, the larger, more colorful *Toxorhynchites* larvae survive as predators on flesh and blood—mosquito flesh and blood, that is. They are mosquitoes that eat mosquitoes.

The appearance of tree holes usually signals that wood borers and fungi have taken up residence in the tree. Woodpeckers investigate and peck at a limb or trunk that promises to have some wood borers inside. Hunting season for caterpillars among the leaves lasts for only a few months each year, but wood-boring insects are a year-round source of food. Although the borers may not be very active on cold days, woodpeckers can still find them as easily in the winter as they can in the summer. The holes that one woodpecker starts as it searches for wood-boring larvae can be enlarged by other woodpeckers and may eventually end up as one big hole for still another animal's nest.

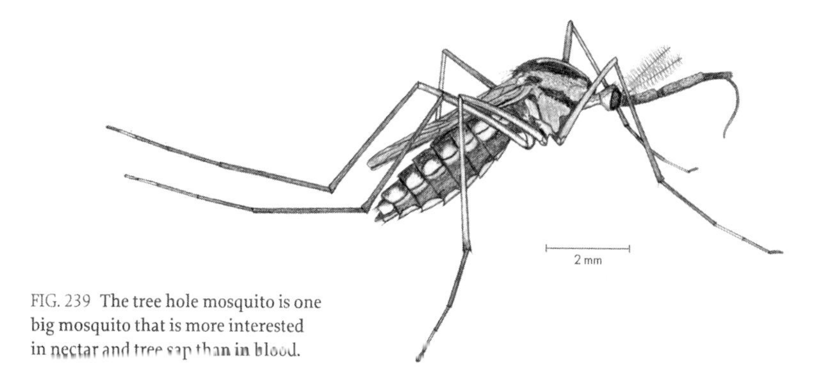

FIG. 239 The tree hole mosquito is one big mosquito that is more interested in nectar and tree sap than in blood.

FIG. 240 A kestrel has claimed a hole in a walnut tree for its nest.

Woodpeckers and wood borers are not the only ones that live in tree holes. Flying squirrels, nuthatches, bats, mice, tree frogs, and even some larger birds settle into secondhand woodpecker holes. Flying squirrels (figs. 46 and 285) and nuthatches (fig. 9) use strips of fine bark, a little grass, and a few leaves to pad their nests. Wood ducks line hollow trees with their down feathers. Great crested flycatchers (fig. 2) line their tree cavities with any number of materials—bark, twigs, hair, feathers. And for reasons known only to these flycatchers, they also habitually add shed snake skins to their nests. Bluebirds and tree swallows add a few grasses and feathers as nest lining. The small hawks known as kestrels (fig. 240) do not bother to add any bedding of their own to the tree holes they choose.

— WOUNDED-TREE BEETLES, FAMILY NOSODENDRIDAE (*noso* = wounded; *dendro* = tree) (67 species; 2 species NA; 4–6 mm).

These rotund beetles and their chunky larvae hang out at tree holes and tree wounds where oozing, fermenting sap provides ideal conditions for the growth of bacteria, fungi, and fly larvae (fig. 241). The beetles congregate where the sap is deepest and wettest. The larvae are noteworthy among beetle larvae in having a short snorkel at their posterior end, a feature usually reserved for strictly aquatic larvae such as rat-tailed maggots (fig. 238). Here in the depths of the sap the beetles are surrounded by abundant nourishment.

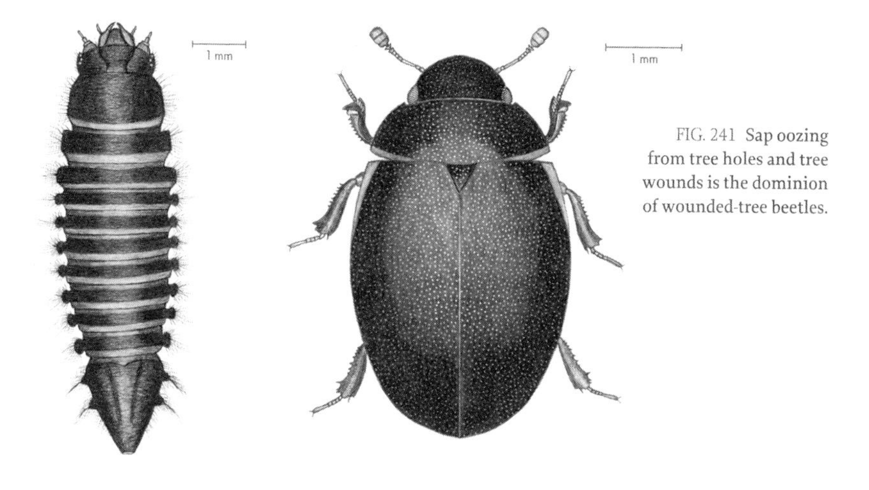

FIG. 241 Sap oozing from tree holes and tree wounds is the dominion of wounded-tree beetles.

BEES AND WASPS OF LITTLE TREE HOLES

While birds, bats, flying squirrels, and rat-tailed maggots use big tree holes, leafcutter bees and mason wasps often move into the vacant tunnels of wood-boring larvae. Once these bees and wasps locate a hollow twig or hole left by a wood borer, they set about remodeling it, each in its own special way. They are usually more fastidious about their nests than the birds are about theirs.

— MASON WASPS, POTTER WASPS, FAMILY VESPIDAE (*vespa* = wasp) (5,000 species; 300 species NA), SUBFAMILY EUMENINAE (3,000 species; 260 species NA; 10–20 mm).

Mason and potter wasps are dark wasps with bright patches of contrasting yellow, white, orange, or red (fig. 242). These artisans use mud to sculpt the chambers

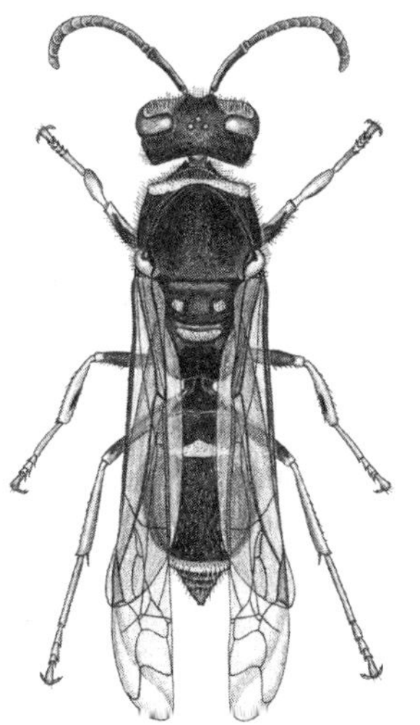

FIG. 242 Potter wasps seek out twig surfaces or hollow twigs for their nurseries, which they refurbish with mouthfuls of mud to accommodate their larvae and the paralyzed insects on which the larvae feed.

FIG. 243 A mason wasp surveys a hollow branch as a possible site for her mud-lined nest.

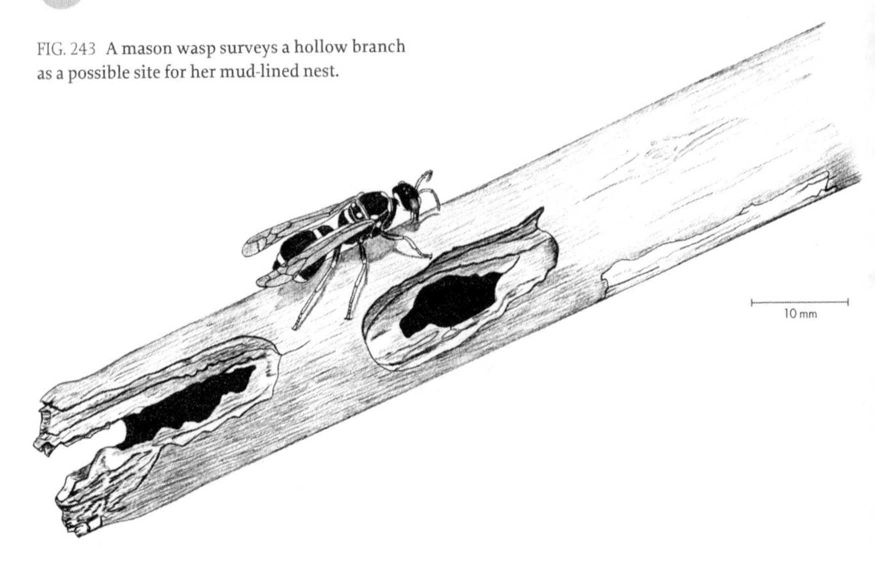

for their brood cells. Some sculpt mud pots on surfaces of leaves and twigs for their nursery chambers, while others use mouthfuls of mud to build chambers in hollow twigs for their larvae (fig. 243). The grass-carrying wasps mentioned in chapter 2 (LeafScape, 21) likewise choose hollow twigs and other small tree holes in which mother wasps use grass fragments rather than mud to fashion and partition chambers for their larvae, which they provision with paralyzed tree crickets. Like other wasps—and unlike bees—these potter and mason wasps provide insects as food for their larvae. A mother mason wasp will stash just enough paralyzed caterpillars in each nursery chamber to nourish a larva from the time it hatches until it undergoes metamorphosis. After the caterpillar and egg are in place, the mother wasp neatly seals off each chamber with a thin, smooth wall of mud.

— LEAFCUTTER BEES, MASON BEES, AND RESIN BEES, FAMILY MEGACHILIDAE (*mega* = large; *chilus* = lip, referring to the long labrum, or upper lip) (4,100 species; 630 species NA; 5–24 mm).

Unlike many bees that carry pollen in special baskets on their hind legs (chapter 5) and nest in the soil or hives, bees in this family carry pollen in dense patches of hairs that cover their bellies (fig. 244). Almost all megachilid bees nest aboveground in various cavities.

Without a ruler or compass, leaf-cutting bees bite off perfectly circular pieces of leaves that are just the right size and shape to build a series of leafy chambers in the tunnels abandoned by wood-boring larvae (fig. 244). In each chamber the bees store a good supply of nectar and pollen before laying an egg. The chambers are then sealed with more round pieces of leaves.

The mason bee and resin bee members of this family respectively use mud and plant resins for constructing their nursery chambers in abandoned passageways,

FIG. 244 A leafcutter bee cuts circular pieces of leaves from redbud leaves to line and seal the nursery chambers it constructs in hollow twigs

— CARPENTER BEES, FAMILY APIDAE, GENUS *XYLOCOPA* (*xylokopos* = wood cutter) (500 species; 9 species NA; 12–26 mm).

Like their cousins the bumblebees and honeybees, carpenter bees are also avid pollinators. However, unlike their cousins in the family Apidae, these bees are avid wood workers, excavating tunnels in limbs and tree trunks with their strong mandibles. They partition each tunnel into typically six to eight nurseries separated by plugs of wood pulp. Through the summer months, the larvae grow up on a rich diet of pollen and nectar, and adult bees spend the winter in suspended animation, awaiting the spring (fig. 245).

Like all male bees, carpenter bee males are not known for their industry. They forgo tunneling and provisioning, leaving those jobs for their mates. However, they are very protective of their spouses and their tunnels. The gentle, stingless males come across as fierce, hovering around the females' tunnels and fearlessly approaching whomever they encounter in the vicinity.

FIG. 245 Inside her wood tunnel, the mother carpenter bee (*left*) has partitioned the space to accommodate a larva (*right*) and an egg (*center*) with ample provisions of pollen and nectar.

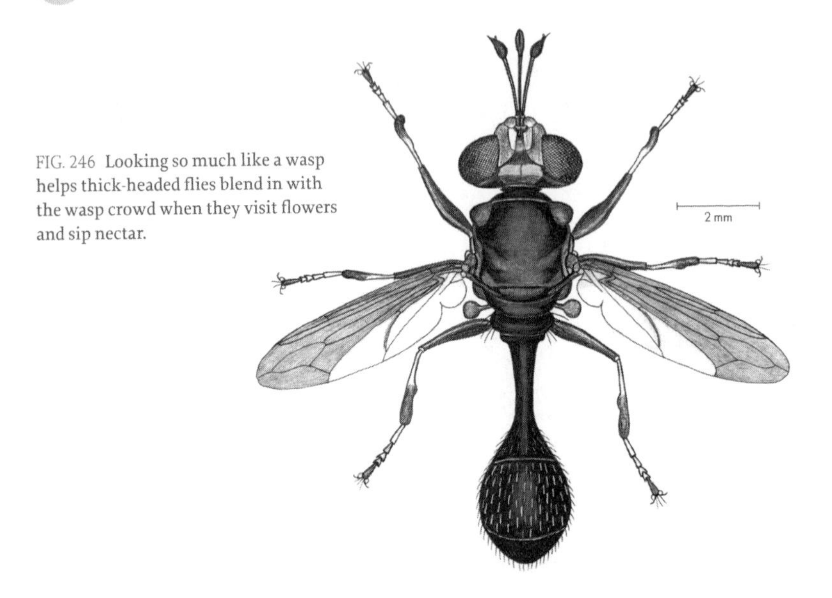

FIG. 246 Looking so much like a wasp helps thick-headed flies blend in with the wasp crowd when they visit flowers and sip nectar.

The wasps and bees of tree holes keep company with parasitic flies whose adults not only look like them but whose larvae also devour them.

— THICK-HEADED FLIES, FAMILY CONOPIDAE (*con* = cone; *ops* = face) (830 species; 70 species NA; 3–20 mm).

Thick-headed flies have heads that appear to be too large for their slender bodies. The head is wider than the thorax, and the long, slender proboscis with which the fly laps up nectar is even longer than its head (fig. 246).

The mother fly deposits one egg at a time by pouncing on a bee or wasp host while in flight. She must move swiftly. She first pries apart two segments of her host's abdomen with her modified ovipositor before inserting her egg through the soft membranes separating the segments.

The pear-shaped fly larva is narrowest at its head end and remains in the spacious abdomen of its host, waiting until it is fully grown before it stretches through the narrow waist that separates the host's thorax from its abdomen. It then devours what little remains of the tissues of its dying host and pupates in the shell of the abdomen. The adult conopid develops inside the dead host's abdomen, having spent all its immature days within the shelter of its host.

BEETLES OF ROTTING WOOD

The most primitive of the beetles and the descendants of the first beetles on Earth happen to be tree inhabitants. The ancestors of these beetles witnessed the early appearance and evolution of conifers and flowering trees. Their ancient lineage is appropriately placed in a beetle suborder of its own—Archostemata (*archo* = ancient; *stemata* = line).

— TELEPHONE-POLE BEETLES, FAMILY MICROMALTHIDAE (*micro* = tiny; *maltha* = pliant) (1 species; 1 species NA; 1.5–2.5 mm).

While most insect groups have at least one claim to notoriety, the micromalthid beetle has many such claims. For one thing, more extinct species than living members have been assigned to this beetle family (Yan et al. 2019). The telephone-pole beetle is the only living species of its family, Micromalthidae (fig. 247). Six extinct species have been described, mainly from amber deposits (Yamamoto 2019). The beetles are found in decaying wood of trees, logs, and even old telephone poles. This native North American beetle has probably been introduced to other continents by traveling as a stowaway in shipments of lumber.

Adult beetles are incapable of reproducing; they appear only very rarely under stressful environmental conditions of heat and drought. The sterile adults are apparently remnants of a time when the life cycle included sexual reproduction. The larvae, by contrast, are relatively common in decaying wood and come in several different forms. Only larvae reproduce (paedogenesis, *paedo* = child; *genesis* = birth), with female larvae giving birth to live larvae that are also all female but different in their form and activity. Mating never occurs, and all these births are virgin births (parthenogenesis, *parthenos* = virgin; *genesis* = birth). When these active female larvae molt, they revert to a more sedentary form again. Love, sex, and metamorphosis have taken unprecedented paths in this nonconformist beetle (Perotti at al. 2016).

Adult males are even rarer than adult females and arise under environmental stress when the sedentary female larva undergoes an extra molt to a third larval form. This mother gives birth to a male larva that spends the next few days feeding on her before pupating and emerging as a rare adult male telephone-pole beetle.

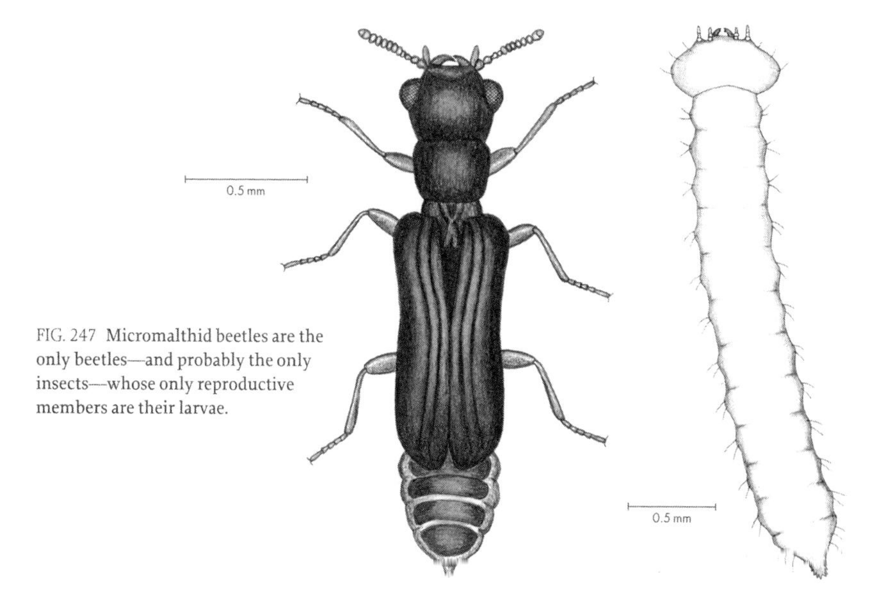

FIG. 247 Micromalthid beetles are the only beetles—and probably the only insects—whose only reproductive members are their larvae.

— RETICULATED BEETLES, FAMILY CUPEDIDAE (*cupedi* = dainty) (30 species; 4 species NA; 10–20 mm).

Rectangular pits aligned in about 50 regular rows and about 20 columns cover the surfaces of the elytra to create a reticulated pattern (fig. 248). These primitive beetles trace their ancestry back about 290 million years, even before the first dinosaurs appeared on Earth. Extinct species outnumber the species presently found on Earth. Adult beetles dwell under bark, and larvae tunnel into the heartwood, feeding on the fungi within the decaying but still solid wood (fig. 249). The adults are often enticed from their dark, sequestered homes under bark to visit evening lights. Male beetles alone, however, are enticed by the odor of laundry bleach. To locate mates among trees in the forest, female beetles apparently emit a pheromone that resembles the odor of bleach. This is one of the many idiosyncrasies of these unusual beetles.

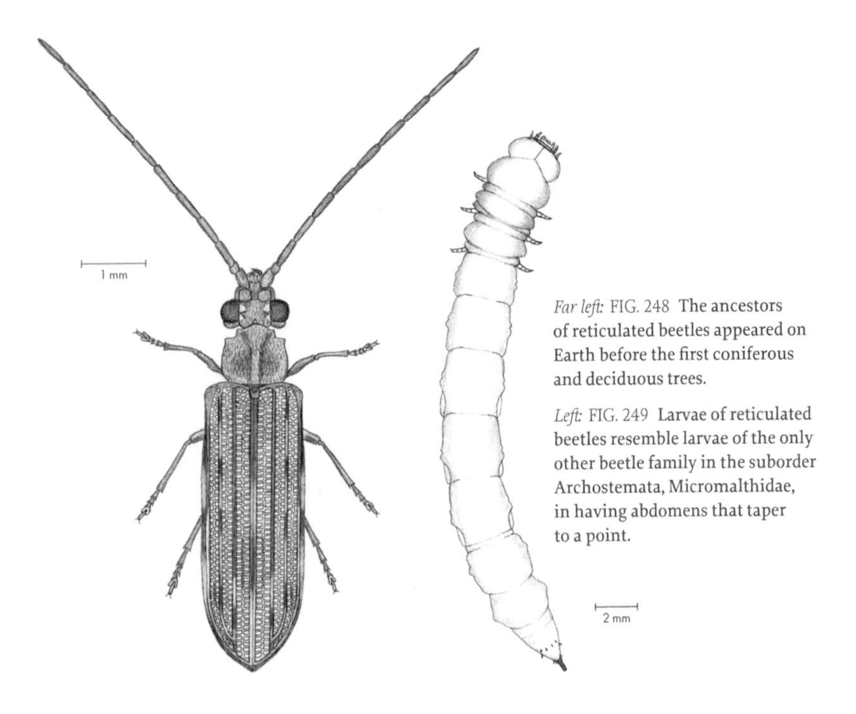

Far left: FIG. 248 The ancestors of reticulated beetles appeared on Earth before the first coniferous and deciduous trees.

Left: FIG. 249 Larvae of reticulated beetles resemble larvae of the only other beetle family in the suborder Archostemata, Micromalthidae, in having abdomens that taper to a point.

— SCARAB BEETLES, FAMILY SCARABAEIDAE (30,000 species; 1,700 species NA; 2–120 mm).

This family of sturdily built beetles has sluggish larvae that feed on roots, dung, or decaying wood. The heftiest, most corpulent of the beetles are members of this family. Adults of beetles such as chafers, June beetles, and Japanese beetles feed on flowers and leaves (chapter 2; fig. 84); over 8,000 species known as dung beetles are decomposers attracted to dung. These scarab beetles efficiently recycle this nutrient-rich resource with the help of their larvae.

— HERMIT BEETLES, SUBFAMILY CETONIINAE (*cetus* = whale, referring to the large size of these beetles), GENUS *OSMODERMA* (*osmo* = odor; *derma* = skin) (9 species; 3 species NA; adults ~ 30 mm long, larvae ~ 50 mm).

The subfamily Cetoniinae not only includes many species (4,000 worldwide) but also contains some of the largest beetles on Earth. The Goliath beetle of Africa has larvae measuring up to 250 mm long and adults measuring up to 120 mm. Among members of this subfamily is the genus *Osmoderma*, a well-known resident of hollow, decaying trees (fig. 250). These beetles are conspicuous to both the eye and nose. The hermit beetle in the genus *Osmoderma* stands out for having the distinct odor of leather and for being a large beetle with an even larger larva. The name "hermit beetle" was probably chosen because of its preference for secluded habitats in old hollow trees. The larvae feed for three years before they are ready to transform to adult beetles. With the loss of forests and large old trees throughout Europe, the hermit beetle has become an icon for endangered insects. In Sweden stamps have been issued with the image of the hermit beetle and its tree habitat. Scientists in Italy have been monitoring the populations of these beetles, even using trained dogs to detect them in hollow trees (Mosconi et al. 2017). These conspicuous beetles are considered umbrella and indicator species for the many other beetles that dwell in hollow trees (Ranius 2002).

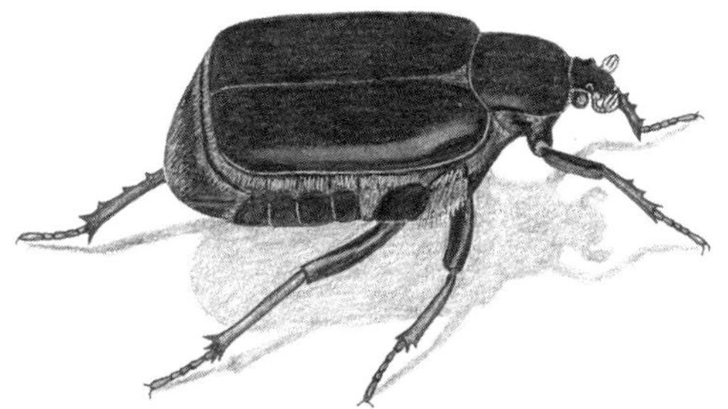

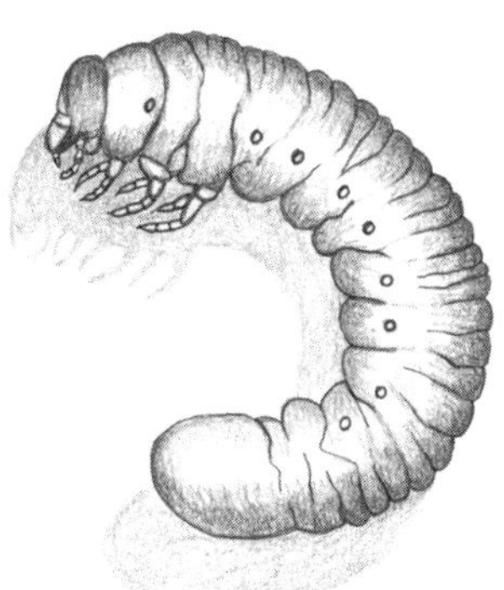

FIG. 250 Hermit beetles and their larvae live their long lives in the confines of hollow trees.

Among these other beetle families and species are some that are about as large and conspicuous as *Osmoderma*, such as stag beetles and patent leather beetles, as well as some that are considered some of the tiniest of beetles. The best way to see the variety of these beetles is to collect decaying wood from a hollow tree in winter or early spring and place it in a large clear glass jar or small aquarium. Over the next few months keep the wood moist, and a progression of adult beetles will emerge.

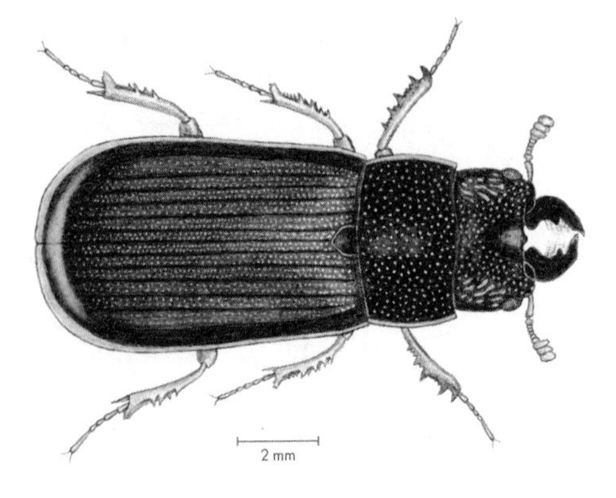

FIG. 251 Adult stag beetles have jaws of impressive proportions.

— STAG BEETLES, FAMILY LUCANIDAE (*lucanus* = kind of beetle) (1,500 species; 38 species NA; 8–60 mm).
— PATENT LEATHER, PEG, OR BESS BEETLES, FAMILY PASSALIDAE (*passalus* = peg) (500 species; 4 species NA; 20–43 mm).

Members of these two beetle families are large and conspicuous, but by beetle standards, the number of species in each family is small. These beetles have substantial jaws (mandibles) and jaw muscles for pulverizing wood. Stag beetles are noted for their often-massive jaws, which male beetles use in competing for female attention (fig. 251). In addition to their muscular mandibles, a prominent curved peg (*passalus* = peg) is perched on the top of each patent leather beetle's or passalid beetle's head like the horn of a unicorn. An encounter with a passalid beetle or a stag beetle is always memorable (fig. 252).

In passalid larvae the last pair of legs is no longer used for walking but is turned 90 degrees to rub against a rough patch of cuticle on the more anterior segment of the thorax. Using their third pair of legs to communicate with sound is important to these larvae. They survive perfectly well ambling along with four legs rather than the usual six (fig. 253). The sound produced by the rubbing of this third pair of legs is part of the discourse between larval and adult beetles in the communities that passalid beetles form in decaying wood. Eavesdropping

FIG. 252 Passalid beetles live in decaying logs as family groups where all life stages are found.

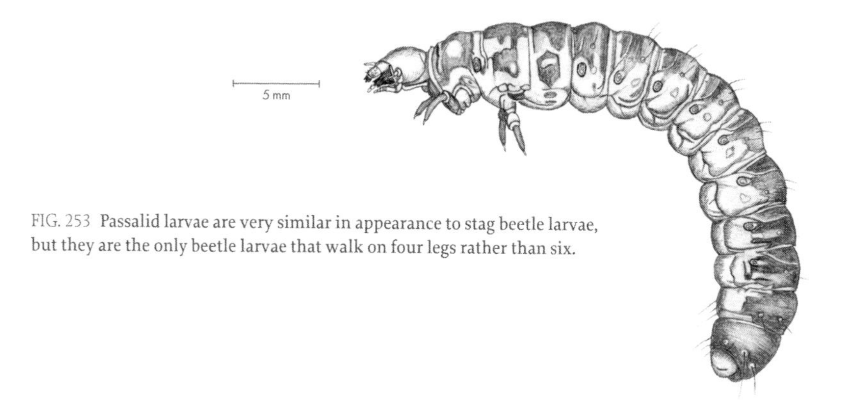

FIG. 253 Passalid larvae are very similar in appearance to stag beetle larvae, but they are the only beetle larvae that walk on four legs rather than six.

on beetle conversations has enabled the recording of several different calls of passalid beetle language. After the eggs of these beetles hatch, the adults stick around to watch over their larvae. These beetles form families. In the close, dark quarters of their fungus-lined chambers, beetles of different generations share rotting wood, fungi, and gut microbes.

In shape and size, stag beetle larvae resemble passalid larvae, but passalid larvae are the only beetle larvae with one pair of legs, which are typically used for walking, converted to sound production. Most larvae are found in decaying wood, but a few have shown up in living roots of trees. All stag beetle and passalid larvae, however, depend on the countless microbes in their guts to facilitate their digestion of the tough wood fibers and to supplement their low-nutrient diets with vitamins and amino acids.

All these beetles are instrumental in recycling the great amount of dead wood that a tree adds to its environment each year. They shred and pulverize the wood as a prelude to the final breakdown of the molecules by fungal and bacterial enzymes. After microbial enzymes finish their work on the pulverized wood, the remnants are water, essential minerals, the undigested carbon of humus, and carbon given off as carbon dioxide. The whole liberating process of converting complex tissues and cells of trees to their simple mineral components is referred to as *mineralization.* Photosynthesis represents a reversal of mineralization; combining carbon dioxide and water with the 15 essential inorganic soil minerals results in their *immobilization* within a tree as the innumerable complex molecules that make up a large tree's estimated 100 trillion cells.

— CLICK BEETLES, FAMILY ELATERIDAE (*elater* = a driver) (10,000 species; 970 species NA; 1–60 mm).

As members of a large family of about 10,000 species, adult click beetles are often referred to as snapping beetles, spring beetles, and skipjacks for their ability to "click," flip, and right themselves after falling on their backs (fig. 254; BarkScape, 8). While adults spend most of their time at flowers, larvae meander through passageways in soil and decaying wood. The larvae have adopted diverse

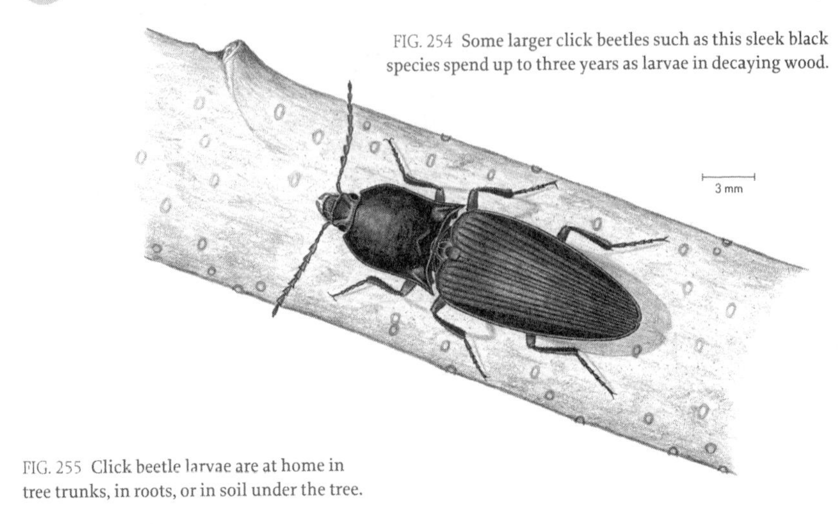

FIG. 254 Some larger click beetles such as this sleek black species spend up to three years as larvae in decaying wood.

FIG. 255 Click beetle larvae are at home in tree trunks, in roots, or in soil under the tree.

occupations in their food webs: as herbivores feeding on roots (fig. 341; RootScape, 26); as predators of other arthropods; or as decomposers that scavenge decaying wood or leaves and the microbes that inhabit them. The larvae are known for their slim, elongate physiques. The hard, sleek cuticle of one of these wood-boring "wireworms" imparts the appearance and texture of a metal wire (fig. 255).

— RARE-CLICK BEETLES, FAMILY CEROPHYTIDAE (*cero* = horn, antenna; *phyto* = plant, apparently referring to antennae that branch like trees) (22 species; 2 species NA; 5–9 mm).

— FALSE CLICK BEETLES, FAMILY EUCNEMIDAE (*eu* = good; *cnem* = leg armor, apparently referring to spines and spurs on legs) (1,700 species; 92 species NA; 2–30 mm).

— THROSCID BEETLES, FAMILY THROSCIDAE (*throsc* = leap, spring) (150 species; 20 species NA; 2–5 mm).

In addition to the "true" click beetles, several other small families of tree-dwelling beetles, related in form and ancestry, can click and flip as adults (figs. 256–258). Larvae of these three beetle families feed in wood infiltrated by fungi. The larvae of false click beetles, unlike the larvae of true click beetles, navigate without legs and have earned the name "cross-cut borers" from their habit of boring across the grain of wood rather than with the grain. The mandibles of eucnemid larvae have their chomping surfaces facing outward rather than inward, and this reverse orientation matches their outlandish manner of boring and feeding. In addition to the larvae of these families being wood borers, one species of throscid feeds on the mycorrhizal fungi of trees, but many mysteries remain about the hidden lives of larvae and adults in these three beetle families.

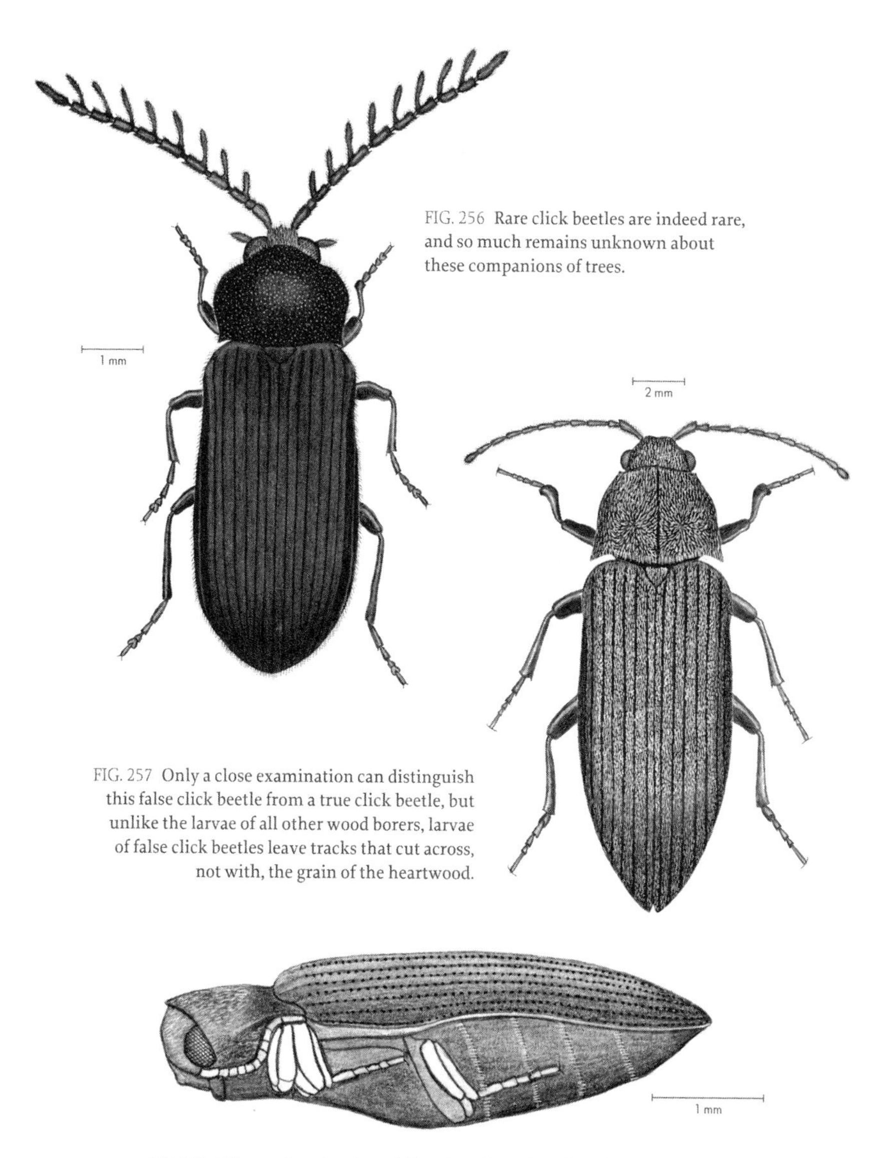

FIG. 256 Rare click beetles are indeed rare, and so much remains unknown about these companions of trees.

FIG. 257 Only a close examination can distinguish this false click beetle from a true click beetle, but unlike the larvae of all other wood borers, larvae of false click beetles leave tracks that cut across, not with, the grain of the heartwood.

FIG. 258 When agitated, a throscid beetle reduces its vulnerability by folding its antennae and legs and then tightly clamping them against its body.

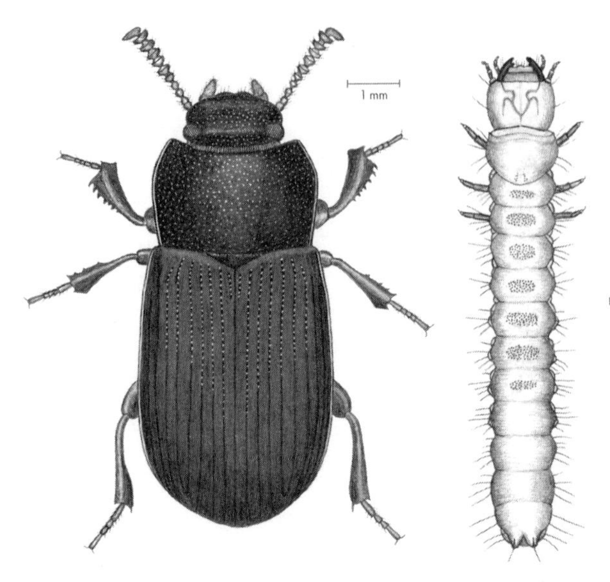

FIG. 259 Darkling beetles are members of a family of diverse forms and habits and are at home not only on trees but also on the ground.

Larvae of false click beetles have the distinction of being the exclusive hosts of an entire family of parasitic wasps, the Vanhorniidae. This group of discriminating parasitoids claims six species worldwide and only one species in North America. This is another wonderful example of the extreme specialization of species that make up the web of life on a tree.

Adult throscid beetles are small and seemingly defenseless, but they have one attribute—in addition to their leaping ability—that frequently spares them from bird beaks and beetle mandibles. On the undersurface of its body, each throscid beetle has grooves whose contours perfectly complement the contours of its antennae and legs. Folding and tucking their appendages into these grooves leaves no loose ends for grasping. With no appendages protruding and impeding its fall from a flower or log, the beetle can freely roll and bounce to cover (fig. 258).

— DARKLING BEETLES, FAMILY TENEBRIONIDAE (*tenebrio* = lover of darkness) (20,000 species; 1,200 species NA; 1–80 mm).

Both adult and larval darkling beetles share habitats in decaying trees, under bark, and on fungi (fig. 259). Some species have taken up residence as squatters in termite and ant nests; others are ground dwellers that inhabit leaf litter or live under rocks and logs. While most members of this large family prefer moist habitats on or under trees, some have colonized deserts and parched sand dunes. Some adult beetles also have reputations as "skunk beetles" that spray foul-smelling odors from the tips of their abdomens to thwart intruders. Tenebrionid larvae can easily be mistaken for larvae of click beetles, which also live under bark and in decaying wood; both beetle larvae are long and cylindrical and have hard cuticles. Since click beetle larvae are referred to as wireworms, their tenebrionid counterparts are often designated false wireworms (fig. 259).

- FALSE DARKLING BEETLES, FAMILY MELANDRYIDAE (*melano* = dark; *dryo* = tree), (1,200 species; 60 species NA; 3–20 mm).
- POLYPORE FUNGUS BEETLES, FAMILY TETRATOMIDAE (*tetra* = four; *tomos* = sections, apparently referring to the reduced four terminal antennal sections that make up the antennal club) (155 species; 30 species NA; 3–16 mm).

Until recently the Melandryidae and Tetratomidae were considered to be the single family Melandryidae, always known for its diverse forms and for being hard to characterize according to size, shape, and color (figs. 260 and 261). To add to the confusion in identifying these two beetle families, members of the Melandryidae are commonly referred to as false darkling beetles because of their superficial

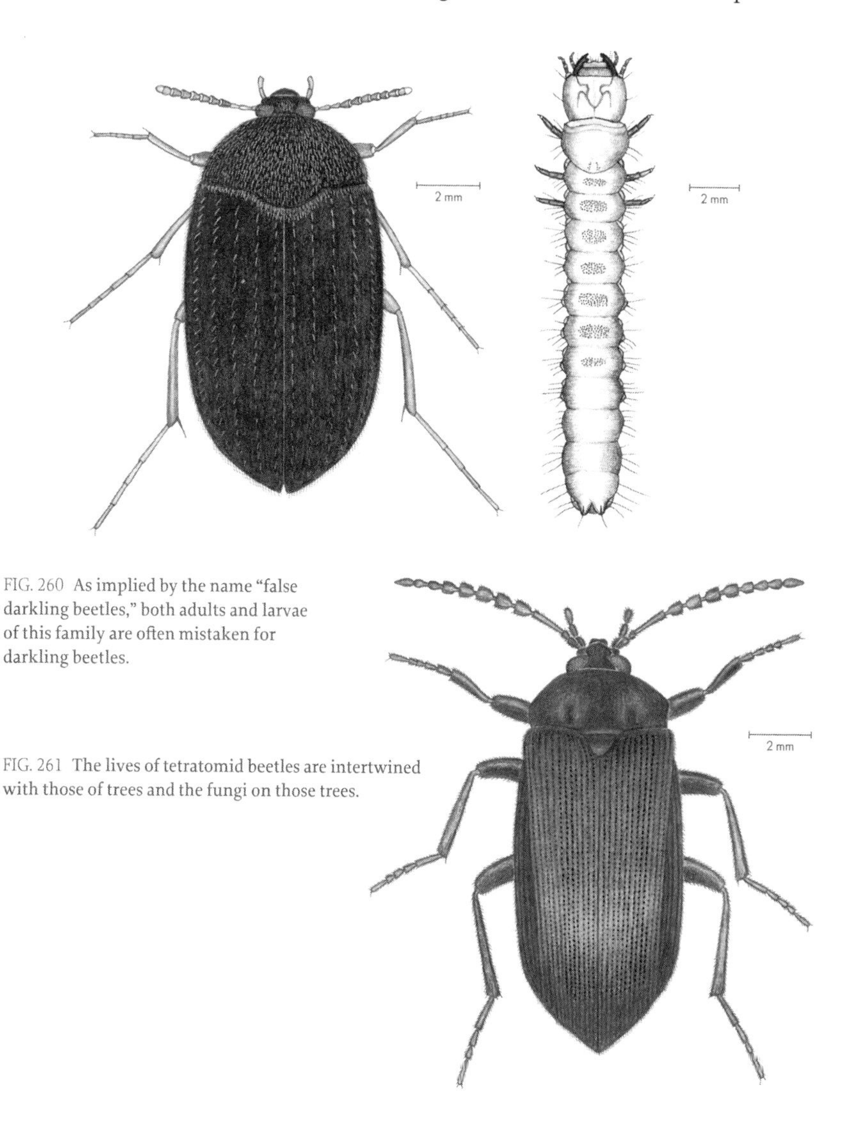

FIG. 260 As implied by the name "false darkling beetles," both adults and larvae of this family are often mistaken for darkling beetles.

FIG. 261 The lives of tetratomid beetles are intertwined with those of trees and the fungi on those trees.

resemblance to true darkling beetles. Sometimes only careful examination of features magnified under a microscope can establish familial connections, making identification of some species quite challenging. All three beetle families including their larvae frequent moist, dark habitats where various fungi grow under loose bark or on bark surfaces. While most larvae have been observed feeding on different fungi, some are reputed to be predators of the fungus feeders.

Tenebrionid, melandryid, and tetratomid beetles are representatives of the many beetles whose livelihoods depend on fungi and bacteria that thrive in moist, decaying wood under bark, in hollow trees, or on bark surfaces. Fungi and bacteria also dwell in the guts of many beetles, and these microbes undoubtedly enhance the quality of their hosts' lives (fig. 262) (Blackwell and Vega 2018). This partnership between beetles and microbes has endured across millions of years. The fossil record suggests that beetles have been dependable transporters and spreaders of fungal and bacterial spores, and in more recent ages, beetles, along with other flying insects, have taken on the additional task of transporting and disseminating pollen of flowering plants.

The partnership between beetles and microbes intersects with the partnership between microbes and trees. Members of all six kingdoms—archaea, bacteria, protists, fungi, plants, and animals—have established alliances aboveground and belowground, many of whose features remain mysteries. Everyone's life, however, seems better for these partnerships.

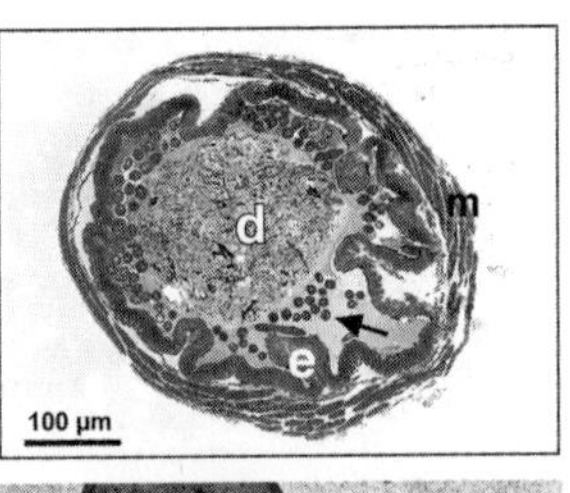

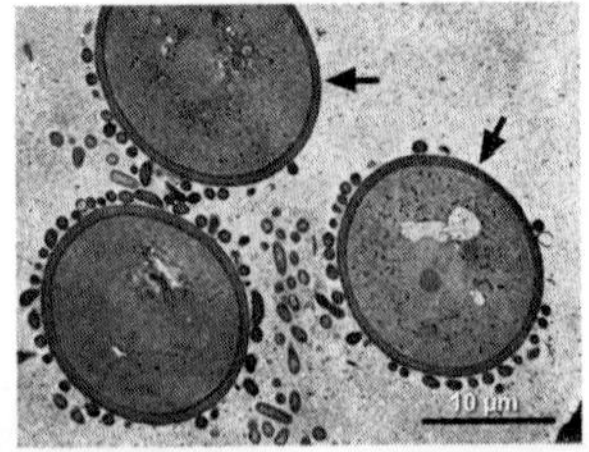

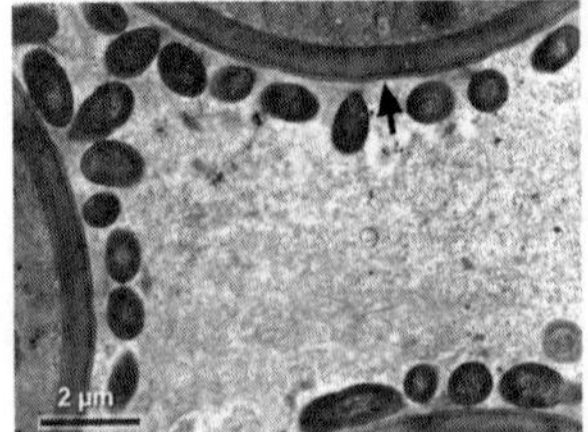

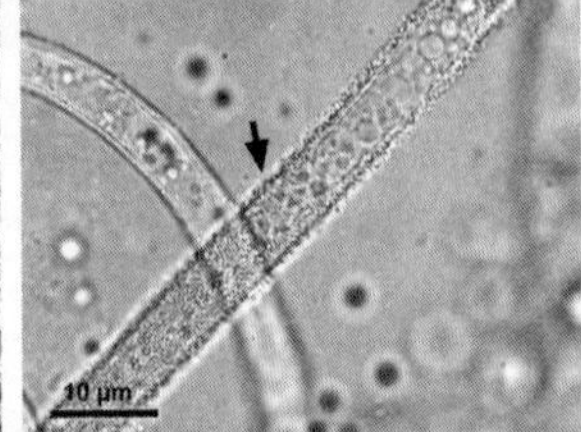

FIG. 262 *Top*: Peering into a cross section of a tenebrionid beetle's gut, we see digested matter (d) in the middle of the gut surrounded by a layer of epithelial digestive cells (e) to which filaments of fungal endophytes (arrow) attach. The outermost layers of muscles (m) control movement of food through the gut. *Middle and lower left*: Close-up sections of the fungal endophytes (marked with arrows) show bacterial companions on their surfaces. *Lower right*: A film of bacteria covers the surface of this whole fungal filament (arrow) in the beetle's gut.

IN THE COMPANY OF FLOWERS AND FRUITS 5

During its lifetime as a member of a tree community, a single insect species can contribute one set of talents early in its life as a larva and another, very different set of talents later as an adult. Not only do insects transform in form and function at metamorphosis, but they also usually take on new roles in the community and settle into new habitats on the tree.

Many beetles, moths, flies, bees, wasps, and butterflies—these creatures of many callings that begin life as herbivores, fungivores, decomposers, even predators or parasitoids—end their days peacefully sipping nectar and pollinating flowers of trees and their plant neighbors. One of the key functions assumed by adult insects—and only the adults—is pollination of flowers. Pollination ensures that trees will bear sweet, juicy fruits, energy-rich nuts and acorns, and the seeds that promise new generations of trees.

TREE FLOWERS AND THEIR INSECT VISITORS

On a walk through the woods on a spring day, you may see carpets of wildflowers covering the ground, but you may not notice other flowers blooming on the branches overhead. The first buds to open and the first colors to appear on many trees are those of their flowers. Some of these early flowers may be tiny and easily overlooked, but others have unmistakable colors, scents, and shapes. Orange, red, and yellow flowers adorn the branches of elms and maples many days before their first leaves appear. On oaks, apples, and willows, flowers and leaves unfold at the same time (figs. 263–265). But the fragrant flowers of tulip trees, black locusts, wild cherries, and basswoods do not open until after their leaves are fully formed.

Each spring the flowers that bloom overhead join forces with insects and wind to bring about pollination. Without pollination of their flowers, trees cannot produce seeds and fruit. With the help of insects and wind, pollen is carried from the male anthers of stamens to the female stigmas of pistils. After landing on a stigma, a pollen grain sends forth a tube that grows down through the pistil until it reaches an ovule at the base. The meeting of pollen and ovule produces a seed,

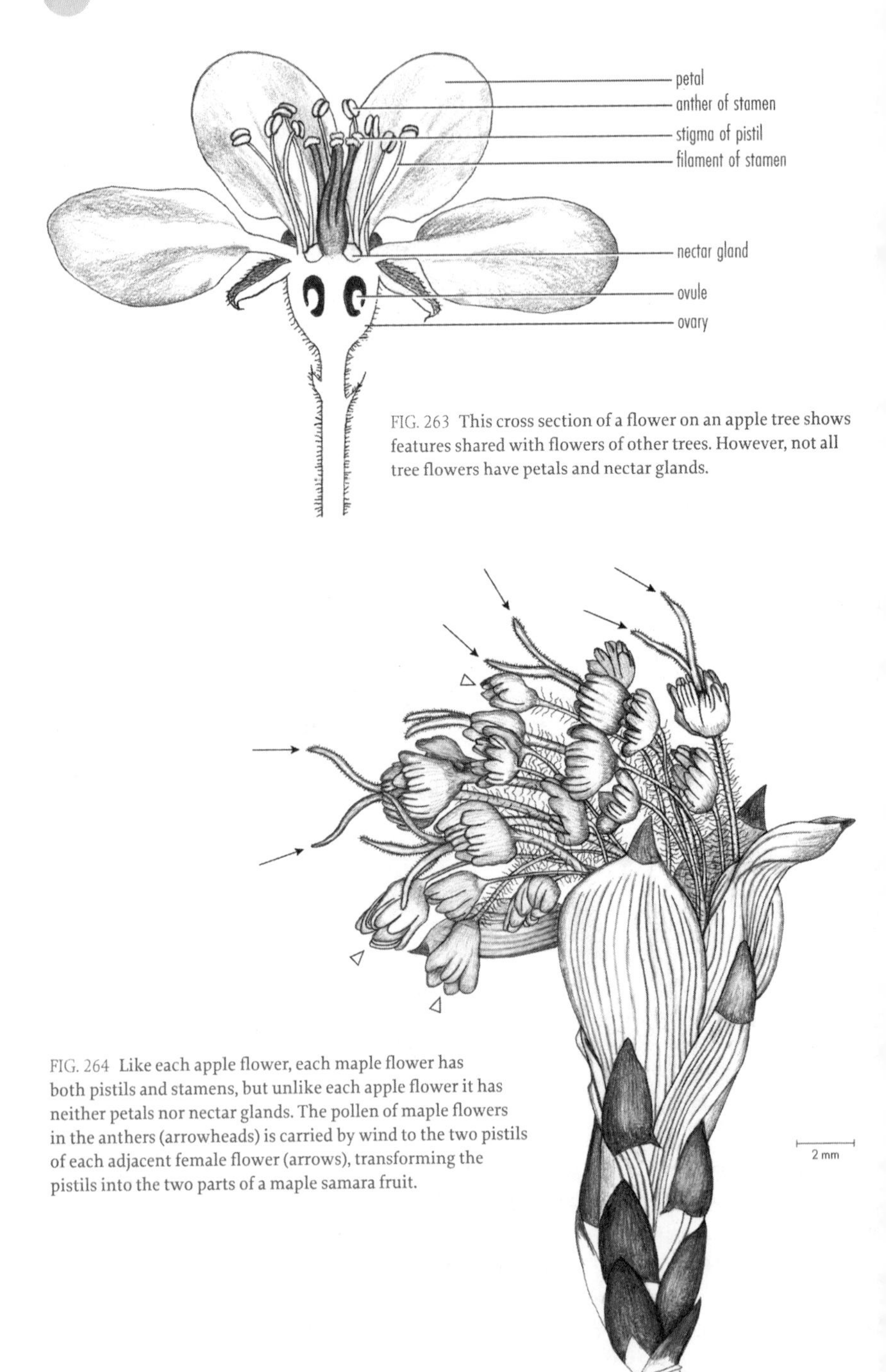

FIG. 263 This cross section of a flower on an apple tree shows features shared with flowers of other trees. However, not all tree flowers have petals and nectar glands.

FIG. 264 Like each apple flower, each maple flower has both pistils and stamens, but unlike each apple flower it has neither petals nor nectar glands. The pollen of maple flowers in the anthers (arrowheads) is carried by wind to the two pistils of each adjacent female flower (arrows), transforming the pistils into the two parts of a maple samara fruit.

FIG. 265 The male flowers (staminate catkins) are separated from the female (pistillate) flowers of white oak.

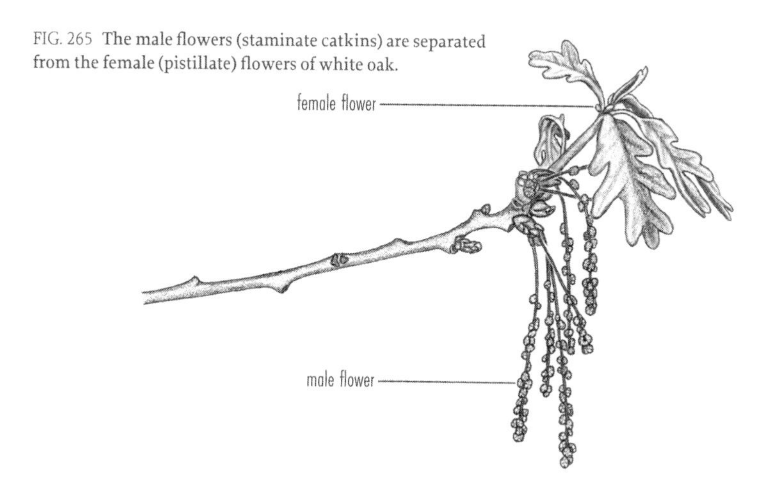

and cells of the ovary that surround the ovule produce a fruit. Some fruits have only one seed; some have many seeds.

The flowers of trees come in a variety of arrangements. Stamens and pistils may be found in the same flower, in separate flowers on the same tree, or in separate flowers on different trees. Flowers having only stamens and anthers are known as staminate or male flowers. Flowers having only pistils are known as pistillate or female flowers. Whatever the arrangement of flowers on a tree, pollen grains somehow find their way to the pistils to ensure pollination and setting of fruit.

Wind is the chief carrier of pollen for many trees like oaks, elms, and maples. You may never have noticed the flowers of these trees, for they are neither showy nor fragrant (figs. 264–266). These trees are extravagant with their pollen, and on a windy day in spring, clouds of pollen can fill the air as the grains are blown from tree to tree. Most of the grains will never reach a pistil of another flower and will settle to earth, but enough will still land on pistils of appropriate flowers to ensure

FIG. 266 The red oak also has separate male and female flowers on the same tree (*center*). Let's take a closer look. Each male (staminate) flower has four or five stamens (*left*). Each female (pistillate) flower has one pistil with three stigmas (*right*). Neither flower has a nectar gland.

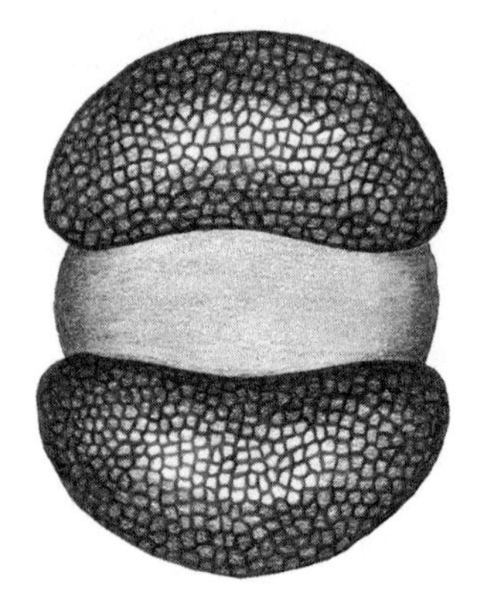
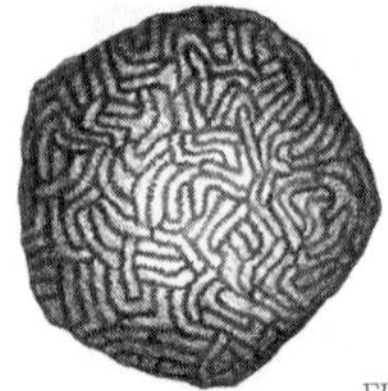
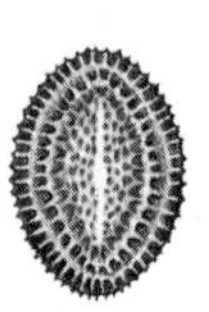

FIG. 267 A sampling of tree pollen shows a range of grain sizes and sculpting: pine (*left*); willow (*lower right*); elm (*upper right*).

a good crop of acorns, elm fruits, and maple samaras. These trees and others like birch, hickory, and walnut do not rely on insects to pollinate their flowers—wind is the main carrier of their pollen. All trees have pollen, but many lack the sweet nectars and colorful petals that attract pollinating insects (fig. 267).

Trees like apple, cherry, catalpa, and willow have flowers that are fragrant or colorful or both, with sweet nectar and protein-rich pollen to entice insects to carry their pollen directly to their pistils. Insect-pollinated flowers store alluring nectar in special glands at their base, and their pollen is generally stickier than the pollen of wind-pollinated flowers. In their search for nectar, insects usually get well dusted with sticky pollen that clings to them as they brush against it. The insects do a thorough job of visiting these flowers and transporting their pollen. Depending on pollination by insects is less chancy than depending on breezes that shift from day to day. Thus, insect-pollinated trees do not need to invest nearly as much energy in producing pollen as do wind-pollinated trees. However, even insect-pollinated flowers produce relatively large numbers of pollen grains. A single apple flower may produce 70,000 grains of pollen even though only 10 of those grains are needed to produce the 10 seeds of an apple. There is obviously a good supply of flower pollen remaining for the use of insect visitors (fig. 268).

Many insects such as beetles, flies, wasps, and butterflies visit flowers and help with pollination while they search for nectar. However, bees can claim most of the responsibility for pollinating flowers. Unlike other insects, bees collect nectar and pollen not only to feed themselves but also to stock their hives and brood chambers. They store the nectar that they do not consume themselves in a chamber of their gut called the crop or honey sack.

Honeybees are well suited for gathering pollen, for every bee's body—even its eyes—is densely covered with hairs that entrap pollen grains. They even have special combs and brushes on their legs for combing the pollen from their coat of hairs into "baskets" on their hind legs (fig. 269). The pollen and partially digested nectar (also known as honey) are transferred to special compartments in the hive

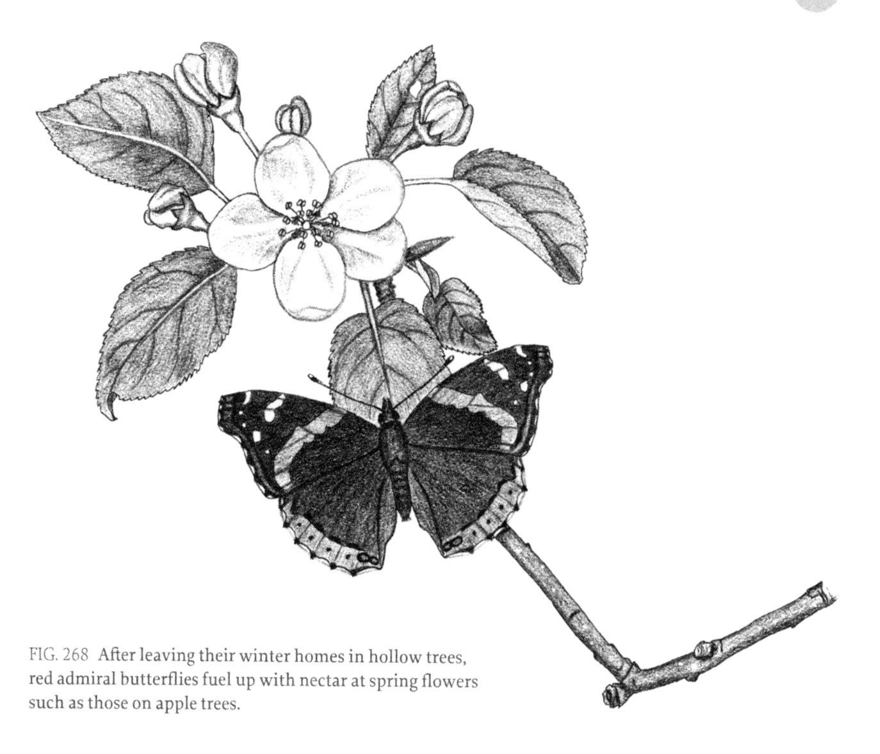

FIG. 268 After leaving their winter homes in hollow trees, red admiral butterflies fuel up with nectar at spring flowers such as those on apple trees.

and are used as needed to feed adults and developing larvae. But honeybees are relatively new arrivals to the North American landscape, having been introduced from Europe around 1600 to join the ranks of the thousands of species of native bees that had traditionally pollinated North American trees.

Bees come in many forms with many habits. Some are social, but most are solitary. While honeybees (fig. 269), carpenter bees (fig. 245), and leafcutter bees (fig. 244) choose hollow trees or tree cavities for their nests, most of their bee relatives—bumblebees, plasterer bees, mining bees, and sweat bees—excavate chambers in

FIG. 269 Honeybees have special baskets on their hind legs for carrying hundreds of pollen grains.

the soil that they provision with nectar and pollen. However, there are a few bees that probably never join in pollinating flowers and that never build nests of their own. Instead they exploit the industry of other bee species by laying their eggs in the nests of these other species, where their larvae dine on provisions intended for others. Like their counterparts in the bird world, these parasites are referred to as cuckoo bees.

— BEES, SUPERFAMILY APOIDEA (*api* = bee) (19,800 species; 3,600 species NA; 2–30 mm).

— BUMBLEBEES, DIGGER BEES, CARPENTER BEES (chapter 4; fig. 245), HONEYBEES, CUCKOO BEES, FAMILY APIDAE (*apis* = bee) (5,700 species, 1,000 species NA; 2–30 mm).

This bee family includes the more familiar bees with the most diversity in forms and habits.

— PLASTERER BEES, FAMILY COLLETIDAE (*collet* = glue together) (2,500 species; 160 species NA; 12–18 mm).

These bees construct their solitary nests in underground tunnels or hollow plant stems. The solitary nests, however, are often grouped in vast cities. Many plasterer bees are among the first bees to appear in spring. They use a secretion to plaster the walls of their nests that dries into a cellophane–like lining described by one observer as "more lustrous than the most beautiful satin." The lining envelops a bee's store of nectar and pollen, ensuring its freshness in damp, underground chambers.

— MINING BEES, FAMILY ANDRENIDAE (*andrena* = type of bee) (3,000 species; 1,250 species NA; 10–20 mm), MELITTID BEES, FAMILY MELITTIDAE (*melit* = honey) (200 species; 32 species NA; 4–22 mm).

These solitary bees all construct underground tunnels for their nests. The life cycles of many of these species are closely coupled to the life cycles of the particular flowers on which they have become dependent for nectar and pollen.

— SWEAT BEES, FAMILY HALICTIDAE (*halic* = salty) (4,300 species; 520 species NA; 6–20 mm).

Most members of this family nest in soil, but sometimes they prefer rotting wood. Some species have a particular fondness for sweat that has earned the common name "sweat bees" for the family. The bright, iridescent members of the family go by the generic names *Augochlorella*, *Augochlora*, and *Augochloropsis* (*augo* = bright; *chloro* = green), which refer to their bright metallic green, blue, or copper colors. A few red-bellied members among the ranks of this family, the cuckoo bees, are parasites of their pollen-provisioning relatives.

— LEAFCUTTER BEES, FAMILY MEGACHILIDAE, are exceptional in many respects and are discussed in chapter 4 (fig. 244).

Willow flowers happen to be some of the first tree flowers of the year that are laden with both nectar and pollen (fig. 270) (Ostaff et al. 2015). They might not be particularly attractive to our eyes or noses, but to bees that have neither seen nor smelled flowers for several months, their nectar and pollen have a strong appeal. In the early spring, wherever willow flowers are blooming, foraging bees soon fill their crops with willow nectar and their pollen baskets with willow pollen.

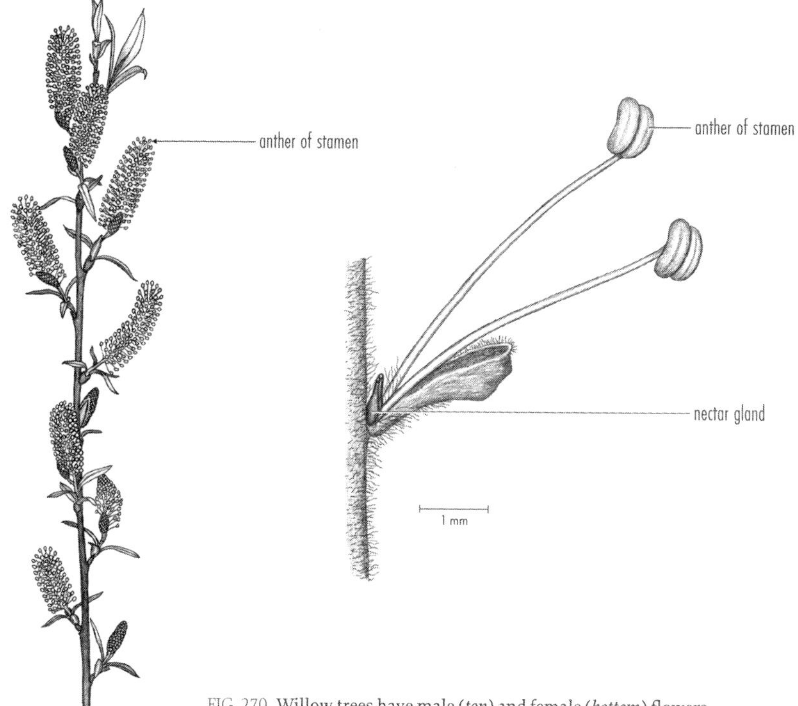

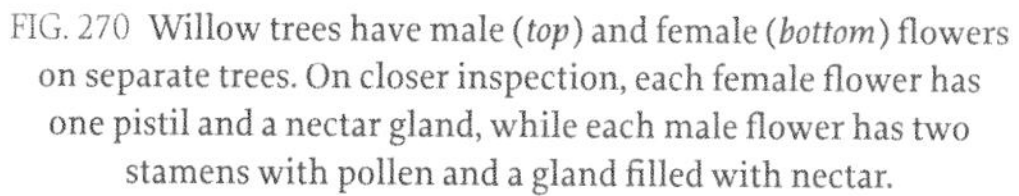
FIG. 270 Willow trees have male (*top*) and female (*bottom*) flowers on separate trees. On closer inspection, each female flower has one pistil and a nectar gland, while each male flower has two stamens with pollen and a gland filled with nectar.

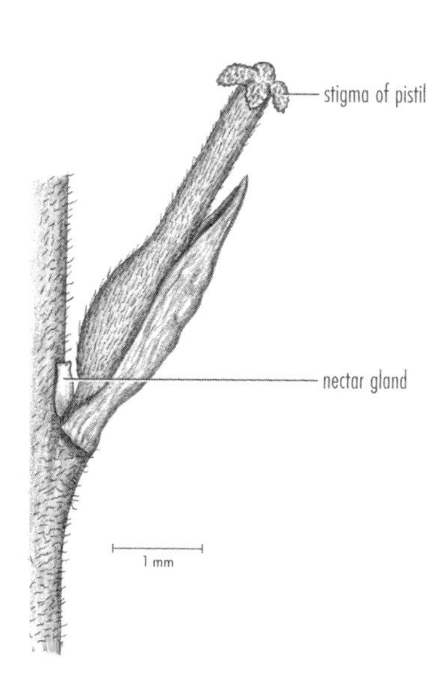

After returning to their colony in the soil, a hollow tree, or a hive box, the bees unload the pollen and nectar they have harvested, provisioning brood chambers or restocking honeycombs whose stores were emptied during the winter.

Different trees arrange flowers in different ways. Willow trees happen to have male and female flowers on separate trees. Male flowers have both pollen and nectar to attract insects. Female flowers have only nectar to offer, but they also attract their share of insects. Female flowers must have some attraction to offer the spring insects; otherwise the insects would visit only the male flowers and pollination would never take place. Any number of arrangements of male and female flowers or parts of flowers are apparently acceptable to insects as long as they are rewarded in some way for their visits to flowers.

— BEE FLIES, FAMILY BOMBYLIIDAE (*bombylius* = buzzing insect) (5,000 species; 800 species NA; 2–40 mm).

Among the other visitors to early spring tree flowers such as those of willows are flies that have recently emerged from their larval homes in the soil. Most people mistake these buzzing flies for buzzing bees. Bee flies not only buzz like bees but also have lush coats of fuzz like bees. Their proboscises for reaching nectar are often longer than those of their bee counterparts (fig. 271). On even closer examination of a bee fly, one observes that these flies do not have pollen baskets; they do not collect and provision pollen for others. Instead, after mating, the mother fly drops her eggs over the soil surface, where the newly hatched larvae are very active and scurry about until they locate an acceptable soil-inhabiting larval host—be it beetle, fly, caterpillar, ant lion, or bee. Charcoal bee flies of

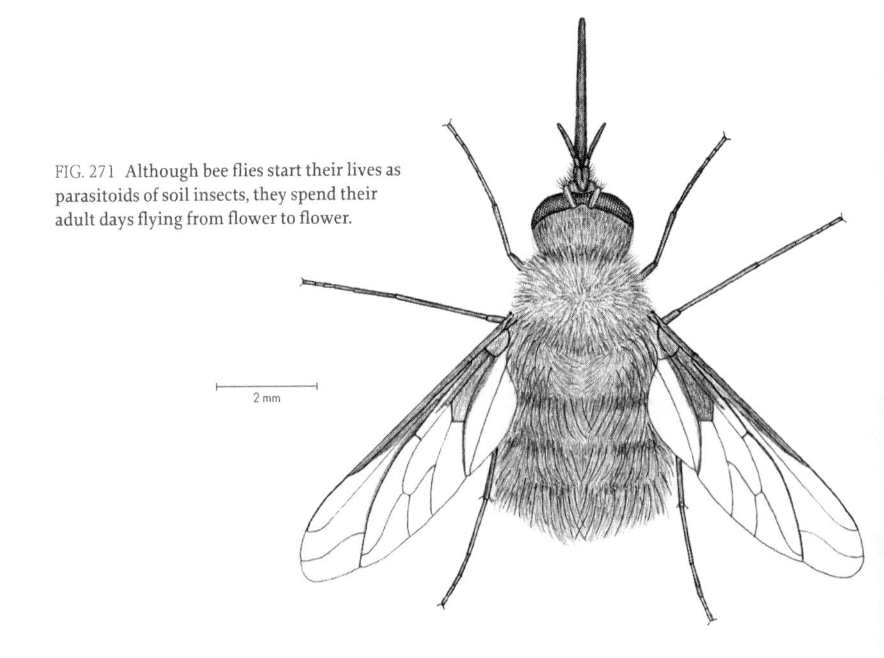

FIG. 271 Although bee flies start their lives as parasitoids of soil insects, they spend their adult days flying from flower to flower.

FIG. 272 The spring azure is one butterfly whose caterpillars feed on the flowers of dogwood trees rather than on the leaves.

summer (fig. 92) have larvae that are parasites of digger wasps and digger bees. The larvae of some species of bee flies that mature in late summer and fall locate eggs that female grasshoppers carefully stash in underground chambers at the end of summer.

Slightly later in the spring, the flowers of dogwood trees host not a pollinator but the hungry caterpillars of one of the first butterflies to emerge from its chrysalis at the end of winter—the spring azure or common blue (fig. 272). Accompanying the caterpillars are different species of ants that ascend the dogwood, not only for the nectar of its flowers but also to savor nutrients from the dorsal honey glands of the obliging caterpillars. The ant escorts fend off threats to their vulnerable charges from foraging birds and predatory insects.

— BLUES AND AZURES, FAMILY LYCAENIDAE, GOSSAMER-WINGED BUTTERFLIES, SUBFAMILY POLYOMMATINAE (*poly* = many; *ommat* = eyes) (1,135 species; 31 species NA; mature larvae 12–20 mm; wingspan 18–38 mm) (Shields 1989).

Even before leaves unfold on the trees, patches of blue flit about the woods, brightening the gray, slumbering landscape of late winter. These sprightly butterflies are among the harbingers of spring, appearing several weeks before the opening of dogwood flowers on which their caterpillars feed. Thoreau celebrated the bluebird that "carries the sky on its back"; spring azures are the insects that carry the sky on their gossamer wings.

As adults, bees, wasps, moths, butterflies, flies, and beetles indiscriminately visit countless flowers for nectar and pollen. A few birds, bats, and other mammals also contribute to the pollination enterprise. Some species might have preferences for which flowers they visit, but many pollinators find the flower nectar and/or pollen of more than one species appealing.

However, a few partnerships of insect pollinators stand out as particularly discriminating. Some insects have established committed relationships with specific floral partners, and these partners can be either herbaceous flowers or tree flowers. Among the best-studied special partnerships between insects and tree flowers are those of fig flowers and wasps, where the wasps' visits to fig flowers ensure pollination and formation of figs. A few fig species are self-pollinating, but for each of the approximately 750 other species, one or at most two species of tiny fig wasps have developed a mutually dependent pollination relationship with them.

— FIG WASPS, SUPERFAMILY CHALCIDOIDEA, FAMILY AGAONIDAE (*agaon* = to adore) (~ 900 species; 13 species NA; ~ 1.5–2 mm).

The match between these figs and their fig wasps is obviously a strong and enduring one and has been traced back about 80 million years. The fig flower is a most unusual tree flower that depends on a most unusual wasp for its pollination. Fig flowers remind me of inside-out flowers. The male and female flowers occupy

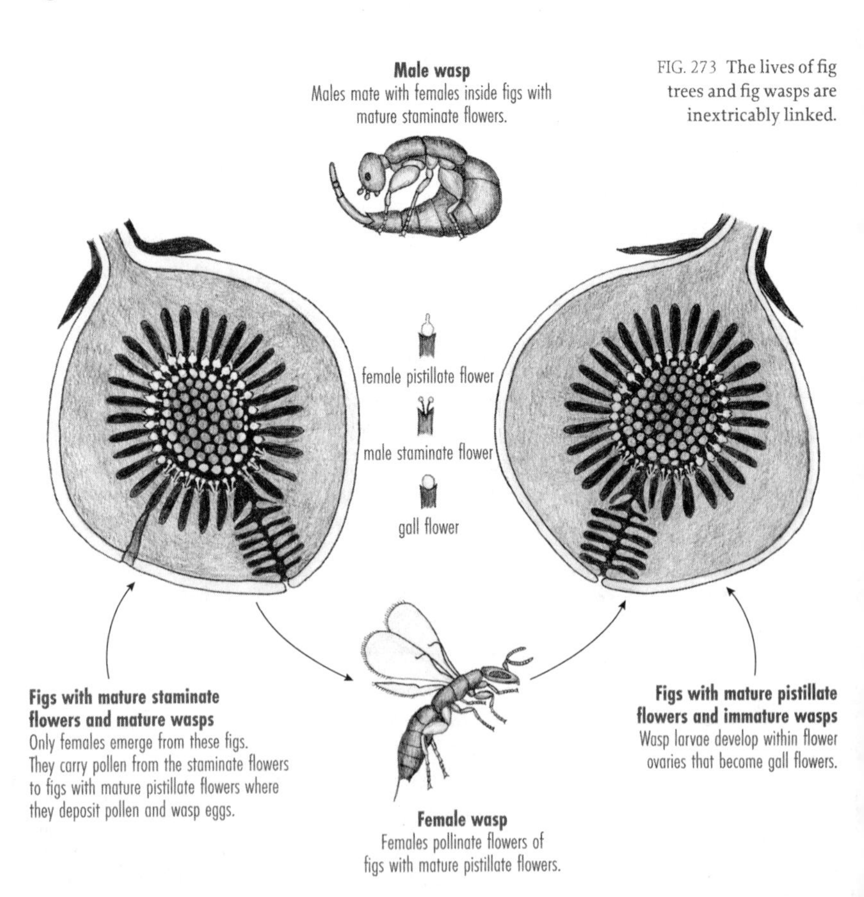

FIG. 273 The lives of fig trees and fig wasps are inextricably linked.

the center of a hollow fleshy structure that will form the future fruit and whose unique form is referred to as a syconium (*syco* = fig). The flowers are arranged inside the syconium with their blooms hidden from the view of bees, beetles, flies, and butterflies. The only external access to the flowers is through a narrow passageway at one end of the syconium, and only a select group of tiny wasps can gain entry as pollinators of these unusual fig flowers. The fig wasps develop in galls that they form inside the flowers (gall flowers), but only the females ever leave their home flowers to pollinate another fig flower. While all the fig flowers on one tree are at the same stage of maturity, those on other fig trees in the neighborhood are at different stages of maturity. The relationship has ensured survival of the wasps, fertilization of the fig flowers, and maturation of the fig fruits (fig. 273).

In the relationship that has developed between wasps and fig fruits, the female wasp is the one that carries pollen from a fig flower on one tree to a mature pistillate fig flower on a nearby tree, ensuring that the enclosed and hidden female fig flowers receive pollen. The wingless, hunch-backed males leave their galls in a mature staminate fig flower and travel only far enough within the flower to find a female with whom to mate. While being wingless is almost always assigned to females in the insect world (e.g., gall wasps, certain moths, and scale insects), in the case of fig wasps, not only are males wingless but their eyes and antennae are small and inconspicuous. The males of a given generation always emerge from their pupal chambers before the females. Each male finds a gall with a female and mates with her even before she emerges from her gall. As the newly emerged female wanders through the mature staminate flower in search of an escape route, she quickly becomes covered with pollen. As his one last act of chivalry, the male chews an escape tunnel in the flower for her. Male wasps never escape and leave home (Frost 1959).

The fertilized female flies off from her home in search of a new unripe fig flower (mature pistillate flower). At just the right time, fig trees with these receptive flowers release odors that attract female wasps. The female picks up their scent and traverses the fig's narrow entryway to enter the chamber of hidden flowers. As she spreads pollen from her birth home among the flowers of her new residence, she also leaves behind her eggs and induces the formation of gall flowers in which the next generation of fig wasps develops. The larval wasps are herbivores on the flowers that the adult wasps pollinate. The rich taste of figs is a result of blending inherent fig sweetness with flavors imparted by male fig wasps that never leave their fig nurseries and female wasps that never leave the figs they pollinate.

POLLINATION IS THE BEGINNING OF THE JOURNEY FROM FLOWERS TO FRUITS AND SEEDS

We know that many trees rely on their animal inhabitants to disperse and plant their seeds (fig. 274). The sweet pulp and sap of ripe fruits produce a variety of volatile compounds that attract creatures—vertebrate and invertebrate—with

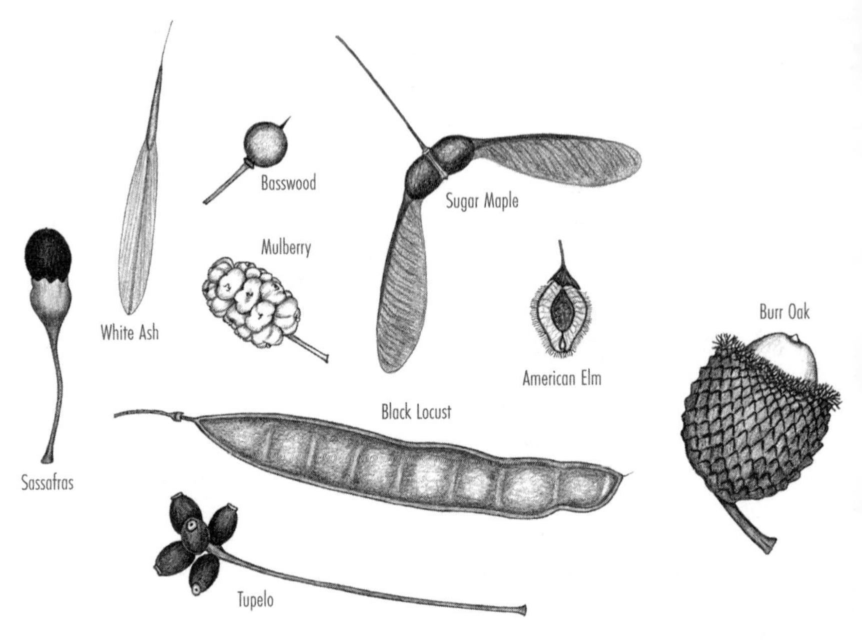

FIG. 274 The seeds and fruits of North American trees in nine different families include several that are sweet, juicy, and filled with energy.

their enticing aromas (Rodriguez et al. 2013). Many small creatures such as ants, butterflies, wasps, and fruit flies simply come to relish the sweet juices that ooze from the fruit's pulp and seeds. Other insect larvae burrow into the fruit, and some, such as caterpillars of the codling moth, travel to the core of the fruit to chomp on seeds as well. None of these insect visitors of fruit consume the whole fruit; however, many birds and mammals enthusiastically participate in the rewards of the fruit harvest along with its insect tidbits.

Moth caterpillars of several families—tortricid moths (Tortricidae), acorn moths (Blastobasidae), twirler moths (Gelechiidae), and fruitworm moths (Carposinidae)—bore into nutrient-rich fruits, seeds, and buds. On their more nutritious diets, these caterpillars mature much faster than their wood-boring relatives; many of the former have from one to four generations a year rather than only one generation every one to four years. Two of these moth families—Tortricidae and Gelechiidae—are discussed in chapter 2 and are only briefly mentioned here because a few members are nonconformists that have taken up the business of boring into fruit rather than rolling or mining leaves like most members of their families.

— TWIRLER MOTHS, FAMILY GELECHIIDAE The relatively small caterpillars in this large, diverse family feed in concealed, unexposed locations on trees. Among their ranks are gall makers; leaf miners; leaf rollers (chapter 2; fig. 125); lichen feeders; borers of seeds, flowers, and fruits; scavengers; and even a few

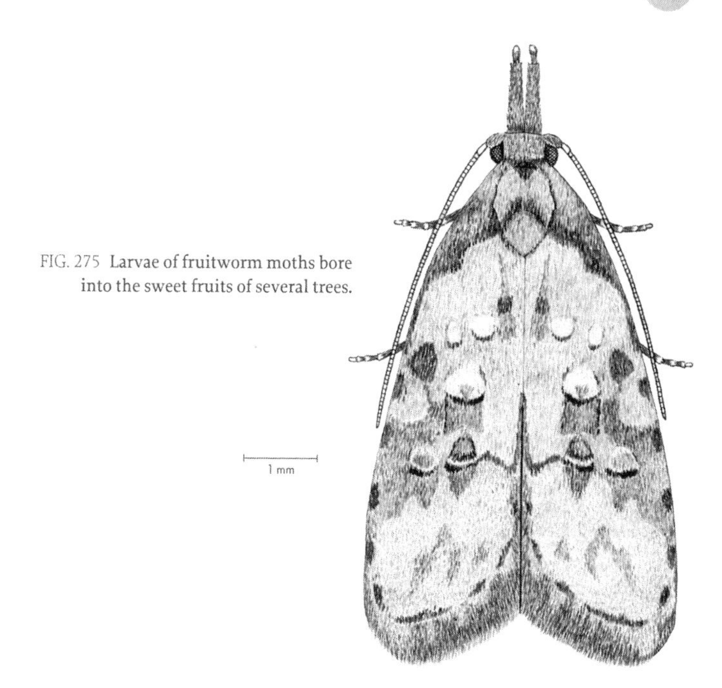

FIG. 275 Larvae of fruitworm moths bore into the sweet fruits of several trees.

predators. A few exceptional species such as the peach twig borer, which was introduced to North America from Europe, have developed bad reputations for being fond of the same fruits that humans enjoy. Although the first year's generation of caterpillars start out as twig borers on peaches, later generations find their way into the rapidly growing peach fruits. The mature larvae of the year's last generation spend the winter under tree bark, ready to pupate and emerge as egg-laying moths in the spring.

— FRUITWORM MOTHS, FAMILY CARPOSINIDAE (*carpo* = fruit; *sina* = damage) (280 species, 20 species NA; larvae 10–12 mm; wingspan 10–40 mm, usually 10–20 mm).

The adult moths are distinguished by their prominent labial palps—a pair of mouthparts that project forward beyond the eyes and antennae (fig. 275). Several species have small but conspicuous tufts of raised scales on their wings. The caterpillars are borers that feed on fruit such as apples and peaches, stems, and even bark. A few leaf-mining caterpillars are also included in this moth family. At the end of their larval days, they spin their cocoons in the soil.

— TORTRICID MOTHS, FAMILY TORTRICIDAE Most members of the Tortricidae are leaf rollers (chapter 2; figs. 112 and 113), but some members of this large family of about 11,000 species are exceptions to this rule and bore into fruits, seeds, stems, and even roots. The codling moth and oriental fruit moth are among the most infamous of these fruit borers. Newly hatched codling moth larvae chew their way through the skin of apples. Once well inside the fruit, the caterpillar seals off its entry point with silk, pieces of fruit pulp, and a few droppings. Now that it is safely ensconced inside the fruit, the caterpillar gravitates to the apple's core and feasts on the developing seeds. Interrupting seed development also interrupts fruit development and induces the fruit's premature ripening. A European tortricid caterpillar—the acorn moth of Europe—feeds on acorns of Old World oaks. Its New World acorn-eating counterpart is a blastobasid caterpillar.

— ACORN MOTHS AND SCAVENGER MOTHS, FAMILY BLASTOBASIDAE (*blasto* = bud; *basi* = base); (430 described species, but possibly several times that number; 70 species NA; larvae 10–15 mm; wingspan 5–35 mm).

The habits of caterpillars in this family are difficult to categorize. Many are opportunistic scavengers both aboveground and belowground. Some have been reported to eat other tree-dwelling insects such as slow-moving beetle larvae, scale insects, or moth pupae. One member of the family, however, has chosen the special path of feeding on energy-rich acorns and chestnuts. This is the acorn moth of North America (fig. 276).

— ACORN WEEVILS, FAMILY CURCULIONIDAE, GENUS *CURCULIO* (*curculio* = weevil) (350 species; 30 species NA; 4–13 mm).

Each autumn thousands of acorns fall from a single oak tree, and each year acorn weevils claim a few of these. Before the acorns start to fall, the female weevil chooses the acorns in which she will lay her eggs. She uses the tiny jaws at the tip of her long snout to drill a hole in each acorn, and then she drops an egg inside.

FIG. 276 Larvae of acorn moths enjoy a share of the acorn harvest and contribute to the flavor of acorns enjoyed by birds and mammals.

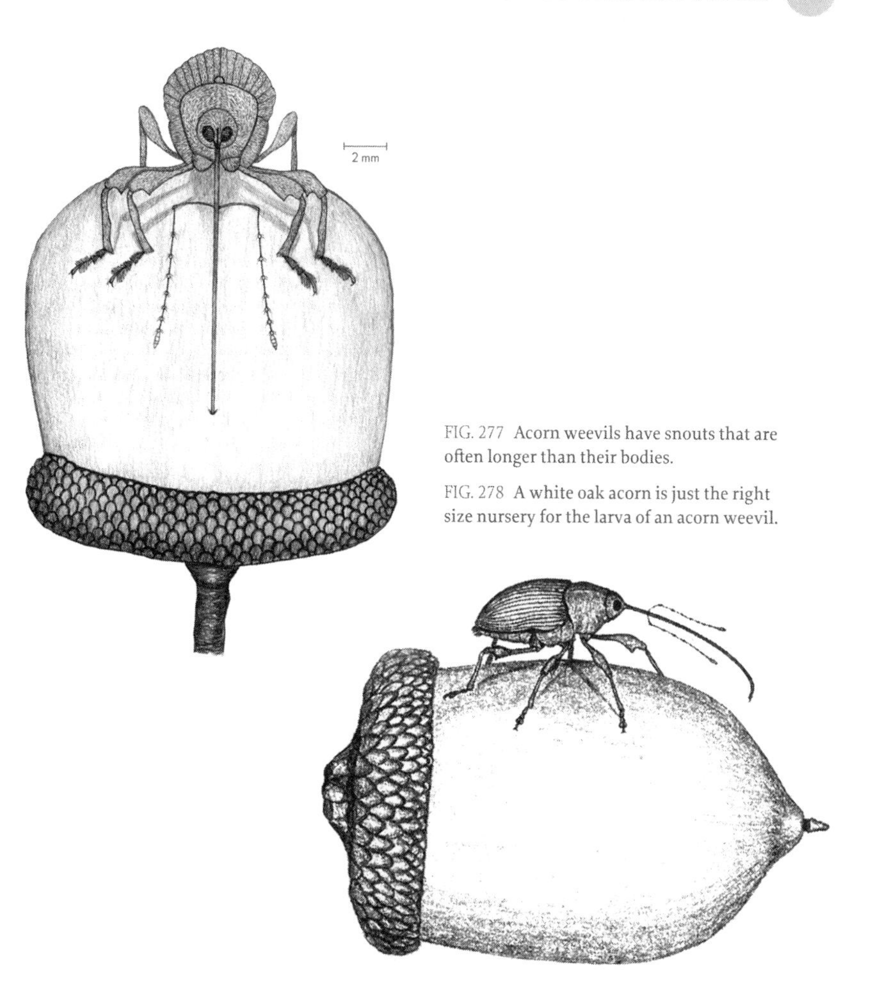

FIG. 277 Acorn weevils have snouts that are often longer than their bodies.

FIG. 278 A white oak acorn is just the right size nursery for the larva of an acorn weevil.

The female weevil's snout is longer than the male's and is often even longer than her body. The white larva that hatches grows fat inside its acorn nursery. When the acorn falls from the oak in an autumn wind, the full-grown larva leaves its nursery and burrows into the ground to pupate and spend the winter (figs. 277 and 278).

Other members of the weevil family in the genus *Curculio* lay their eggs on sweeter, fleshier fruits such as plums, apples, and cherries. The snouts of these other mother weevils are conspicuous but nowhere near as striking in length as the snout of an acorn weevil. Mother weevils emerge from the ground litter in spring to lay the eggs of the next generation on the developing fruit. On nutrient-rich fruit, two generations of weevils mature each year; mature larvae of the second generation emerge from their homes of ripe fruit and crawl into the soil, where they will spend the winter and pupate in the spring.

Weevil and caterpillar larvae share the acorn harvest with many birds and mammals. The harvest of an oak tree is so bountiful that weevils, blue jays, turkeys, mice, and squirrels can all claim a share of acorns. Take chipmunks, for example. Chipmunks that shell the acorns they gather selectively consume those with larvae inside; the noninfested acorns are the ones they prefer to hoard for later consumption. However, chipmunks never reclaim many of these acorns, and they go on to become oak seedlings. By reducing the population of acorn weevils and acorn moths in their woodland territories, chipmunks not only benefit from the extra nutrition provided by the proteins and fats of insect larvae but also increase germination of the remaining acorns, benefiting both themselves and oak trees (Yang et al. 2018). A web of connections embraces many insects, birds, and mammals with oak trees and their acorns (Ostfeld et al. 1996). After all creatures have taken their share of the acorn harvest, plenty will remain to sprout.

Think about all the oak trees that have gotten their start in this world thanks to those industrious squirrels. Every autumn, squirrels rustle among the fallen leaves, gathering and burying acorns for the cold days that lie ahead (fig. 279). Somehow, a portion of those buried acorns never get retrieved. Many will sprout and grow into oaks.

Blue jays and scrub jays also hoard acorns during the autumn harvest (fig. 280). And, like squirrels, the jays sometimes fail to retrieve some of their stash of acorns buried miles away in a forest clearing or an old pasture. They unintentionally find new homes for oak trees and reforest the landscape.

FIG. 279 Squirrels are especially industrious during the harvest of acorns and walnuts.

FIG. 280 Blue jays are one of many animals that love the taste of acorns, often transporting them to distant locations and helping to reforest the landscape.

Other birds like the acorn woodpecker also harvest vast numbers of acorns. Each acorn is stored in one of thousands of acorn-sized holes chiseled out of branches, trees, posts, or tree trunks. These communal granaries are places where a troop of these birds can hoard thousands of acorns in one location. Squirrels or other woodpeckers may pilfer some acorns, but members of the woodpecker community take turns standing guard over their winter larder. These birds may not be very efficient planters and dispersers of oaks, but undoubtedly they inadvertently drop part of their harvest in places where unclaimed acorns can sprout and thrive.

Walnuts and squirrels also have a good partnership. Wind and birds may scatter seeds of many trees, but squirrels are usually the only ones who scatter tough, bulky walnuts. For all these hard-shelled nuts that squirrels manage to chew into, they probably bury just as many in the soil each autumn. Each spring, walnut seedlings appear, often far from the trees from which they fell.

Another creature helps squirrels with their walnut harvest. When the walnuts fall from a tree, each is covered with a firm, green husk. This husk and the hard shell of the walnut present obstacles to squirrels trying to chew their way to the tasty nut within. But this obstructing husk is just what appeals to the walnut husk fly (fig. 281).

— PEACOCK FLIES, FRUIT FLIES, FAMILY TEPHRITIDAE (*tephros* = ash gray; *-ite* = a part) (5,000 species; 300 species NA; 2.5–10 mm).

Male and female flies meet, court, and mate on the surface of a walnut husk. The maggots that hatch from eggs left behind by the females gnaw through the husk with their sharp mouth hooks until it almost disintegrates and readily falls away. The maggots then crawl out and burrow into the soil, where they wait out the winter months. Now, thanks to the maggots, the walnuts can be easily husked, saving the squirrels some time and effort as they chew through the shells and prepare for winter.

The fruit flies of genetic fame (family Drosophilidae) may be featured in far more scientific articles, but they cannot claim the aesthetic charm of the tephritid fruit flies. These flies are unsurpassed for their ornate wing markings—spots, stars, and stripes arranged in myriad combinations. The flies are apparently aware of their good looks, for they constantly strut about flashing and waving their wings. Peacocks raise and fan their tails to display their gorgeous patterns, and peacock flies not only raise their wings but also turn them at all possible angles to ensure they are exposed to their best advantage. They are not bashful about

FIG. 281 Walnut husk flies rendezvous on the surface of a walnut husk.

showing their finery. While most peacock flies feed on a wide range of fruits on trees and shrubs—dogwood, apple, walnut, orange, cherry—a few are leaf miners and gall makers (chapter 2).

Each fall before the persimmon harvest, waves of migrating birds sweep across the country as their instincts to migrate are awakened by the longer nights and cooler days. During the long flight southward, each bird must meet the energy needs of its demanding travel schedule. With the arrival of autumn, the ripening fruits of trees are now a more plentiful source of energy than the few insects that are left at the end of summer. Fruits of dogwood, sassafras, and tupelo are rich in fats and provide birds with more calories for their demanding flights than sweeter tree fruits. Mammals like possums and raccoons prefer the sweet fruits of persimmons, apples, and wild cherries, leaving the fat-laden fruits for the migrating birds (fig. 282). As the fatty fruits ripen, the leaves on these trees turn bright orange and red as though to advertise their locations to birds flying high overhead (Kricher and Morrison 1988). Migrating birds relish the fruits and harvest them all in a matter of days. The few fly larvae that develop inside dogwood fruit add to the fat calories that are so critical for the completion of the birds' arduous migrations. Where bird droppings fall, new trees will someday sprout and nourish future generations of birds that pass

FIG. 282 Raccoons, possums, and foxes eagerly celebrate the harvest of sweet autumn tree fruits such as persimmons.

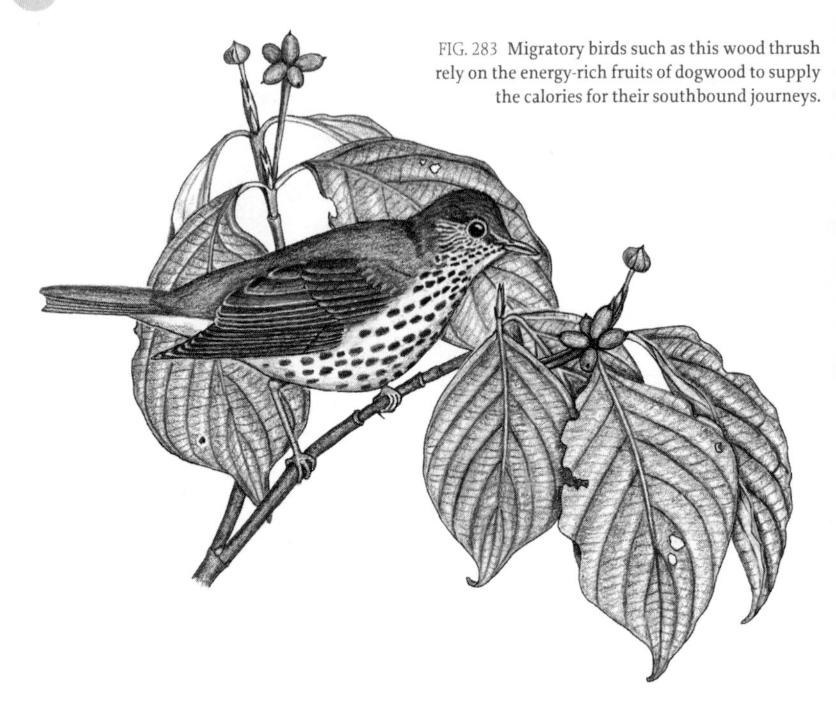

FIG. 283 Migratory birds such as this wood thrush rely on the energy-rich fruits of dogwood to supply the calories for their southbound journeys.

the same way. Over 70 percent of Earth's flowering plants depend on birds to disperse their seeds (Bello and Barretto 2021) (fig. 283).

Try sampling fruits from a variety of trees. Sweet fruits like apples, cherries, persimmons, and plums are rich in sugars. Fruits of dogwood, tupelo, and sassafras may be rather bitter, so take only a nibble from them. Although these fruits are neither sweet nor juicy, they are excellent sources of energy. Gram for gram and ounce for ounce, they provide more energy from their fats than the sweet fruits ever provide from their sugars. Migrating birds take full advantage of this energy value.

Invisible connections link the birth of trees with the insects, mammals, and birds that find their fruits so enticing. In the tropics, tapirs and elephants have earned the title "gardeners of the forest." Both consume a great variety of fruits and seeds that they gather with flexible, versatile snouts that serve for both sniffing and grasping. These roamers wander from forest patch to forest patch and along the way transport many seeds in their droppings, dispersing plants and reforesting the landscape during their wide-ranging travels. The mammals and birds that eat fruit prepare seeds for sprouting by passing them in their droppings. Many seeds sprout much better after their seed coats have been scraped and scratched a bit by passage through the gizzards and enzymes of animal digestive tracts. Many of the fruits and seeds that are not immediately eaten are often hidden and stored in places where they can sprout and thrive. These animals are important contributors to reforesting the planet. All forest residents benefit

FIG. 284 The benefits of a good harvest ripple through the tree community, from insect lives to the lives of large predators such as barred owls.

from good harvests and the efforts of the many "gardeners of the forest" (fig. 284) (Fricke et al. 2022).

HIDDEN COMPANIONS OF A TREE'S VERTEBRATE INHABITANTS

While distributing fruits and seeds from place to place, however, these birds and mammals are mostly oblivious to the small arthropod passengers that they also inadvertently transport (fig. 285). These arthropods are considered companions of trees by virtue of their intimate associations with the birds and mammals of trees; they include a family of flies, along with many fleas, lice, and mites.

After considering all the insect parasitoids that choose insects and a few other invertebrates as hosts, one would be remiss in not mentioning these insect and mite parasites that live at the expense of their vertebrate hosts but nevertheless spare the lives of their hosts so that future generations of parasites can share the same animal homes.

— PARASITIC MITES AND TICKS, CLASS ARACHNIDA, SUBCLASS ACARI, SUPERORDER PARASITIFORMES (*parasiti* = parasite; *formes* = forms) and SUPERORDER ACARIFORMES (*acari* = mite; *formes* = forms) (48,300 species currently

FIG. 285 The bird and mammal inhabitants of trees like this flying squirrel can host several different arthropod companions.

described, well over 1,000,000 species estimated; at least 10 percent of these are probably found in North America; 0.15–16 mm).

This vast estimated number of mite species includes mites of all descriptions and destinies. Mites are at home just about everywhere and have taken on just about every occupation—on all parts of trees, throughout the soil under trees, and on most animal companions of trees. The subset of mite species that lives as parasites is hard to estimate, but considering all the vertebrate and arthropod species that could possibly serve as hosts for these mites, the parasitic mites probably represent the majority of all mite species. Many parasitic mites, unlike their free-living relatives in the leaves and soil, have their claws replaced by circular pads referred to as ambulacral (*ambula* = walk; *acra* = at the tip) disks. Most are

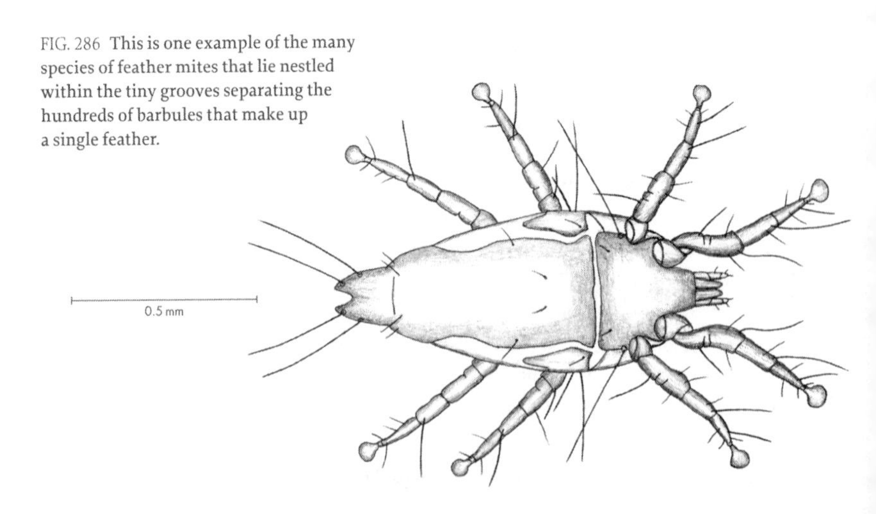

FIG. 286 This is one example of the many species of feather mites that lie nestled within the tiny grooves separating the hundreds of barbules that make up a single feather.

exceptionally discriminating in their choice of host. Mites have found homes on all vertebrate animals that also call trees home—mammals, birds, lizards, snakes, and even frogs (fig. 286). Bees, beetles, wasps, and other arthropods of trees have their own mites specially adapted to their bodies and dependent on them for transport from place to place. Feather-dwelling mites stake out specific habitats on and inside bird feathers, cleansing the feathers of fungal spores and apparently often conferring more benefits than detriments to their hosts. As arthropod and vertebrate companions, mites can help their hosts by eliminating other species that can harm or compete with their hosts, or by eliminating debris from their hosts; however, while some just go along for the ride as hitchhikers, others are actual parasites and feed on blood, lung tissue, or secretions from skin glands.

— LICE, ORDER PHTHIRAPTERA (*phthir* = lice; *aptera* = wingless) (5,500 species; 1,000 species NA; 0.5–10 mm).

These parasitic relatives of bark lice (chapter 4; fig. 205) forsook their free-living lifestyle on bark, lichens, and nests for the warmth and shelter of fur and feathers. Changes in form accompany changes in lifestyle. While bark lice have rotund bodies and bulging eyes, parasitic lice have flat bodies that scuttle between closely packed hairs or feathers; their eyes are absent or reduced to tiny specks. The claws of lice oppose each other in the manner of human opposable thumbs. Unlike their insect relatives whose two claws on each foot never meet, lice have opposable claws that allow them to grasp and manipulate hairs or the barbs and barbules of feathers. Every louse species spends its entire life on one bird or mammal, and each species has adopted one specific bird or mammal species as its host. The bird louse illustrated here (fig. 287) was found on a wood thrush (fig. 283). Lice always feed among the hair of mammals or the feathers of birds, either chewing the hair or feathers or piercing the skin and drawing blood. For some inexplicable reason, bats are the only tree-dwelling mammals that have been shunned by lice.

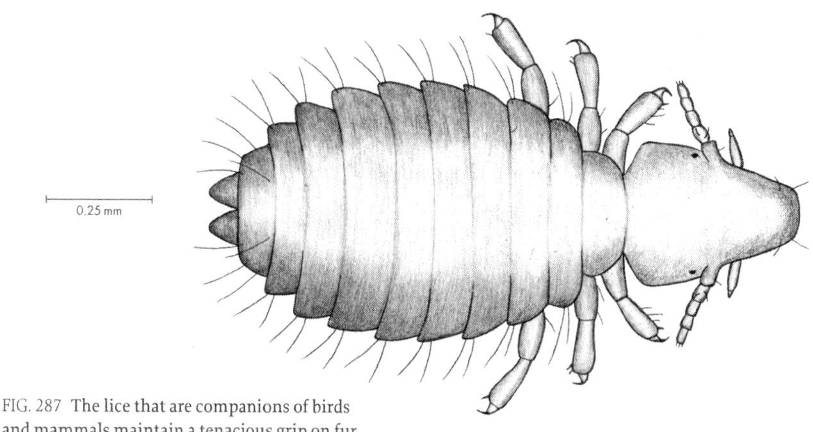

FIG. 287 The lice that are companions of birds and mammals maintain a tenacious grip on fur and feathers with specially modified toes.

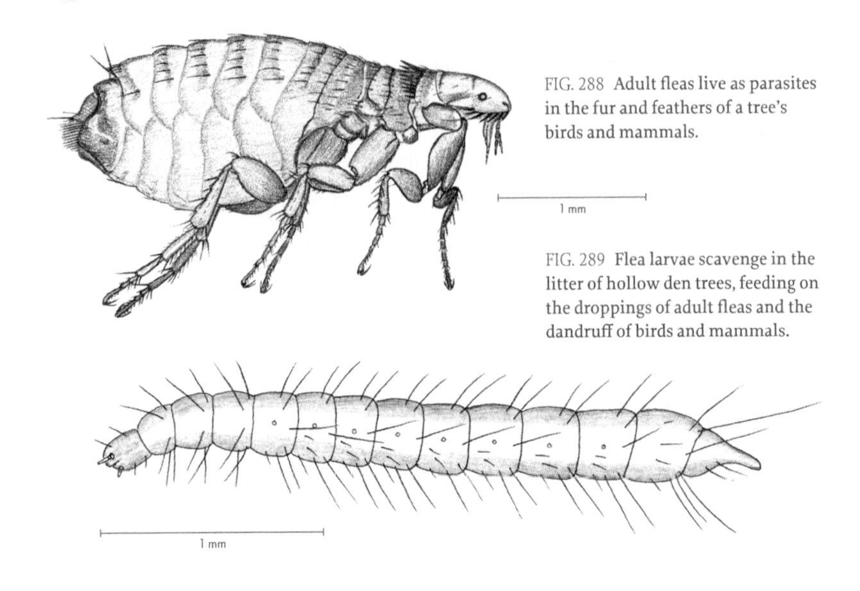

FIG. 288 Adult fleas live as parasites in the fur and feathers of a tree's birds and mammals.

FIG. 289 Flea larvae scavenge in the litter of hollow den trees, feeding on the droppings of adult fleas and the dandruff of birds and mammals.

— FLEAS, ORDER SIPHONAPTERA (*siphon* = tube; *aptera* = without wings) (2,500 species; 325 species NA; 1–10 mm).

The litter at the base of a hollow den tree inhabited by bats, squirrels, chipmunks, or raccoons is the place to get an idea about who shares not only the hollow den tree but also its furry residents. Adult fleas tenaciously cling to fur with strategically placed bristles and claws as they readily avail themselves of blood meals provided by their warm-blooded hosts. A few fleas have taken up residence with birds, but most feed only on mammals, and some are very particular about which mammals. The members of one flea family are found only on bats. A flea's taste for blood is acquired only as an adult, for flea larvae begin life crawling and feeding in the debris that accumulates in den trees, including such things as fur, dandruff, dead insects, and the droppings of their parents (figs. 288 and 289).

— BOT FLIES, WARBLE FLIES, FAMILY OESTRIDAE (*oestro* = gadfly) (180 species; 42 species NA; larvae 15–30 mm).

Only very close inspection of these insects reveals that they are flies and not bees. Most flies are covered with bristles and spines, but these stout, handsome flies are covered with patches of soft hairs (fig. 290). Many larval members of the family are parasites of deer, horses, and other hoofed mammals, and one tropical species has chosen humans as hosts. However, the larval lives of tree-dwelling bot flies (genus *Cuterebra*: *cut* = skin; *ereb* = darkness) are spent as parasites feeding beneath the skin of squirrels, chipmunks, and other small rodents. They leave their warm-blooded hosts in anticipation of their metamorphosis, drop to the ground, and pupate in the soil. Although the larvae have strong, rasping mandibles, the adult flies completely lack mouthparts and do not eat.

Pollen and nectar, fruits and nuts, bark, wood, and leaves attract and nurture the hidden invertebrate company of trees along with their hidden predators and

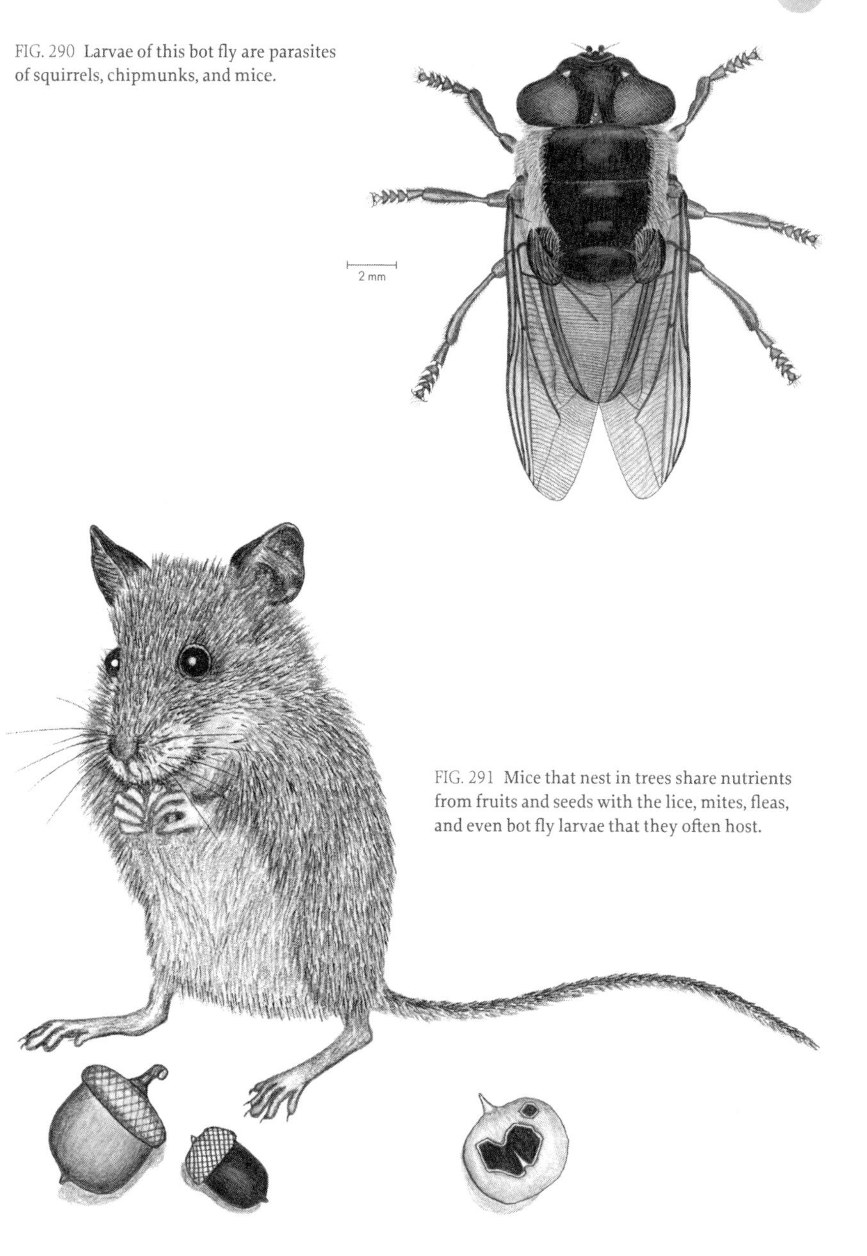

FIG. 290 Larvae of this bot fly are parasites of squirrels, chipmunks, and mice.

FIG. 291 Mice that nest in trees share nutrients from fruits and seeds with the lice, mites, fleas, and even bot fly larvae that they often host.

parasites. The larger vertebrate companions of trees—birds, mammals, even some frogs, lizards, and snakes—also have their share of hidden invertebrate companions that still indirectly enjoy the benefits of the tree's gifts (fig. 291).

All these creatures of the treetops along with the pollen, nectar, fruits, nuts, bark, wood, and leaves that nourished them directly or indirectly eventually settle to the soil to nourish the decomposers beneath the tree. The decomposers set about the essential job of recycling all this plant and animal matter that rains down from above. The decomposers eventually liberate the mineral nutrients in simple chemical forms that trees can take up from the soil and use to support new generations of trees and tree companions.

New generations of trees begin from seeds and acorns that survive the gauntlet of birds, squirrels, weevils, caterpillars, and other larvae. Every mighty oak begins from a tiny oak embryo nestled at one end of an acorn (fig. 292). The first act of the embryo is to send forth a root and seek out essential mineral nutrients from the soil. The two massive energy-rich seed leaves, or cotyledons, of the acorn nourish the embryo until it has established a firm roothold in the earth and has begun taking up essential nutrients for growth. Only then does the embryo send forth its first shoot and first leaves toward the sky and begin capturing energy from the sun.

The next chapter will explore the underground world that nurtures these first days as well as all succeeding days of every tree's life.

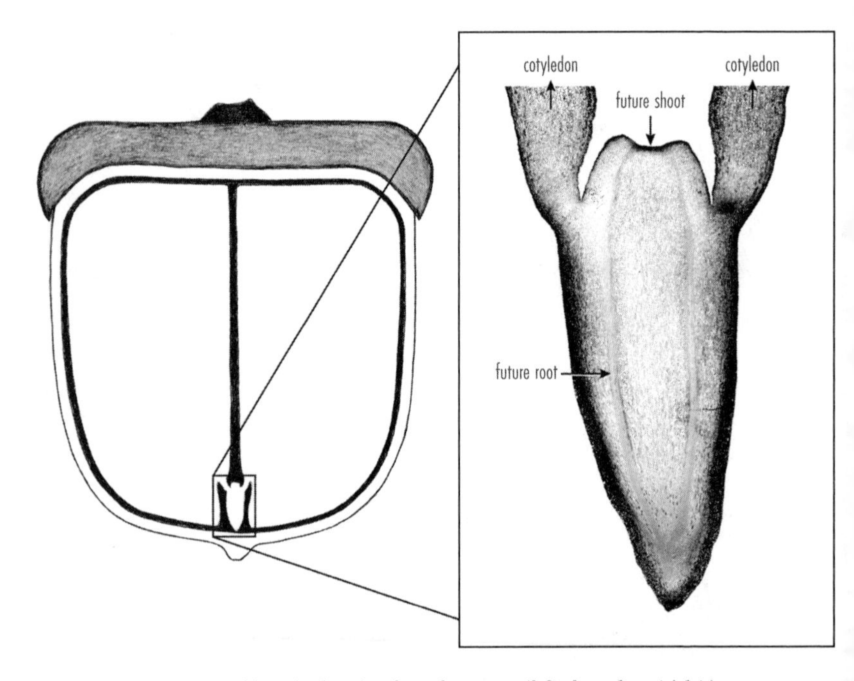

FIG. 292 In this sagittal section through an acorn (*left*), the embryo (*right*) is nestled between two nutrient-rich cotyledons that will supply the energy for its future root to emerge from the acorn and begin its first journey into the soil.

Three different habitats of a tree (fig. 1) and their representative companions are illustrated on the following six pages. Microscopic views of the smaller creatures are placed in circles. Each creature in each of the three regions—LeafScape, BarkScape, and RootScape—is assigned a number and identified in the accompanying legends.

LEGEND, LeafScape

1. Red-eyed Vireo
2. Rough Green Snake
3. Spiny Oakworm Moth (*Anisota stigma*)
4. Orange-striped Oakworm Caterpillar (*Anisota senatoria*)
5. Red-bordered Emerald Moth (*Nemoria lixaria*)
6. Walkingstick
7. Long-horned Beetle, Red Oak Borer (*Enaphalodes rufulus*)
8. Oak-leaf Skeletonizer (Bucculatricid Moth larva)
9. Tachinid Fly
10. Apple gall of Gall Wasp
11. Yellowjacket Wasp
12. Female Tree Cricket
13. Lacewing larva
14. Galls of Cecidomyiid Midge (*Polystepha pilulae*)
15. Gracillariid Moth, Solitary Oak Leafminer (*Cameraria hamadryadella*)
16. Leaf mines of Oak Blotch Miner (Gracillariid Moth)
17. Larva of Oak Sawfly
18. Twig gall of Gall Wasp
19. Caterpillar of the Tiger Moth (*Halysidota tessellaris*)
20. Caterpillar of Banded Hairstreak Butterfly
21. Grass-carrying Wasp (*Isodontia*)
22. Leaf-rolling Weevil
23. Thief Weevil
24. Leaf roll nursery of Leaf-rolling Weevil
25. Robber Fly
26. Ladybird Beetle
27. Caterpillar of White-marked Tussock Moth
28. Ladybird Larva
29. Inchworm
30. Parasitic Torymid Wasp
31. Gall Wasp
32. Red Spider Mite (*Tetranychus urticae*)
33. Oak Treehopper (*Platycotis vittata*)
34. Mirid Leaf Bug
35. Red-eyed Green Leafhopper
36. Brown Lacewing
37. Cecidomyiid Midge
38. Common Oak Aphid (*Tuberculatus annulatus*)
39. Actinopelte leaf spot fungus (*Tubakia dryina*)
40. Actinomycete (*Microbispora*)
41. Actinomycete (*Thermoactinomyces*)
42. Actinomycete (*Streptomyces*)

LEGEND, BarkScape

1. Eastern Towhee
2. Brown Creeper
3. Eastern Plains Garter Snake
4. Eastern Chipmunk
5. Tree Frog
6. Snail
7. Daddy Longlegs
8. Click Beetle
9. Carpenter Ant
10. Underwing Moth
11. Long-horned Beetle (*Prionus*)
12. Christmas Fern
13. Fragile Fern (*Cystopteris*)
14. *Hepatica*
15. Apron Moss
16. Leafy Liverwort (*Frullania*)
17. Lichen
18. Rotifer
19. Nematode
20. Tardigrade
21. Protozoa
22. Crane Fly larva (*Liogma*)
23. Thrips
24. Bark Louse
25. Jumping Bristletail
26. *Brochymena* Stink Bug
27. Lichen-carrying Lacewing larva
28. Pseudoscorpion
29. Springtail
30. Oribatid Mite

LeafScape: Its landscape of leaves and branches hosts companions seeking food and refuge.

12
11
13
9
10
8
7
6
19
4
37
5
3

29
30
28
25
26
27
23
24
22
8
7
5
9
10
14
3
11

BarkScape: Its bark and trunk support tiny forests of mosses and lichens and are the gateway to its heartwood.

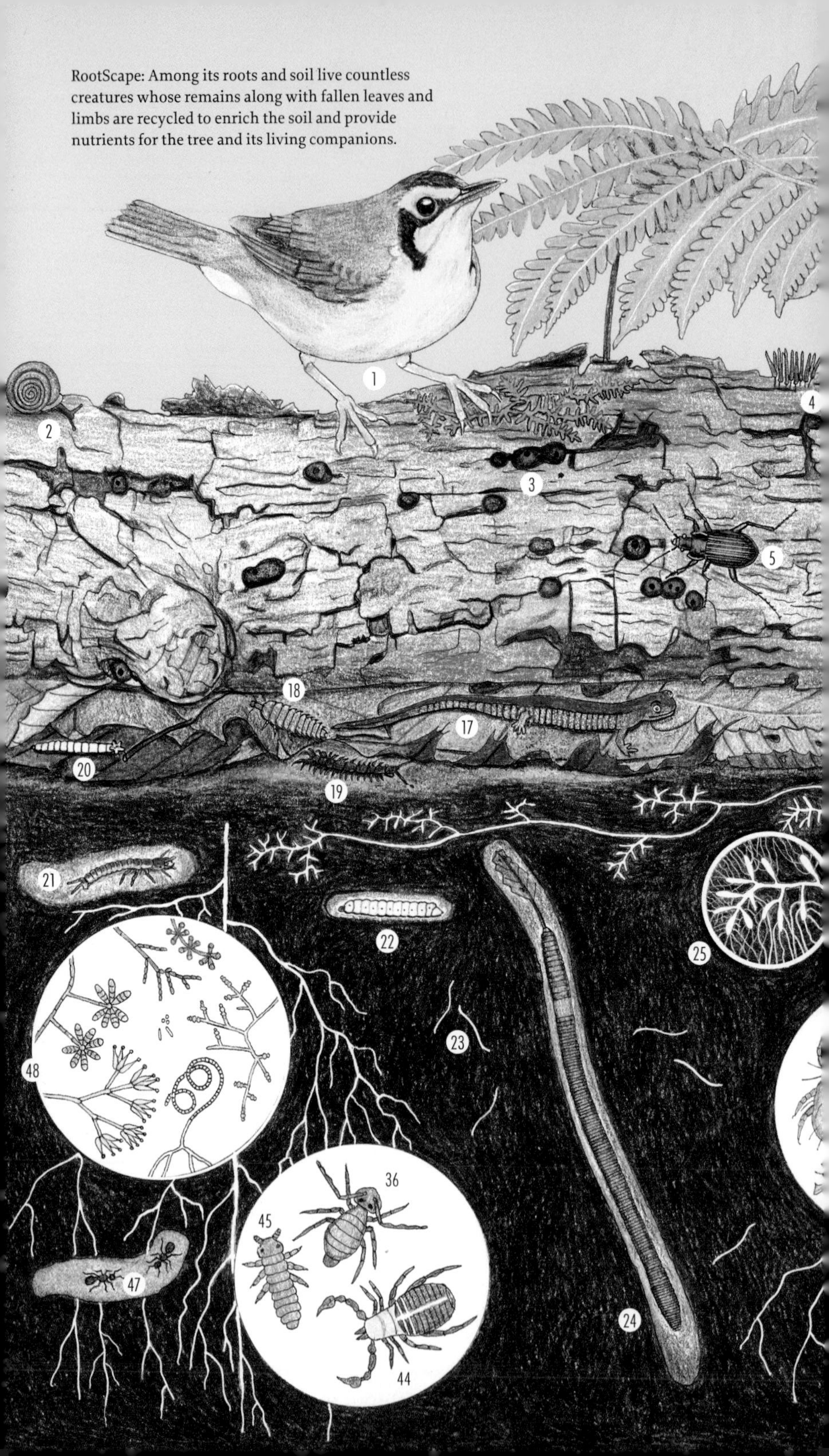

RootScape: Among its roots and soil live countless creatures whose remains along with fallen leaves and limbs are recycled to enrich the soil and provide nutrients for the tree and its living companions.

9
8
10
6
7
15
11
16
13
14
12
27
30
26
43
28
29
39
40
34
35
36
33
37
32
31

LEGEND, RootScape

1. Kentucky Warbler
2. Snail
3. Mushroom, Carbon Balls (*Daldinia*)
4. Slime mold, (*Stemonitis*)
5. Ground Beetle
6. Handsome Fungus Beetle
7. Mushroom (Old Man of the Woods)
8. Pleasing Fungus Beetle
9. Crane Fly
10. Mushroom (Indigo Milky)
11. Daddy Longlegs
12. Soldier Fly larva
13. Millipede
14. Spider (*Dysdera*)
15. Rove Beetle
16. Woodlouse
17. Redback Salamander
18. Firefly larva
19. Stone Centipede
20. Crane Fly larva
21. Ground Beetle larva
22. Robber Fly larva
23. Potworms
24. Earthworm
25. Close up of mycorrhiza
26. Click Beetle larva
27. Dipluran (Japygidae)
28. Soil Centipede
29. Cicada nymph
30. Scarab Beetle larva
31. Soil algae
32. Nematode
33. Tardigrade
34. Protozoa (Amoeba)
35. Protozoa (Ciliate)
36. Rotifer
37. Protura
38. Symphylan
39. Pauropod
40. Dipluran (Campodeidae)
41. Ceratopogonid Midge larva
42. Oribatid mite
43. Gamasid mite
44. Pseudoscorpion
45. Smooth Springtail
46. Globular Springtail
47. Ants
48. Actinomycetes, Bacteria

By autumn and winter, many of the tree's companions have sought refuge under bark, in leaf litter, and in the soil.

THE WORLD BENEATH A TREE

6

THE WEB OF AFFAIRS AMONG ROOTS AND LEAF LITTER

Trees and the creatures that share the soil with them carry on an elaborate underground intercourse. On the tree and under the tree, predators, parasites, herbivores, and fungivores find full employment in their respective professions. The decomposers of the food web concentrate their efforts under the tree, recycling whatever plant and animal remains fall to earth. However, some decomposers such as mites and springtails (BarkScape, 29, 30; chapter 4) also carry out recycling aboveground. On many trees, especially those in the tropics and temperate rain forests, enough soil accumulates from dust in the air in addition to humus and nutrients from the recycling efforts of these decomposers to nurture forests of mosses, lichens, and flowering epiphytes that have established rootholds on tree trunks and branches. Soil—whether under a tree or on a tree—supports similar arrays of creatures carrying out equally elaborate interchanges.

Many of the tree's aboveground companions pupate in the soil or retire for the winter and join the permanent residents in the dark recesses of the tree's underground world. Their complex web of interactions defies a facile understanding of how all these lives are intertwined. And the web becomes even more tangled when one realizes that some arthropods can be decomposers as larvae belowground yet graduate to being herbivores, pollinators, or even predators as adults aboveground. Some fungi such as the nematode-destroying fungi discussed in chapter 1 are decomposers at one stage in their lives but can transform into carnivores under certain environmental conditions. Nature's webs are dynamic and fluid, changing with the ebb and flow of their environments.

Beneath the trees, animal and microbial decomposers liberate essential nutrients from decaying animal and plant remains, providing recycled nutrients for the trees. While armies of microbes—invisible without the aid of a microscope—pump out enzymes to digest the litter under each tree, countless larger decomposers such as snails, earthworms, potworms, and legions of arthropods are busy shredding, chewing, digesting, and defecating the same litter. They consume many microbes as they carry out their recycling and expand the surface area of the litter on which the remaining microbes and their enzymes can work. Many microbes pass through digestive tracts of larger decomposers unperturbed by their journey, continuously multiplying in numbers and secreting enzymes along their way down alimentary canals (Nardi 2007). At the same time, these larger decomposers burrow and move about in the soil, helping to mix organic matter with mineral particles, improving the soil's structure so that air and water can freely penetrate.

All the cells and tissues of the countless inhabitants and the leaves, twigs, flowers, and fruits that fall from the tree provide raw material for the recycling work of the decomposers. Everyone feeds the decomposers, and decomposers feed everyone the nutrients that they recycle. Throughout its life, each decomposer's droppings are also ingested, digested, and recycled—sometimes multiple times—by a group of decomposers referred to as coprophages that clean up whatever dung is left behind in the grazing grounds. Finally, at the end of its life, the tissues and cells of each decomposer are in turn recycled by its fellow decomposers.

The decomposers have one big job to do—convert decaying plant and animal matter into essential minerals that tree roots can use and into humus that holds these nutrients and water within reach of tree roots. In technical jargon, this conversion and release of essential minerals is known as mineralization of the plant and animal remains. Each decomposer contributes its share to the liberation of nutrients that can be readily taken up by roots, and to the formation of the complex organic matter known as humus that stubbornly resists digestion by the decomposers. Decomposers generate the humus and mix it with the soil's mineral particles, constantly improving the structure of the soil. With this improved structure, the soil holds water and mineral nutrients where they are readily accessible to tree roots.

Among the countless decomposers, roots, and fungi of the underground dwell legions of those that feed on them—predators, parasites, root feeders, fungus feeders. These include the seven classes of arthropods found almost exclusively in the soil. While most people have heard of millipedes (class Diplopoda), centipedes (class Chilopoda), and springtails (class Collembola), very few are acquainted with the other four arthropod classes that usually go only by their scientific names—Protura, Diplura, Pauropoda, and Symphyla. Each of these classes has its own idiosyncrasies that make it stand out in the crowded world of the underground.

In the densely inhabited underground world, decomposers, predators, and parasites are in a constant search for food that brings them into constant encounters. Predators and parasites count on speed and subterfuge, highly penetrating mandibles, ovipositors, and piercing stylets. Decomposers and those predators that must avoid larger predators depend on such defenses as impenetrable and protective cuticles or secretion of repellent, toxic chemicals.

DECOMPOSERS AND THEIR PREDATORS

Mites and springtails are by far the most numerous of the arthropod decomposers. Sampling the population of a fistful of leaf litter using a Berlese funnel (chapter 7) quickly reveals who's who in the world beneath a tree.

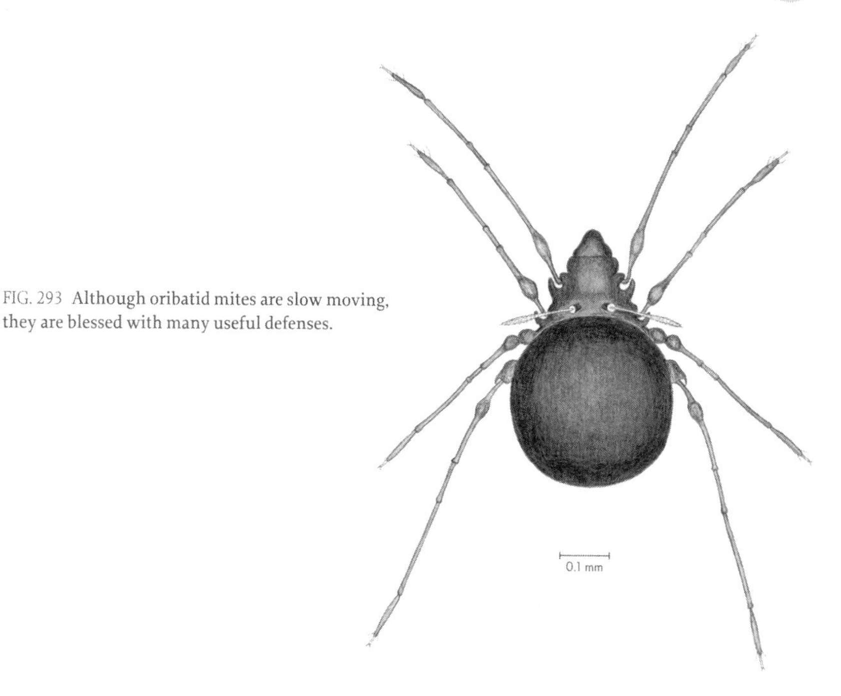

FIG. 293 Although oribatid mites are slow moving, they are blessed with many useful defenses.

— ORIBATID OR BEETLE MITES, CLASS ARACHNIDA, SUBCLASS ACARI, ORDER ORIBATIDA (*oribat* = mountain roaming) (10,000 described species, estimated 110,000 species; unknown number species NA; 0.2–1.4 mm) (Maraun et al. 2007).

Down in the leaf litter beneath a tree, thousands of oribatid mites labor at their recycling job—anywhere from 50,000 to 500,000 mites per square meter. These slow-moving mites encounter many fast-moving predators such as rove beetles and gamasid mites but are well prepared with many defenses. These mites have hard, stout exoskeletons, often sporting sturdy cuticular flaps under which they can tuck their appendages out of reach of predator jaws (fig. 293; BarkScape, 30; RootScape, 42). Toxic secretions from opisthonotal (*opistho* = back part; *nota* = the back) glands on dorsal surfaces of these mites can ward off advances from even the largest predators. Tropical poison tree frogs with reputations for being highly toxic can attribute their alkaloid toxins not to their own synthetic activities but to chemicals acquired from the oribatid mites they consume in tree canopies (Saporito et al. 2007).

— GAMASID MITES, CLASS ARACHNIDA, SUBCLASS ACARI, ORDER MESOSTIGMATA (*meso* = middle; *stigmata* = spots, referring to the single pair of spiracles on the sides of their bodies) (11,000 species; unknown number species NA; 0.2–2 mm) (Marchenko and Bogomolova 2015).

While a few species of gamasid mites have taken up residence as parasites on vertebrates and invertebrates (chapter 5), most species are free-roaming predators

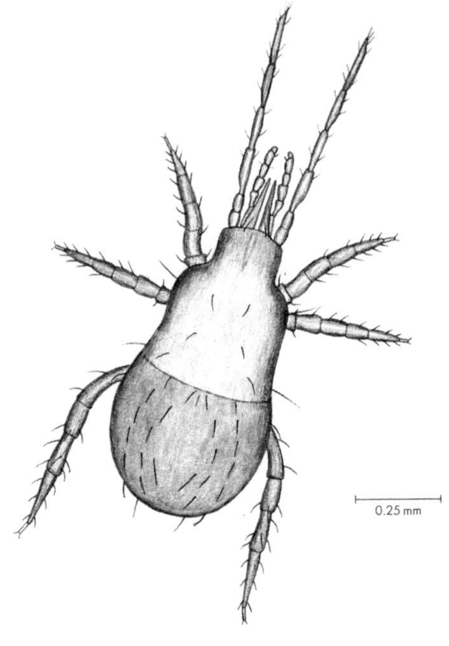

FIG. 294 Gamasid mites have rich hunting grounds in their underground world.

of the soil. These restless, fast-moving mites are constantly searching the leaf litter for nematodes and arthropod prey. They can be spotted not only by their rapid movements but also by their elongate, often pointed snouts or chelicerae (*cheli* = claw; *cera* = horn) with which they stab and drain their victims (fig. 294; RootScape, 43). By contrast, the slow-moving oribatid mites use their short, stout chelicerae to gnaw at soil litter.

— SPRINGTAILS, CLASS AND ORDER COLLEMBOLA (*coll* = glue; *embol* = peg) (8,800 species; 840 species NA; 1–5 mm) (Turnbull and Stebaeva 2019).

Next to the soil mites, springtails are the most abundant arthropod decomposers in the soil, numbering around 50,000 individuals per square meter of soil under a tree. The 8,800 species of springtails described so far come in many forms: slender, stout, and rotund; some with long springs, some with short springs, some with no spring; some with eyes, some without (fig. 295; BarkScape, 29; RootScape, 45, 46). The signature attribute of each springtail is its forked "spring tail," or furcula (*furca* = fork), whose fork latches to a small projection on its belly when the tail is at rest. When alarmed, the springtail flings its tail down and back, sending the springtail leaping forward. The springtails of the soil surface have the largest tails, the most colors, and the best eyesight. The springtail species that inhabit deeper, darker layers of the litter and soil have less color, poorer eyesight, and shorter furculae. In the deepest recesses of the soil, springtails are eyeless and colorless and have completely lost their springs.

Without the hard, thick, almost impermeable cuticles of most mites, springtails must resort to other schemes to avoid desiccating in a hot, dry environment. Some of them are covered with insulating, overlapping scales like those on moth and butterfly wings that help retain body water, but many springtails have thinner, more vulnerable cuticles. All springtails, however, have extensible organs that project from their ventral surface and continually monitor the surrounding humidity. This water-conserving strategy is best observed as these animals cross the inside of a clear, moist glass container; the springtail repeatedly taps the glass surface with its ventral tube, stopping to take up moisture with the tube whenever needed. While the springtail's common name is based on its springing ability, the name Collembola (*coll* = glue; *embol* = peg) refers to this ventral appendage that monitors and maintains normal levels of life-sustaining water.

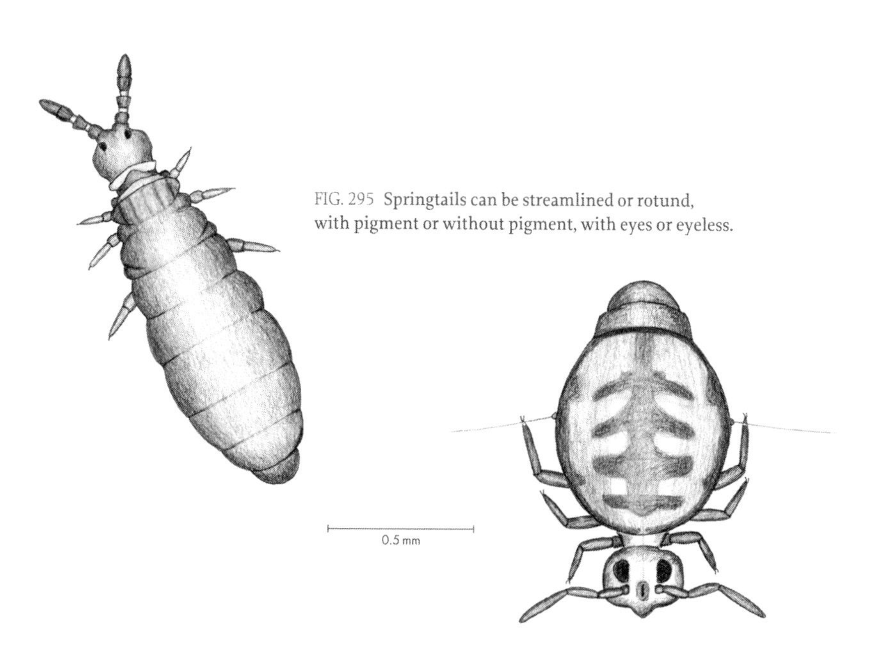

FIG. 295 Springtails can be streamlined or rotund, with pigment or without pigment, with eyes or eyeless.

— PSEUDOSCORPIONS, CLASS ARACHNIDA, ORDER PSEUDOSCORPIONES (*pseudo* = false; *scorpio* = scorpion) (3,600 species; 500 species NA; 2–8 mm). These arachnids scurry about in the litter under a tree waving their large pincers as they survey their habitat for springtails, mites, and tiny insects (fig. 296; BarkScape, 28; RootScape, 44). Like their larger relatives the spiders, these arachnids also spin silk, but unlike spiders, which spin silk from their posterior end, pseudoscorpions spin silk from chelicerae at their anterior end. Rather than using silk to entangle prey, pseudoscorpions weave silken shelters for themselves that look like diminutive igloos. They retreat into these shelters during winter days and whenever they outgrow their old exoskeletons and must shed them for new, larger ones. While its new exoskeleton is thickening and hardening, the newly molted pseudoscorpion lies low in its silken chamber.

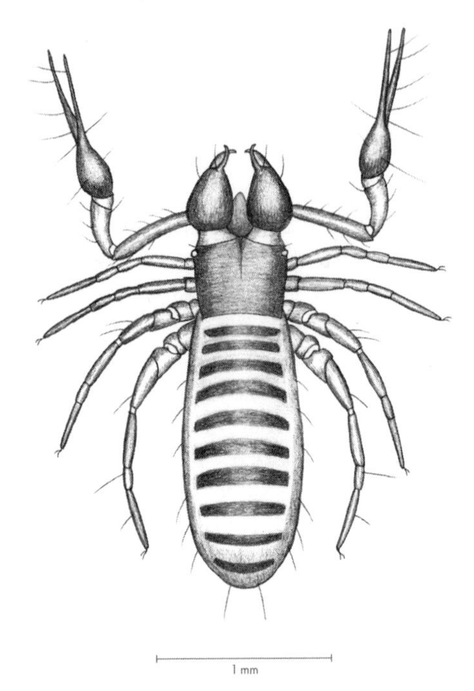

FIG. 296 These fierce-looking but stingless relatives of true scorpions are tiny predators that take a toll on springtails and mites in the soil.

— WOODLICE, CLASS CRUSTACEA, ORDER ISOPODA (*iso* = equal; *poda* = feet), SUBORDER ONISCIDEA (*oniscus* = woodlouse) (4,000 species; 100 species NA; 5–14 mm).

Sowbugs, pillbugs, chucky pigs, cheesy wigs, slater bugs, roly-polies, armadillo bugs, potato bugs—these are just some examples of the numerous common names for these very common but exceptional arthropods (fig. 297; RootScape, 16). They are exceptional by virtue of being among the very few members of the class Crustacea that have forsaken an aquatic existence. Vestiges of their aquatic origins, however, persist in their unparalleled ability to conserve water, their excretion of ammonia, and their use of gills. Woodlice have inconspicuous channels on their cuticles that take up water from their surroundings by capillary action while simultaneously passing this water over their gills. Water is excreted along with ammonia. While this precious water is conserved and recycled, the toxic ammonia rapidly evaporates. You can see for yourself how much ammonia woodlice expel by smelling the contents of a small, closed jar that has been occupied for about an hour by several woodlice. Woodlice are not only efficient at water recycling but are also among the most numerous of the recyclers, constantly consuming the endless supply of plant litter that falls to the ground.

Like their vertebrate counterparts the marsupials, woodlice also protect and brood their eggs in a ventral pouch—a marsupium (Latin for pouch)—until the baby woodlice hatch. Even after her brood hatches and leaves her pouch, the mother woodlouse hovers around her offspring for weeks.

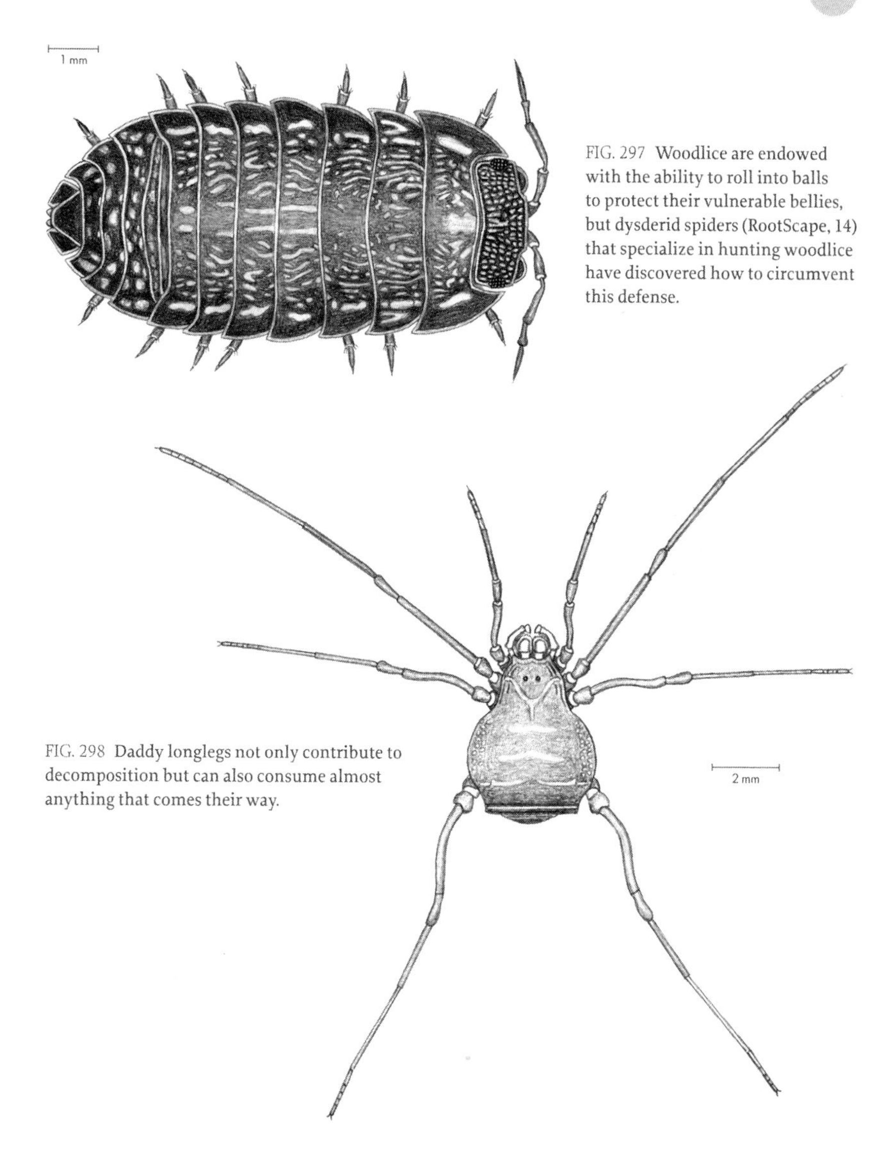

FIG. 297 Woodlice are endowed with the ability to roll into balls to protect their vulnerable bellies, but dysderid spiders (RootScape, 14) that specialize in hunting woodlice have discovered how to circumvent this defense.

FIG. 298 Daddy longlegs not only contribute to decomposition but can also consume almost anything that comes their way.

— DADDY LONGLEGS OR HARVESTMEN, CLASS ARACHNIDA, ORDER OPILIONES (*opilio* = shepherd; referring to an old practice of shepherds walking on stilts to oversee their large flocks) (6,650 species described, but the number of species is probably around 10,000; 65 species NA; body length 2–10 mm, with legs measuring 50 mm).

Legs dominate the anatomy of daddy longlegs, and they are continually grooming these legs with their chelicerae to ensure that these appendages perform optimally (fig. 298). They navigate over foliage and across the ground with their long strides, waving their second pair of legs to sample sensory information about the environments through which they pass (BarkScape, 7; RootScape, 11). With large surface areas for attachment and with their capacity for long strides, the long, skinny legs offer hitchhiking services for tiny and far less mobile creatures of the soil such as mites and pseudoscorpions.

With their long, gangling legs, daddy longlegs appear fragile and vulnerable; however, they have been gifted with chemical defenses that compensate for their physical frailty. At the base of the front legs are glands that are sources for repellent chemicals. Spiders, ground beetles, and birds that try to subdue a daddy longlegs are soon frantically wiping a disgusting stench and flavor from their jaws. A close encounter with one or two daddy longlegs confined to a jar affirms the potency of their odor.

These relatives of spiders and mites do not spin silk, weave webs, or secrete venom. As implied by their common name "harvestmen," they are encountered most frequently under trees during the harvest season of late summer and early autumn. Unlike most spiders, which have four pairs of eyes, the single pair of eyes on harvestmen is poised like a periscope on top of their bodies. They can be predators like spiders, and they can also appreciate fungi and fresh woodland fruits such as berries. However, scavenging on dead creatures and leaf litter is just as acceptable for these omnivores.

— COCKROACHES, CLASS INSECTA, ORDER BLATTODEA (*blatta* = cockroach) (4,400 species; 75 species NA; adults 4–70 mm).

With its head tucked under its thorax and with its sleek, flattened body, a cockroach is well suited for living between fallen leaves under trees or beneath tree bark (fig. 299). Many species never develop wings, which could obstruct

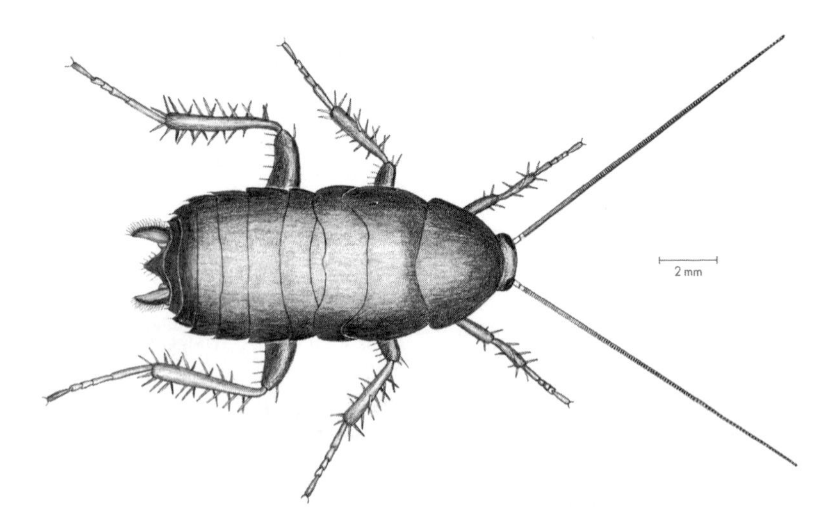

FIG. 299 Cockroaches are covered with sensory bristles that constantly supply them with information about the dangers and opportunities in their environments.

their movements in these tight quarters. To discourage any close encounters or attacks from the rear in these confined spaces, cockroaches have many glands at their posterior ends that exude sticky, repellent compounds. Hundreds of highly sensitive bristles are strategically distributed over each cockroach's legs, antennae, and body, providing it with thorough surveillance of its environment.

Adult and immature cockroaches are well endowed with both defenses and senses for environmental surveillance. The passive, more vulnerable egg cases of cockroaches, however, are the preferred hosts for members of an entire family of parasitic wasps (Evaniidae). Mother hatchet or ensign wasps (fig. 177) scrupulously scrutinize cockroach habitat for their egg case nurseries.

Cockroaches are great scavengers that eat just about anything that comes their way and contribute mightily to clearing debris and preparing it for new uses. Like other noteworthy recyclers such as their close relatives the termites (chapter 4), cockroaches are assisted in their recycling efforts by the armies of microbes that inhabit their digestive tracts and provide additional enzymes for breaking down even the toughest and most recalcitrant plant and animal remains. Also, like their more social termite relatives, many woodland cockroaches establish small family units in which the adults pass on these microbes to the next generation as they tend the younger members of their families. The small fraction of unwelcome cockroach species that share our homes give these important recyclers undeservedly bad reputations.

Insects are not the only six-legged animals. Three other arthropod classes are included in this group of arthropods known as the hexapods (*hexa* = six; *poda* = feet). Even though they may have several hundred legs as adults, baby millipedes hatch from their eggs with only six legs. So, for a brief period in their lives even millipedes have six legs. The three other hexapod classes—Collembola, Protura, and Diplura—may have many attributes not shared with insects, including mouthparts that do not remain exposed but retract into a pouch within the head, but they do have the one feature of leg number in common.

— CONEHEADS OR PROTURANS, CLASS AND ORDER PROTURA (*prot* = first; *ura* = tail) (800 species; 80 species NA; 0.2–2.5 mm).

Much about proturans remains a mystery. In 1907 the Italian naturalist Filippo Silvestri described the first proturan in the scientific literature. In the years since their discovery, biologists have learned about some of the idiosyncrasies of these unusual, minuscule creatures (fig. 300; RootScape, 37).

At hatching, baby proturans have abdomens with 9 segments, and they add one segment at each molt until they attain adulthood with 12 abdominal segments. However, the other hexapods are born with all the abdominal segments they will have as adults: a maximum of 11 for insects, 10 for diplurans, and 6 for collembolans. Throughout their lives, however, with only 9 abdominal segments or with as many as 12 segments, proturans feed on fungi and decaying leaves, contributing to the decomposition of plant litter and enriching the soil.

Proturans have neither eyes nor antennae, but they compensate for the lack of these visual and olfactory structures by using sensitive receptors on their first

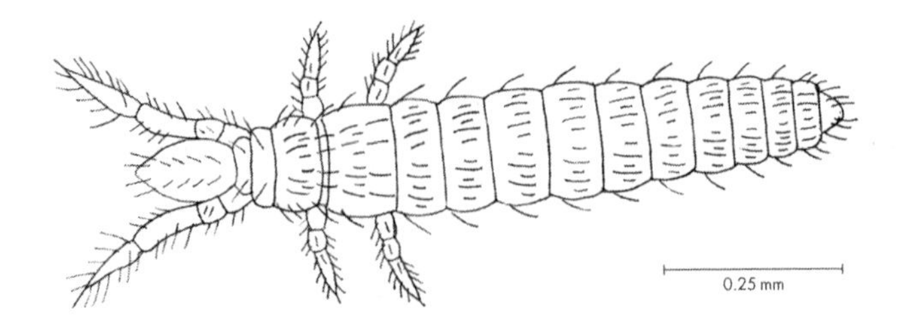

FIG. 300 The lives of even the diminutive proturans influence the lives of trees that tower above them.

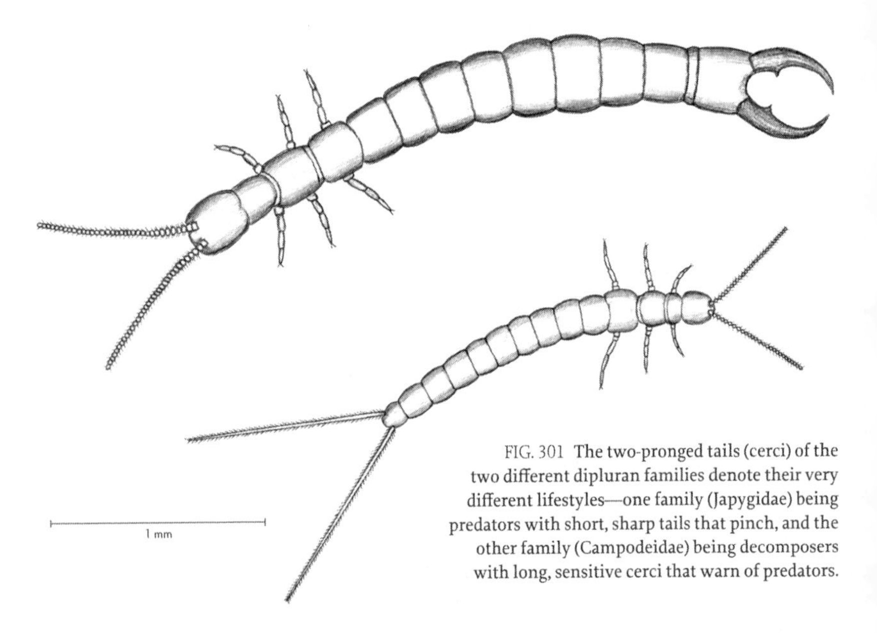

FIG. 301 The two-pronged tails (cerci) of the two different dipluran families denote their very different lifestyles—one family (Japygidae) being predators with short, sharp tails that pinch, and the other family (Campodeidae) being decomposers with long, sensitive cerci that warn of predators.

pair of legs to interpret their surroundings. Proturans walk on their second and third pairs of legs while waving their first pair of legs about to pick up essential sensory information about their dark, confined, and silent habitats.

— TWO-PRONGED BRISTLETAILS, CLASS AND ORDER DIPLURA (*diplo* = two; *ura* = tails) (900 species; 170 species NA; 2–50 mm), FAMILIES JAPYGIDAE (*japyg* = strong wind) and CAMPODEIDAE (*campo* = bending, flexible).

Tails are a conspicuous part of the anatomy of diplurans (fig. 301; RootScape, 27, 40). The paired tails at their rear ends are known as cerci (*cercus* = tail) and are the posterior counterparts of the anterior antennae. The two prominent dipluran families, however, use their cerci in very different ways. The predatory japygids have cerci modified as sturdy pincers for capturing prey as well as providing sensory information from the rear. The equally fast-moving but frailer members of the Campodeidae are decomposers that probably also consume smaller arthropods when these are available. Their longer, more delicate cerci mirror the appearance of their anterior antennae. Both cerci and antennae sweep in wide arcs to detect possible food and foes.

While only a few of the 10 classes of arthropods dwell on trees, all of them inhabit the soil and litter under trees. So far six of the arthropod classes—the insects, arachnids, crustaceans, springtails, proturans, and diplurans—have been mentioned as companions of trees.

Four entire—and the remaining—classes of the phylum Arthropoda are grouped in the SUBPHYLUM MYRIAPODA (*myria* = many; *poda* = legs) (16,000 species; 1,700 species NA; 0.5–300 mm) (Langor et al. 2019).

Myriapodans, known for their abundant legs, include these four classes of soil arthropods: centipedes, millipedes, symphylans, and pauropods. The number of pairs of legs ranges from 8 in the class Pauropoda and 12 in the Symphyla to as many as 177 in some centipedes and even 653 in some millipedes.

— CENTIPEDES, CLASS CHILOPODA (*chilo* = thousand; *poda* = legs) (3,000 species described, but almost three times this number are estimated to exist; 150 species NA; 10–300 mm).

Centipedes are known for their legs and always have an odd number of pairs, ranging from 15 to 177 (fig. 302; RootScape, 19, 28). The long, thin, sinuous soil centipedes hatch from eggs with all pairs of legs formed. However, the shorter, broader stone centipedes sprout more legs after hatching until they attain their maximum number of 15 pairs. Centipedes are unique in having their first pair of appendages modified as fangs with poison glands. These fearsome fangs contain a concoction of neuroactive and inflammatory compounds that are quite effective in dispatching their prey. At their posterior ends stone centipedes have glands that secrete a sticky material that gums up jaws of predators and obstructs attempts to subdue these centipedes. Soil centipedes coil and twist if attacked, releasing a noxious secretion from their ventral surfaces that discourages any further advances from predators.

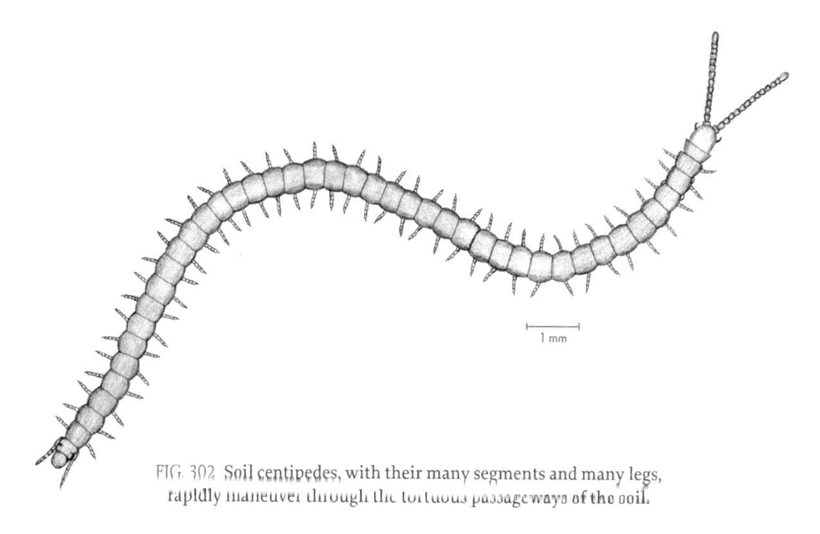

FIG. 302 Soil centipedes, with their many segments and many legs, rapidly maneuver through the tortuous passageways of the soil.

— MILLIPEDES, CLASS DIPLOPODA (*diplo* = double; *poda* = legs) (11,000 species, 80,000 estimated; 1,400 species NA; 3–300 mm).

Millipede jaws contribute to the fragmentation of forest litter. As they meander through the litter, millipedes move their many legs slowly and rhythmically, always managing to put the right feet forward. They achieve a remarkable feat in coordinating so many feet, having as few as 26 legs to as many as 1,306 on a recently discovered Australian millipede that lives 60 meters underground (Marek et al. 2021). However, mastery of the graceful coordination of their leg movements is achieved only after multiple molts. Each millipede is born with only three pairs of legs. When baby millipedes hatch from their eggs, they can easily be mistaken for insects. At each molt several pairs of legs are added until the millipede attains its full length (fig. 303; RootScape, 13).

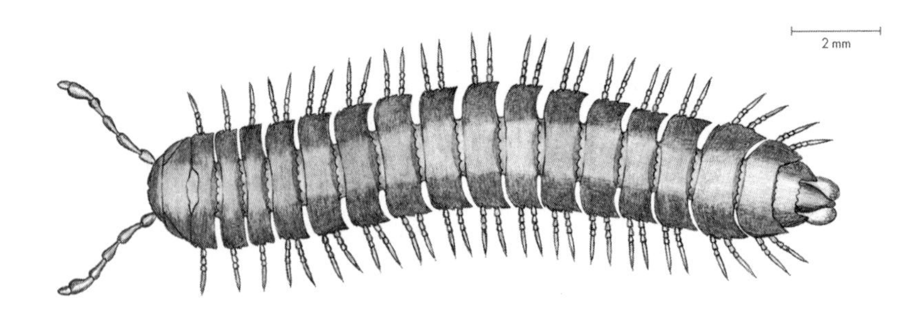

FIG. 303 The countless microbial partners that occupy their digestive tracts help millipedes digest some of the tougher items in forest litter.

If millipedes cannot escape unpleasant encounters on foot, they rely on their potent chemical defenses to ward off unwanted advances. Pairs of glands on most body segments produce these noxious substances. Different families and species of millipedes have perfected many different chemical syntheses in their segmental glands, with chemicals ranging from irritants and repellents to toxic hydrogen cyanide (Shear 2015).

But even all these chemical defenses of millipedes cannot deter some beetle predators. A larva of the glowworm beetle will run alongside a millipede until it is able to fling itself around the millipede, coil around its head, impale its neck with its sharp mandibles, and inject a protein toxin that quickly paralyzes it (fig. 304). The beetle larva's potent digestive enzymes then emulsify the soft tissues of the millipede on which the larva feeds, leaving behind only the exoskeletal remains.

Glowworm larvae are most often observed at night as collections of glowing spots that move across the leaf litter and soil. The bioluminescence of these larvae is reminiscent of lights shining from the windows of railroad cars on a dark night, with a glowing spot on each side of each segment. Someone noted this resemblance to railroad cars and coined the name railroad worms, which has been used ever since.

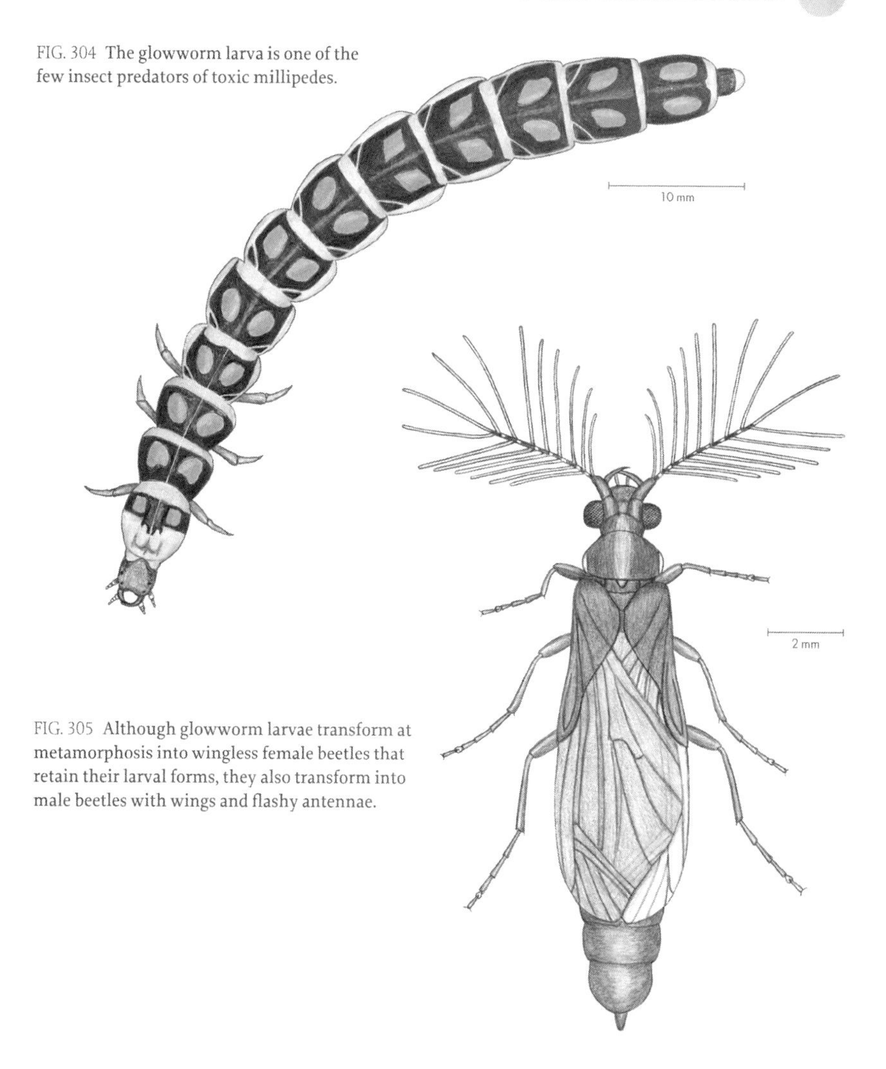

FIG. 304 The glowworm larva is one of the few insect predators of toxic millipedes.

FIG. 305 Although glowworm larvae transform at metamorphosis into wingless female beetles that retain their larval forms, they also transform into male beetles with wings and flashy antennae.

— GLOWWORM BEETLES, FAMILY PHENGOGIDAE (*phengo* = light) (250 species; 25 species NA; larvae 40–65 mm).

Millipedes are spared from attacks by adult glowworm beetles, which do not feed after their molts to adulthood. Nevertheless, the adult beetles still retain the prominent, menacing jaws of their youth. The adult female beetle also retains her larval form and can easily be mistaken for a larva. Only the adult female, however, has conspicuous compound eyes like her male partner. The flightless, earthbound females emit fragrant pheromones that are picked up by the many branches of the males' antennae, each endowed with abundant sensitive olfactory receptors. Few insect antennae can rival the flamboyance of the adult male's feathery antennae (fig. 305).

— PAUROPODS, CLASS PAUROPODA (*pauro* = small; *poda* = feet) (830 species; 100 species NA; 0.5–2 mm)

Pauropods were unknown to science until one day in 1860 when an amateur naturalist in England realized he had discovered a novel arthropod in his garden. When Sir John Lubbock was not occupied with business in the British Parliament, he spent many of his free hours cataloging the life in his garden. Lubbock's passion for natural history is just one of many examples of how amateurs have contributed to our knowledge and appreciation of life on Earth.

All adult pauropods have distinctive three-pronged antennae and between 8 and 11 pairs of legs. They move about quickly and abruptly, often suddenly halting, hesitating, and changing course. Like newly hatched millipedes, baby pauropods start life with only three pairs of legs. Although pauropods are eyeless, they are certainly sensitive to light, and the long sensory projections distributed over their bodies keep them well informed about their encounters and their environments (fig. 306; RootScape, 39).

— SYMPHYLANS, CLASS SYMPHYLA (*sym* = together; *phyla* = guard) (195 species; 36 species NA; 1–10 mm).

Symphylans hatch from their eggs with 6 pairs of legs, but they add an additional pair of legs after each molt, attaining adulthood with 12 pairs (fig. 307; RootScape, 38). Spinnerets on the last segment of a symphylan's body can discharge silk

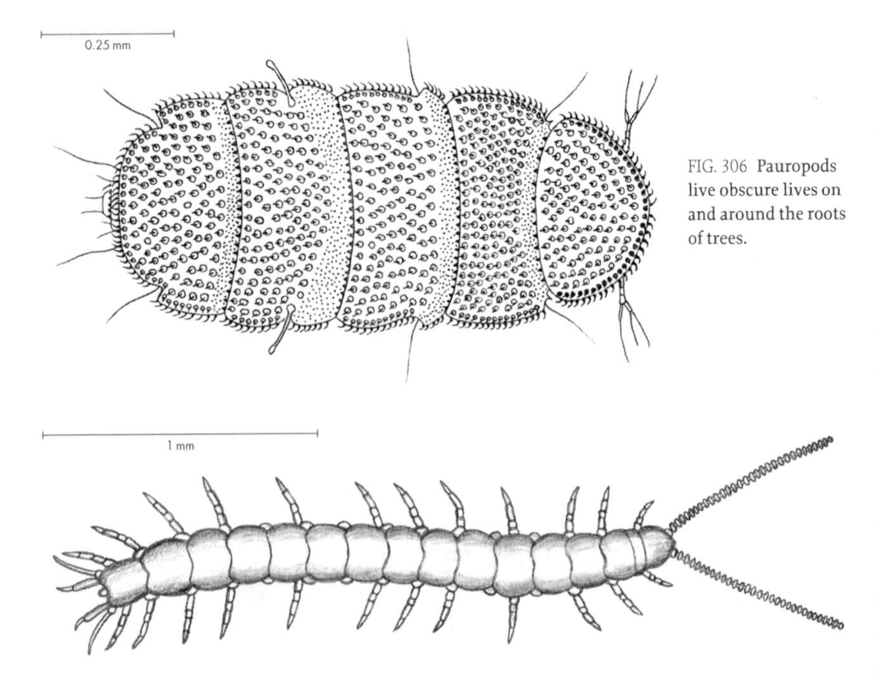

FIG. 306 Pauropods live obscure lives on and around the roots of trees.

FIG. 307 The superficial resemblance of a symphylan to a centipede has earned it the common name garden centipede, but all true centipedes have 15 or more pairs of legs.

strands that can tangle and gum up the jaws and antennae of another arthropod threatening a symphylan from the rear. These arthropods are very sensitive to moisture in the soil and can swiftly travel through pores and passageways in the soil as they respond to their subterranean climate.

Symphylans feed on fungi and decomposing leaves, but they can also consume living tree tissues of mycorrhizae and root hairs. Their sex lives are noteworthy for their peculiar approach to courtship and mating. Males deposit packets of sperm on tiny stalks that the female picks up in her mouth and deposits in special cheek pouches. When she lays her eggs on a surface in the soil, she smears the sperm over them.

A vast number of beetles and beetle species contribute to the decomposition of plant and animal litter. These beetles and their larvae probably devour fungi along with the decaying leaves and wood on which the fungi are growing. The smaller the beetles, the more numerous they are in the litter. They make up in their vast numbers for their small sizes.

— ROUND FUNGUS BEETLES, FAMILY LEIODIDAE (*leio* = smooth; *didi* = distribute) (3,800 species; 350 species NA; 1.5–10 mm).

These small, dark beetles that labor in the litter have few distinguishing characters to unaided human eyes, but I have learned to spot them quickly with the help of a stereomicroscope from their having a constriction in the eighth section, or antennomere, of each antenna. This eighth antennomere stands out from the three larger antennomeres at the end of the antenna and the seven antennomeres at its base (fig. 308).

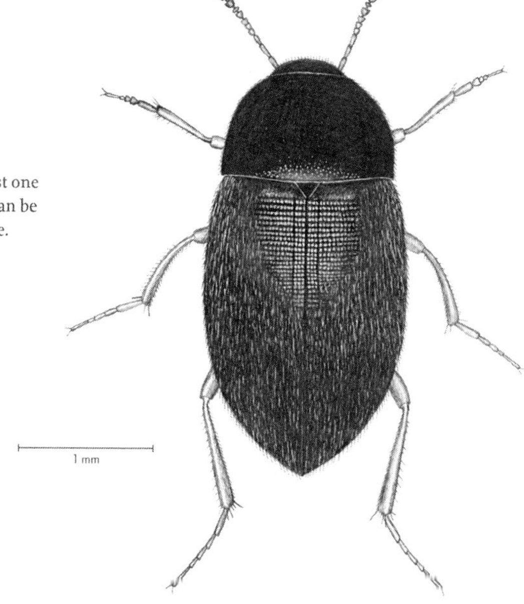

FIG. 308 Round fungus beetles are just one family of fungus-loving beetles that can be very common in the litter under a tree.

— **SAP BEETLES, FAMILY NITIDULIDAE** (*nitid* = shiny, handsome; *ul* = little) (4,500 species; 170 species NA; 2–6 mm).

Sap beetles help with decomposition belowground and aboveground, dealing with decaying fruit, leaves, fungi, and bark (fig. 309). These beetles love the scent of anything fermenting; a concoction of malt, molasses, and a pinch of yeast dissolved in an equal volume of water proves irresistible. One of the most common sap beetles—the shiny black and yellow picnic beetle—tends to show up wherever people gather outdoors with pickles, sweets, and other food.

— **FEATHERWING BEETLES, FAMILY PTILIIDAE** (*ptilon* = feather) (600 species; 120 species NA; 0.3–4.0 mm).

The smallest of all beetles are members of this family (fig. 310). Sometimes they can be almost as numerous in woodland soil as mites and springtails. As is so

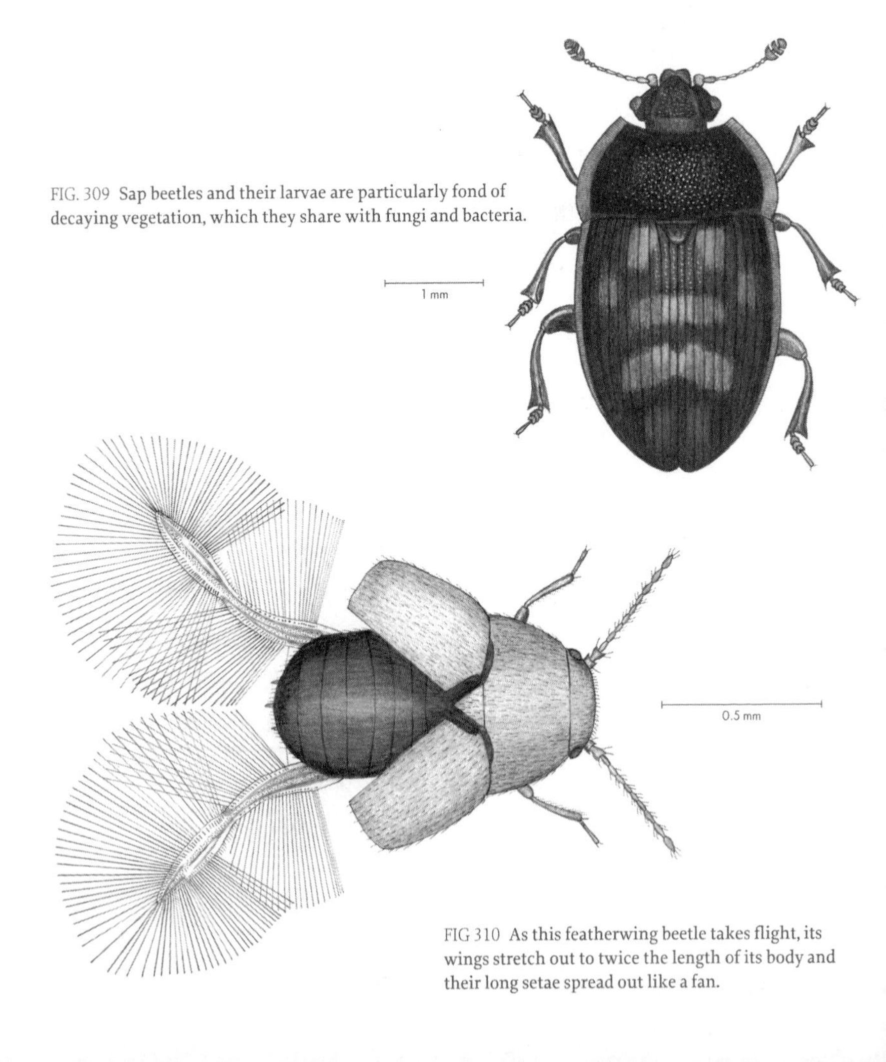

FIG. 309 Sap beetles and their larvae are particularly fond of decaying vegetation, which they share with fungi and bacteria.

FIG 310 As this featherwing beetle takes flight, its wings stretch out to twice the length of its body and their long setae spread out like a fan.

often the case with tiny insects like featherwing beetles and fairy wasps (fig. 334) that live in the close confinement of the soil, their wings are reduced to thin, feathery appendages or tiny stubs, or they are completely absent. Sex is also often simplified for these tiny beetles; for some species, no males have ever been observed. While most insects lay many small eggs, each of the many featherwing beetle females invests much of her energy in one really large female egg and embryo.

— MINUTE FUNGUS BEETLES, FAMILY CORYLOPHIDAE (*cory* = helmet; *lophi* = crown) (200 species; 65 species NA; 0.5–2.3 mm).

The Greek family name of this cheloniform (*chelon* = turtle; *form* = shape of) beetle refers not only to the shield over the head of the larva but also to the forward extension of the adult beetle's thorax, which when viewed from above covers and conceals its head.

While most larvae of these fungus-feeding beetles remain undescribed and unknown, one beetle larva is exceptionally memorable despite its small size. The larva has expanded dorsal plates on its thorax and abdomen in the form of a pancake-shaped shell that almost completely covers its head so that it resembles an insect turtle. Its round carapace is fringed with delicate scales that resemble those on butterfly wings. This round, flattened larva is a fine example of how the practically two-dimensional environment between layers of fallen leaves has shaped its form (fig. 311).

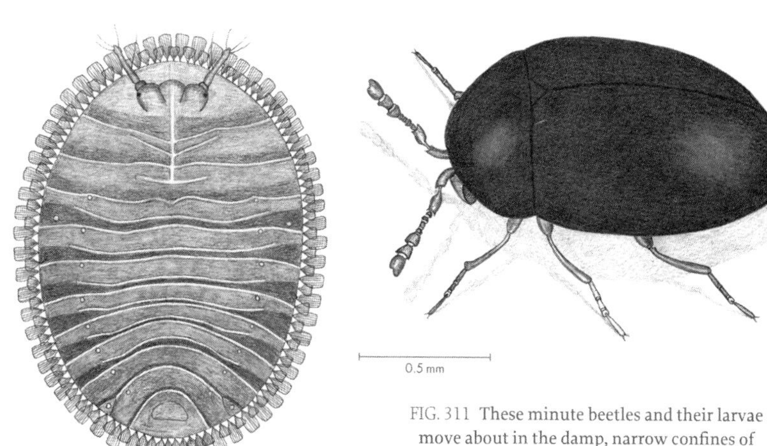

FIG. 311 These minute beetles and their larvae move about in the damp, narrow confines of layered, decaying leaves on the forest floor.

— CARRION BEETLES AND BURYING BEETLES, FAMILY SILPHIDAE (*silpha* = kind of beetle) (200 species; 30 species NA; 12–20 mm).

The countless decomposers that labor to recycle the remains of plants and animals under each tree divide their work so that one group specializes in recycling plant matter and another group specializes in recycling animal remains. The recycling of plant litter requires a much larger crew of workers.

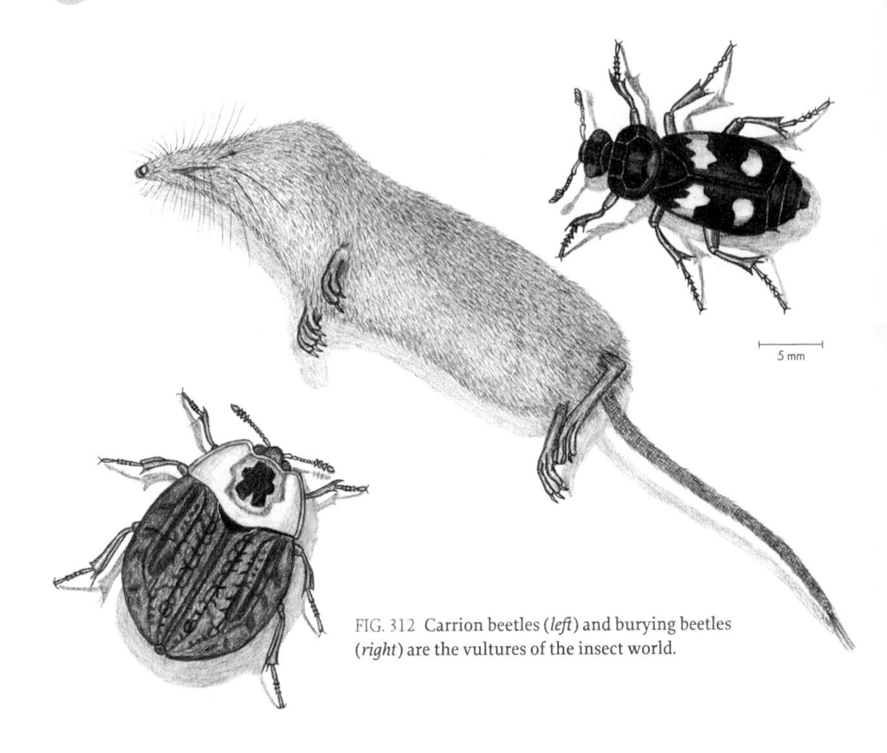

FIG. 312 Carrion beetles (*left*) and burying beetles (*right*) are the vultures of the insect world.

Dead animals are usually in much shorter supply, and the recyclers of carrion face fierce competition for this limited resource (figs. 312 and 313). Fly maggots are these beetles' main competitors, and the adult beetles include fly larvae as well as decaying matter in their meals. Hister beetles are predators frequently found under bark (fig. 220; chapter 4), but certain members of this diverse family also gather on decaying animals to feast on the congregation of fly eggs and maggots. These beetles and their larvae dive into the mass of decaying meat in search of prey and help reduce the populations of maggots that compete with them for this limited resource. Both hister beetles and silphid beetles carry mites as passengers and partners. Upon the beetles' locating a dead meal, the mites disembark and begin searching for fly eggs and eliminating many potential competitors.

Burying beetles are longer and sleeker than the more rotund carrion beetles in the same family. While any large or small animal body—deer, raccoon, mouse, or sparrow—will suffice for carrion beetles, burying beetles choose only the small ones for themselves and their larvae. With bodies about mouse size, a mated pair of beetles can mold and roll the well-decomposed remnants into a soft, compact ball that they bury just beneath the soil surface. The female beetle lays her eggs on the decaying mass, which will serve as a nursery for the larvae. After the larvae hatch, the parent beetles invest much time and energy in guarding and nurturing their sedentary nestlings. Carrion beetles do not prepare nurseries for their young

FIG. 313 Larval carrion beetles feed on the same carcasses as their adults.

but lay their eggs on whatever dead animal they choose. Adult carrion beetles and their larvae both roam freely over their decaying homes.

Although most caterpillars of moths and butterflies grow up on diets of crisp green leaves, the caterpillars of one subfamily of erebid moths and some members of one subfamily of lycaenid butterflies prefer dry, dead leaves that they help recycle to enrich the soil under trees.

The unorthodox taste for leaf litter among caterpillars of these erebid moths accounts for the name litter moths. The moths in this group take on the form of the Greek letter delta whenever they come to rest and hold their wings out flat against bark and leaves. Many have conspicuous snouts formed by a pair of mouthparts known as labial palps (fig. 314). After spending their larval days down

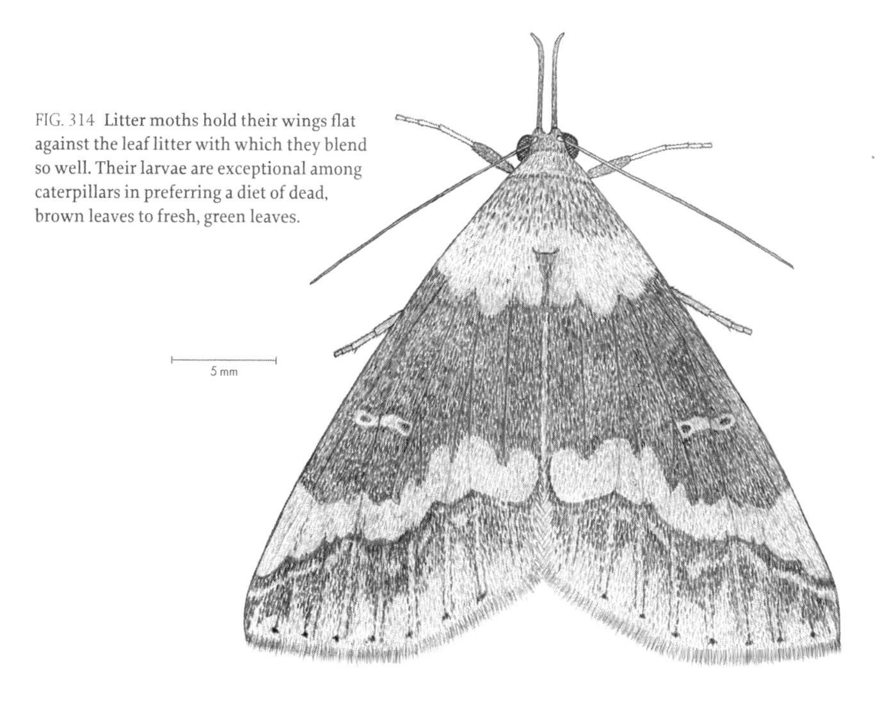

FIG. 314 Litter moths hold their wings flat against the leaf litter with which they blend so well. Their larvae are exceptional among caterpillars in preferring a diet of dead, brown leaves to fresh, green leaves.

in the leaf litter, the adults show up on summer evenings at lights and at sugar baits (chapter 7) that moth lovers use to entice moths and other insects from their hiding places. Many of the insects attracted to lights and sugar baits, including litter moths, have eyes that reflect the lights and eerily glow back at you.

— LITTER MOTHS, FAMILY EREBIDAE, SUBFAMILY HERMINIINAE (*hermos* = beautiful) (~ 8,000 species; 85 species NA; mature larvae 15–25 mm; wingspan 18–48 mm).

— HAIRSTREAK BUTTERFLIES, FAMILY LYCAENIDAE, SUBFAMILY THECLINAE (*thecla* = glory of God) (547 species; 98 species NA; mature larvae 15–18 mm; wingspan 18–38 mm).

The caterpillars of an estimated 200 species of butterflies feed on leaf litter, and most of these exceptional butterflies are hairstreaks (Duarte and Robbins 2010). A tropical member of the group has a caterpillar that feeds on the fungi of leaf litter. The best-known representative of these litter-feeding butterfly caterpillars is the red-banded hairstreak of eastern North America. Its red band lights up the lower surfaces of its forewings and hind wings, which are prominently displayed whenever the butterfly comes to rest (fig. 315).

All fly larvae are legless, but many are decked out with appendages of various lengths and widths that allow them to gain traction in the passages they traverse in leaf litter and soil. They shuffle among dead leaves chomping small chunks of leaves into even smaller chunks, preparing the way for smaller decomposers to break down the dead leaves even more. Microbes move in and multiply in

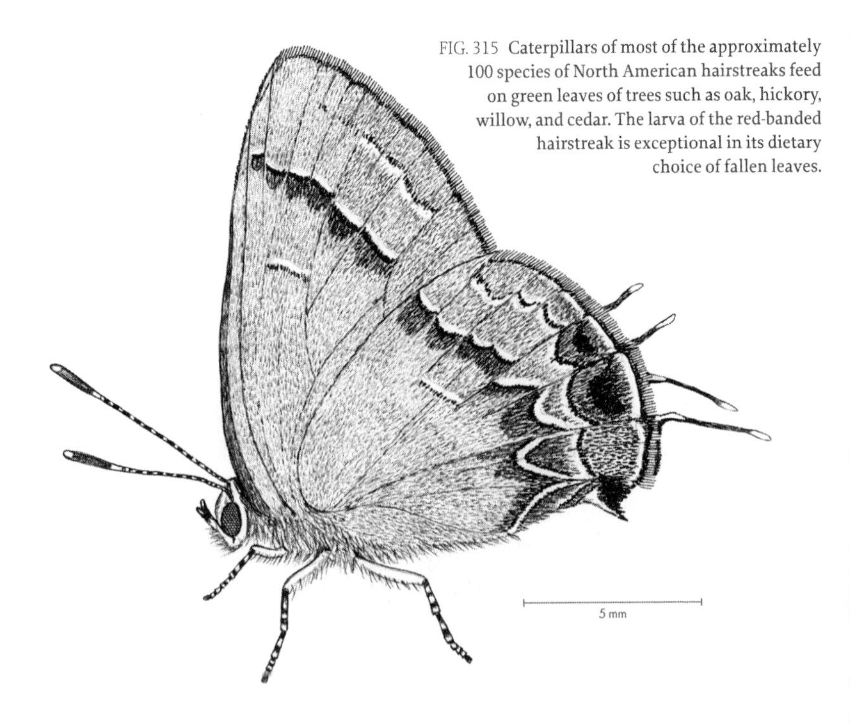

FIG. 315 Caterpillars of most of the approximately 100 species of North American hairstreaks feed on green leaves of trees such as oak, hickory, willow, and cedar. The larva of the red-banded hairstreak is exceptional in its dietary choice of fallen leaves.

this new territory. They release enzymes that ultimately break these leaf chunks down into humus, molecules, and minerals.

The members of these fly families spend all their larval hours eating decaying plant matter, contributing immeasurably to the liberation of nutrients and to generation of the humus that holds these nutrients and soil moisture within reach of tree roots. Each summer day growing trees transport many gallons of nutrient-laden water up their trunks and into their branches, twigs, and treetops thanks to the labors of so many fly larvae.

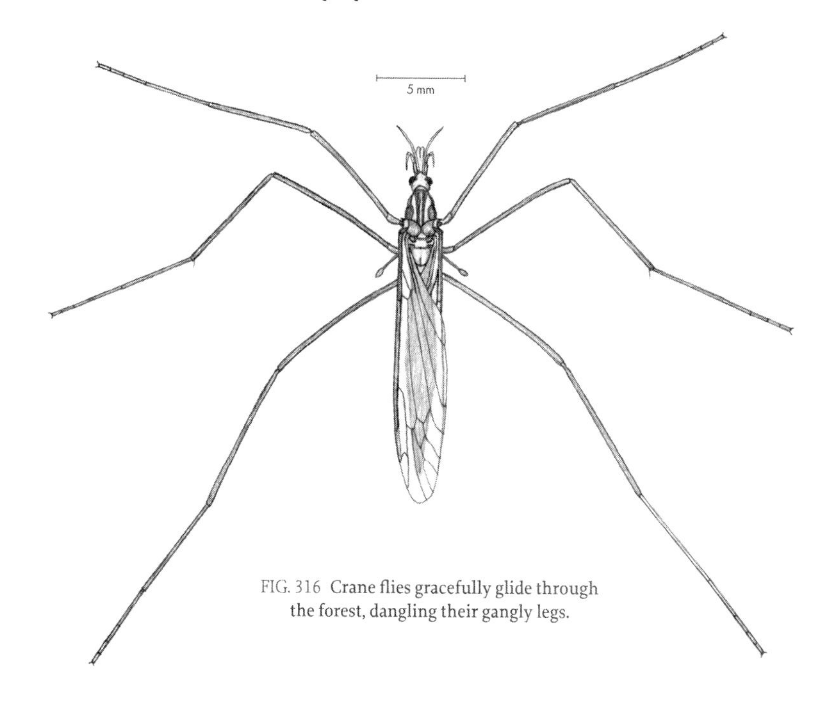

FIG. 316 Crane flies gracefully glide through the forest, dangling their gangly legs.

— CRANE FLIES, FAMILY TIPULIDAE (*tipula* = spider) (15,000 species; 1,600 species NA; 7–35 mm).

Because of their long, spindly legs and similar flight patterns, crane flies often pass for oversized mosquitoes. They represent the largest family of flies, and among the over 15,000 species are members that, at some stage in their life cycles, occupy about every habitat on or under a tree (figs. 316 and 317; RootScape, 9, 20). Many species spend their larval days in aquatic habitats. The larvae, whose tough exteriors have earned them the name leatherjackets, are abundant as recyclers under trees or under the bark of dead branches. At least one species is a leaf miner (chapter 2), and some species feed on living roots. Those larvae that live in mosses on tree trunks blend imperceptibly with their surroundings by sporting green lobes and flaps that match so well in size and shape the leaflets in their mossy home (BarkScape, 22). As adults, crane flies feed innocuously at flowers or do not feed at all.

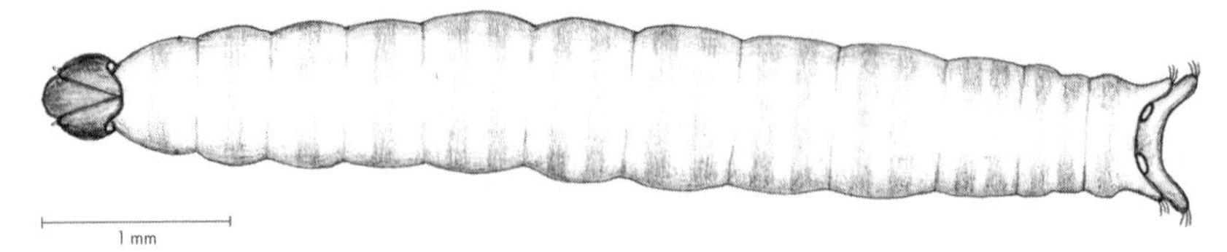

FIG. 317 Larval crane flies are among the most abundant larval decomposers of the forest litter.

Larvae of soldier flies and wood soldier flies, in the families Stratiomyidae and Xylomyidae, which are found under the bark of dead limbs and dead trees (chapter 4; figs. 211 and 212), are also decomposers in the leaf litter under trees.

— MARCH FLIES, FAMILY BIBIONIDAE (*bibio* = type of insect) (1,100 species; 70 species NA; 4–10 mm).

March flies are gregarious as adults and as larvae (figs. 318 and 319). Adults not only emerge in unison from their larval homes in the soil but also form large swarms for mating. Adults devote an inordinate amount of their short lives to mating and copulating; their amorous antics have earned them the colloquial name "lovebugs." With their muscular forelegs, mated females scoop out depressions in the litter in which they will lay several hundred eggs. Their larvae never wander too far from their birthplace and form family gatherings underground.

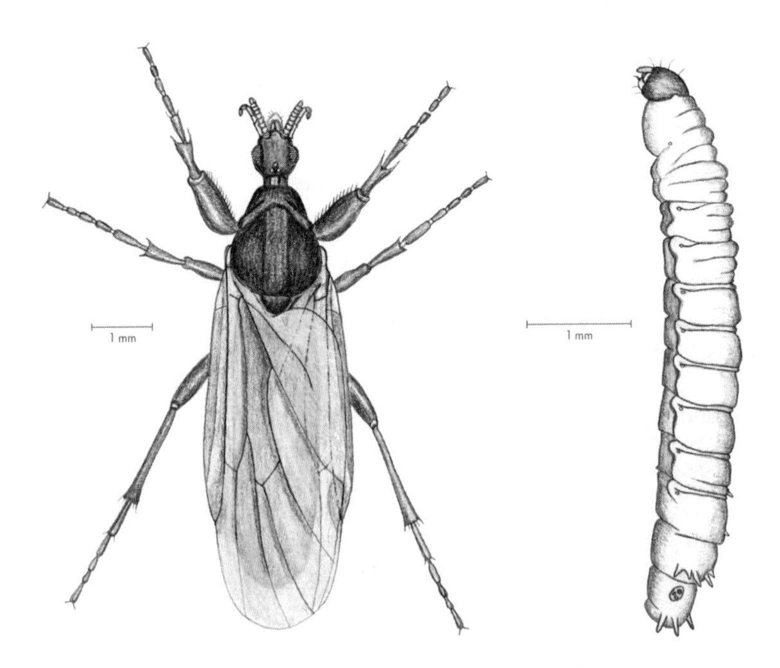

FIG. 318 March flies, like many bee flies, emerge from their larval homes in the soil to visit the early flowers of spring.

FIG. 319 Larvae of March flies chew and shred leaf litter, preparing it for the jaws and enzymes of smaller decomposers.

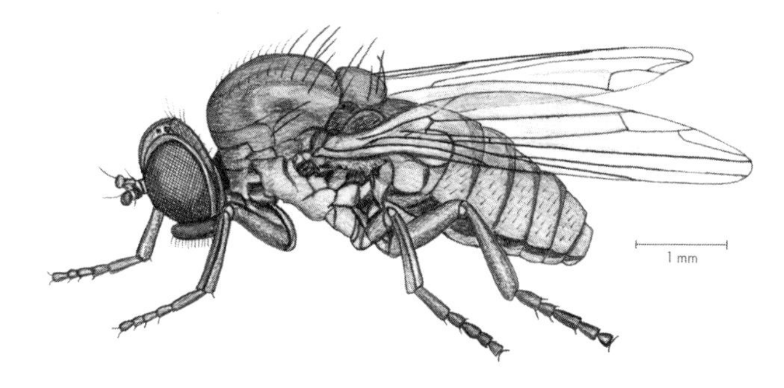

FIG. 320 Like the related hump-backed flies, platypezid flies are small flies with hunched backs, but they also have singularly flat feet.

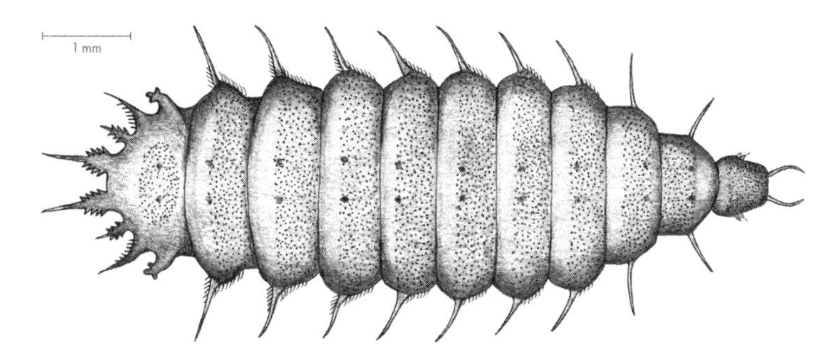

FIG. 321 Fly larvae are always legless, but many, like this platypezid larva, nevertheless have many lateral appendages with which to scoot about in the leaf litter.

— FLAT-FOOTED FLIES, FAMILY PLATYPEZIDAE (*platy* = flat; *pez* = foot) (280 species; 70 species NA; 2–6 mm).

With their big, flat hind feet, the adults probably help spread fungal spores around a forest. These flies have a strong propensity to form swarms, especially during mating, and some members of the family—the smoke flies—have an inexplicable fondness for wood smoke. If you have ever noticed these dark little flies flocking around your campfire, you have probably shared your camping experience with platypezid smoke flies. Perhaps the smell of smoke triggers them to anticipate the imminent flush of fire-loving fungi that erupt on the charred remains of trees after a forest fire. The broad, flat, spiky larvae are fond of fungi of all sorts (figs. 320 and 321).

— HUMP-BACKED FLIES, SCUTTLE FLIES, FAMILY PHORIDAE (*phor* = thief) (4,200 species; 380 species NA; 0.5–6 mm).

With a distinct arch to its thorax and a preference to escape by rapidly scuttling away rather than taking flight, this fly has a gait and physique that have earned it the names scuttle fly and hump-backed fly (fig. 322). These close relatives of flat-footed flies also feed on fungi and help recycle plant litter under trees. However, as members of a much larger fly family, they have taken on far more occupations

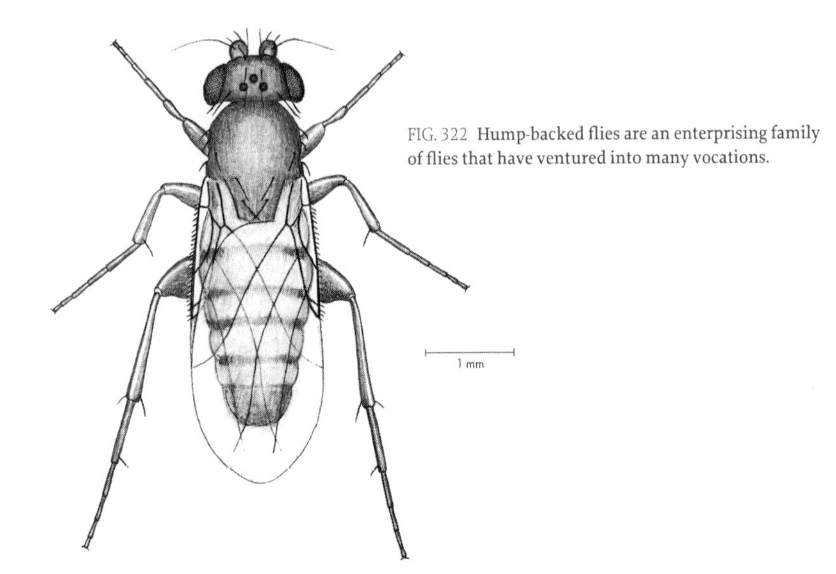

FIG. 322 Hump-backed flies are an enterprising family of flies that have ventured into many vocations.

in the tree's food web as parasitoids and predators of many other arthropods, including ants, millipedes, centipedes, fireflies, soldier beetles, digger bees, and bumblebees. Some species have established themselves as freeloaders and scavengers in the nests of ants and small mammals.

Many of these soil-inhabiting hump-backed flies spend their entire lives underground, surviving on decaying plant matter. Wings are not only superfluous for their subterranean lifestyle but also an impediment as they scuttle about in dark recesses among fallen leaves.

— MOTH FLIES, FAMILY PSYCHODIDAE (*psyche* = moth, butterfly; *-odes* = like) (3,000 species; 120 species NA; 3–10 mm larvae).

Larval moth flies dwell in damp, often wet, habitats and often breathe through snorkels (fig. 323). With their scale-covered wings and long, feathery antennae, the adult flies could at first glance easily pass for diminutive moths (fig. 324). The surfaces of larval moth flies often gather chunks of surrounding soil, probably conferring extra protection from the many predators in their neighborhood, among which are the larvae of many other fly species.

Being a multitalented group, flies of the soil and their larvae contribute not only as decomposers in recycling massive amounts of plant and animal remains but also as predators and parasites in controlling populations of other arthropods.

— STILETTO FLIES, FAMILY THEREVIDAE (*therium* = wild beast) (1,600 species; 150 species NA; 2.5–15 mm).

Narrow abdomens that taper to a stiletto point are a distinctive feature of these rather drab flies. The long, serpentine larvae rapidly slither through the soil and litter as they track down their quarry. Their sleek physiques and extra body joints allow them to twist and contort in the tortuous passageways of the soil. These

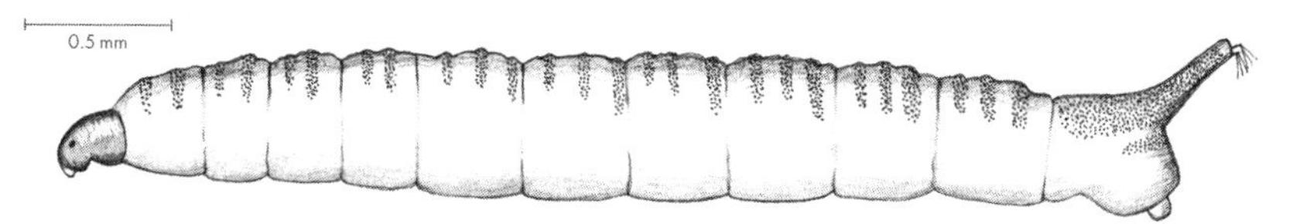

FIG. 323 Larvae of moth flies thrive in damp, dark places of decay and decomposition.

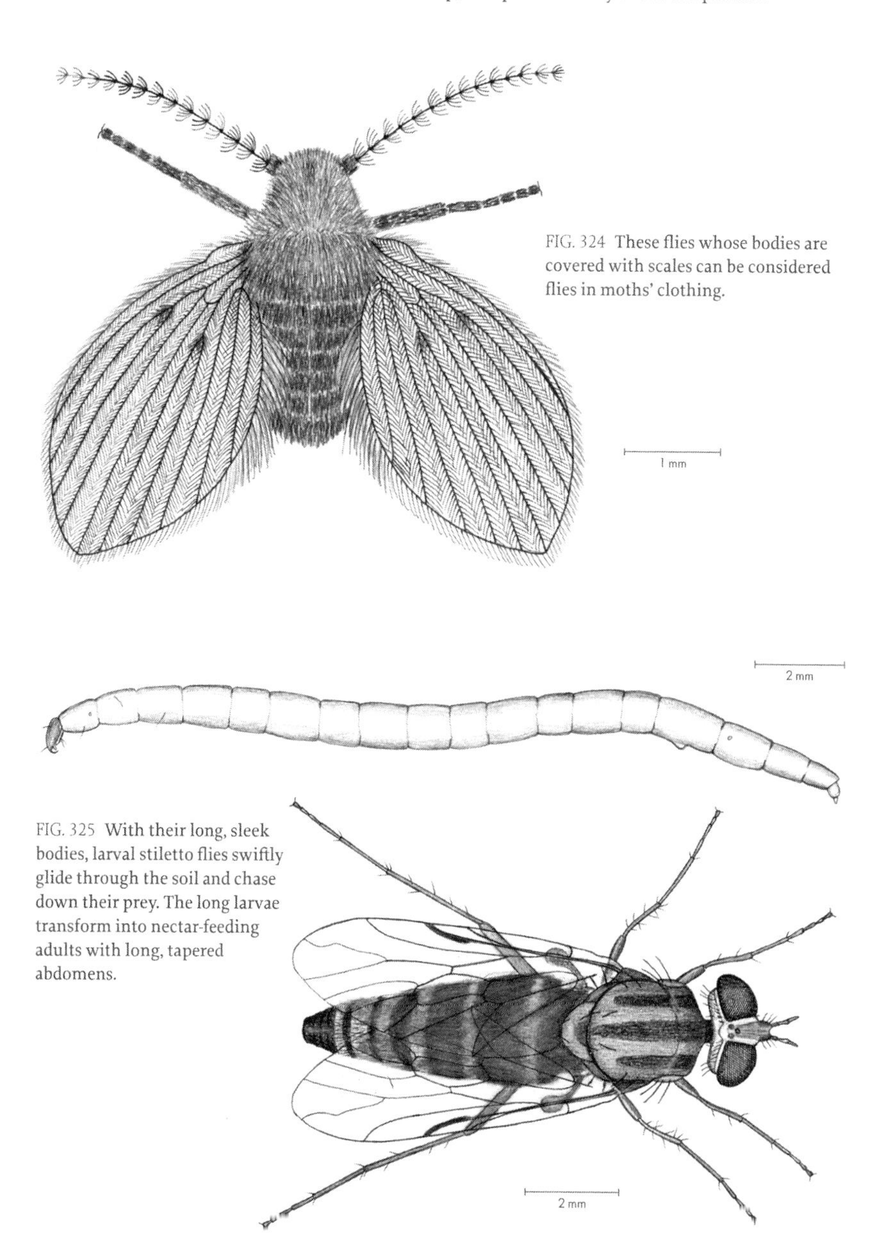

FIG. 324 These flies whose bodies are covered with scales can be considered flies in moths' clothing.

FIG. 325 With their long, sleek bodies, larval stiletto flies swiftly glide through the soil and chase down their prey. The long larvae transform into nectar-feeding adults with long, tapered abdomens.

larvae are more likely to be mistaken for potworms or small earthworms than for insects (fig. 325).

Larvae of ROBBER FLIES AND LONG-LEGGED FLIES, FAMILIES ASILIDAE AND DOLICHOPODIDAE (fig. 326; RootScape, 22).

While most adult flies are predators or pollinators, many of their larvae contribute to the decomposition of plant litter or animal dung. Flies that are predators as larvae in the soil usually remain predators as adults. Robber flies and long-legged flies (fig. 24) are good examples of adult insects carrying on the traditions and occupations of their larvae, but aboveground.

While most flies leave the soil at metamorphosis, some beetles never leave their larval homes, spending all their life stages as predators underground. Other beetles begin life in the soil but then take flight to pollinate flowers or feed on vegetation.

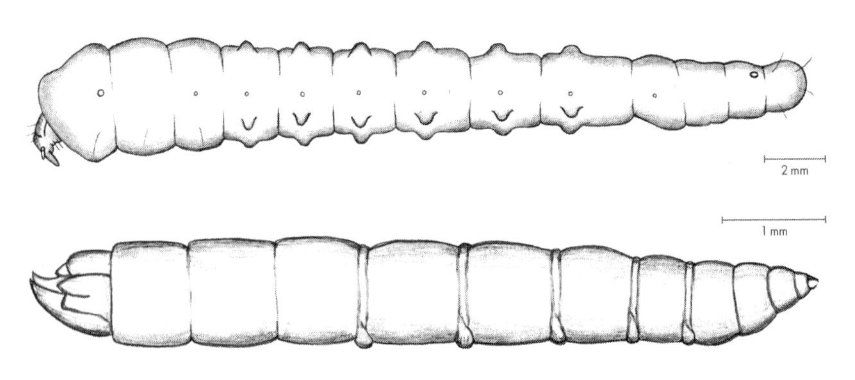

FIG. 326 While predatory larvae of robber flies (*top*) and long-legged flies (*bottom*) are commonly encountered in soil, some species live under bark and feed on wood-boring insects.

— ROVE BEETLES, FAMILY STAPHYLINIDAE (*staphylinus* = kind of beetle) (63,000 species; 5,000 species NA; 1–35 mm).
— GROUND BEETLES, FAMILY CARABIDAE (*carabus* = kind of beetle) (40,000 species; 2,440 species NA; 1–65 mm).

With few exceptions, rove beetles and ground beetles are earthbound throughout their entire lives. They are fast-moving beetles with formidable-looking jaws. Most species are dark brown or black to match the soil and leaf litter, but many are attired in iridescent shades of green, blue, copper, or orange (figs. 29, 327–329; RootScape, 5, 15, 21).

Almost all are insect, snail, or slug predators; a few are seed or fungus predators. A few larval carabids are parasitoids on other immature beetles. One subfamily of rove beetles contains members whose larvae are parasitoids of fly pupae. As larvae, ground beetles and rove beetles are hard to distinguish from each other. They are all long and sleek, with powerful jaws and rapid movements; however, their tails and tarsi offer distinguishing features. Rove beetle larvae have a single

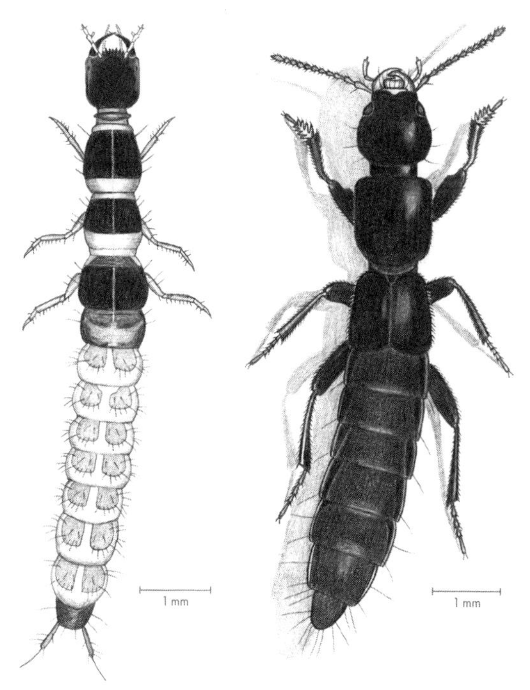

FIG. 327 The saber-like jaws of both larval and adult rove beetles hint at their fierce reputation.

Below left: FIG. 328 Ground beetle larvae have a striking resemblance to rove beetle larvae.

Below right: FIG. 329 Most of the approximately 40,000 species of ground beetles hunt on the ground under trees; however, a few species stake out hunting grounds on leaves and branches aboveground.

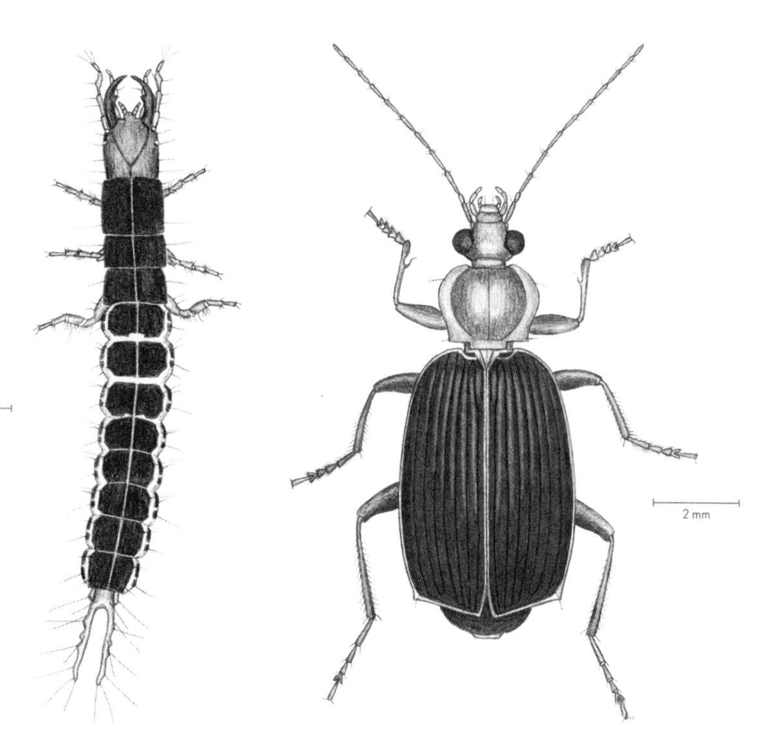

tarsal claw, while ground beetle larvae have two tarsal claws. Rove beetle larvae have two jointed tails (urogomphi) that are articulated at their bases on the ninth abdominal segment; but larval ground beetles have two unjointed, stiff tails extending from the ninth segment.

Rove beetles and ground beetles move with astounding speed, and few prey organisms can outrun these fleet-footed hunters. When threatened by others, these beetles can rely on the potent stench from their abdominal stink glands to quickly halt any unwanted advances. As an additional forewarning before releasing their defensive odors, rove beetles raise their tails as skunks do before spraying and as scorpions do before stinging. Among the ranks of the larger, more voracious ground beetles are the stunning iridescent caterpillar hunters in the genus *Calosoma* (*calo* = beautiful; *soma* = body), which are known for consuming forest caterpillars with bad reputations such as those of tussock moths. The local name for a British rove beetle, devil's coach horse beetle, conveys the intimidating deportment of these dark, ravenous beetles.

Among the ranks of the much smaller, but still very voracious, rove beetles are the ANT-LIKE STONE BEETLES, FAMILY STAPHYLINIDAE, SUBFAMILY SCYDMAENINAE (*scydmaen* = angry) (4,600 species; 500 species NA; 0.5–3 mm).

These tiny beetles are predators that are undaunted by the defenses of well-armored oribatid mites (fig. 330).

FIG. 330 The two jaws (*arrows*) of the ant-like stone beetle act like a can opener, piercing and opening the exceptionally tough shells of oribatid mites.

— FIREFLY BEETLES, FAMILY LAMPYRIDAE (*lampyri* = shining) (2,200 species; 170 species NA; 5–25 mm).

Fireflies and lightningbugs are neither flies nor bugs but beetles that take to the air on summer evenings to fill the skies with dancing flashes of light (fig. 331). These displays, enchanting to human eyes, are the signals exchanged by fireflies during their evening courtship. After their evening courting, the females lay eggs in the soil, where their larvae will patrol the endless passageways for insects, potworms, earthworms, snails, and slugs. After a summer of feeding on other creatures of the soil, the larvae will put on at least one final light show. On certain autumn evenings before they retire underground for the winter, the mature larvae or glowworms can dot the surface of the soil with hundreds of their glowing lamps (fig. 332; RootScape, 18).

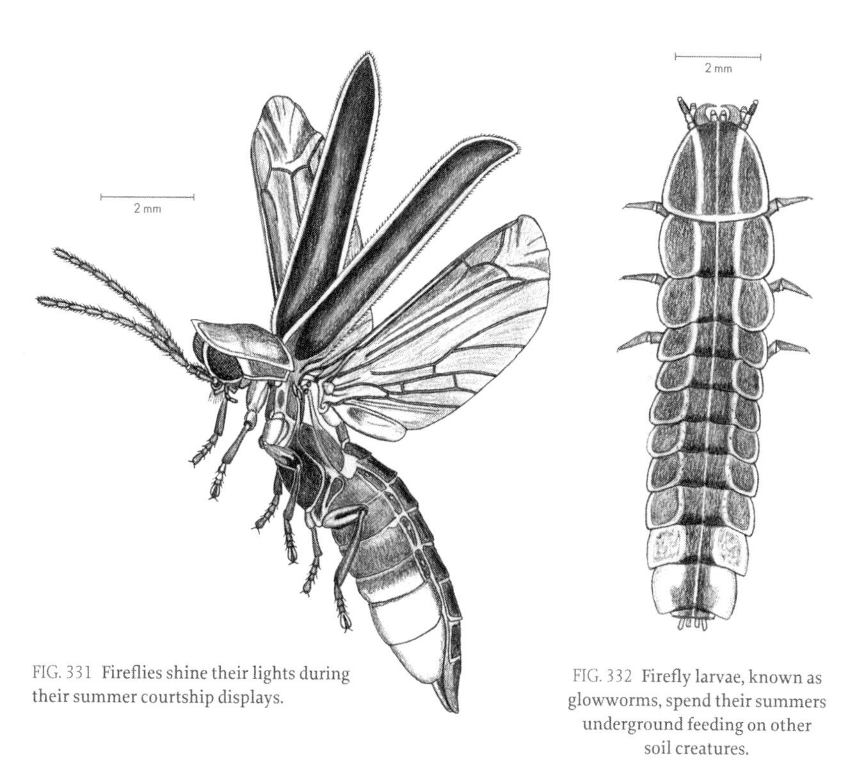

FIG. 331 Fireflies shine their lights during their summer courtship displays.

FIG. 332 Firefly larvae, known as glowworms, spend their summers underground feeding on other soil creatures.

Beneath the flashy exteriors of fireflies lies a blend of toxins very similar in structure and mode of action to toxins from toad skins and foxglove flowers. Firefly flashiness conveys a warning about these beetles' edibility.

— SOLDIER BEETLES, FAMILY CANTHARIDAE (*cantharis* = kind of beetle) (5,100 species; 500 species NA; 5–15 mm).

The name "soldier beetle" is derived from the mix of colors on these beetles that evoke images of the uniforms worn by soldiers (fig. 333). Soldier beetles, like their

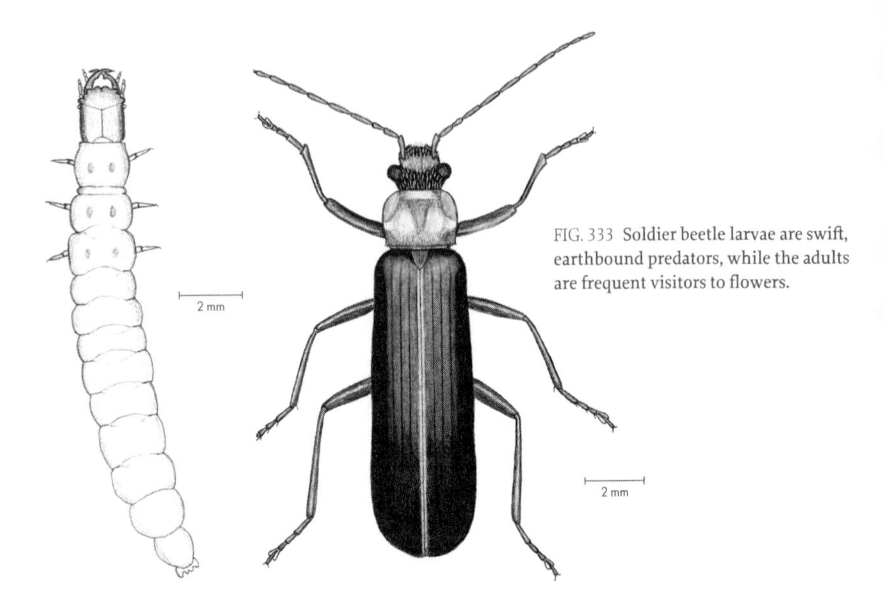

FIG. 333 Soldier beetle larvae are swift, earthbound predators, while the adults are frequent visitors to flowers.

relatives the firefly beetles (figs. 331) and net-winged beetles (fig. 198), have soft, leathery wing covers. To compensate for this vulnerability, however, each beetle family has its own special toxins to repel predators. Fireflies use toxic steroids; soldier beetles and net-winged beetles employ toxic fatty acids. The fatty acid found in soldier beetles is a compound known as dihydromatricaria acid (Haritos et al. 2012), which is also found in certain fungi and many plants in the aster family that are often visited by soldier beetles and to whose repellent chemicals the beetles are apparently immune.

Predatory soldier beetle larvae, like firefly larvae, move rapidly not only with their anterior six legs but also with the last (10th) segment of the abdomen. This surrogate foot, the pygopod (*pygo* = rump; *pod* = foot), projects downward and pushes the larva off from its rear end. On the dorsal surface of each larval abdominal segment are a pair of inconspicuous pores from which perturbed larvae expel a defensive odor that is especially repugnant. A multitude of very fine sensory bristles covers the larva's body and gives it a distinct velvety appearance and texture. Like so many of their fellow soil dwellers, cantharid larvae are endowed with highly refined sensory nervous systems and exceptional senses of touch (fig. 333).

PARASITIC INSECTS OF THE SOIL AND THEIR HOSTS

The insect species that have adopted a parasitic lifestyle number well over 160,000. Considering the high density of arthropod species and individuals of each species that dwell underground, these parasitic insects of the soil have abundant host choices.

Larvae of bee flies, in the family Bombyliidae, are parasites of other soil insect larvae such as those of beetles and other flies or the large clutches of eggs that mother grasshoppers deposit in underground chambers. The over 5,000 species of bee flies worldwide seem to find subterranean larvae of all sorts to be acceptable hosts—from many orders and from all vocations. Even though they all begin their larval lives as parasitoids belowground, as adults bee flies (figs. 92 and 271) are important pollinators aboveground.

— FAIRY WASPS, SUPERFAMILY CHALCIDOIDEA, FAMILY MYMARIDAE (*mymar* = spot, blot) (1,400 species; 200 species NA; 0.15–1.2 mm).

These parasitic wasps are among the smallest of insects (fig. 334). They are also some of the loveliest, having long, narrow wings adorned with long sensory bristles on their fringes. Their singular beauty made them popular subjects of Victorian naturalists and artists. The hosts chosen by mymarids are commensurate with their sizes: the only recorded hosts are eggs of other insects. After depositing an egg in a larger egg of a suitable host, the female wasp lays claim to her nursery by marking the egg's surface with her ovipositor. This mark is detected by the probing ovipositors of subsequent female visitors and enables them to discriminate occupied from unoccupied egg hosts.

A number of other families of parasitic wasps, parasitic flies, and parasitic beetles have discovered the rich hunting grounds of the underground. Many flies and beetles begin life in the soil, where their larvae and pupae offer a wide assortment of eligible host choices for parasitoids. In addition to innumerable immature beetles and flies acting as hosts, other soil inhabitants such as spiders, centipedes, snails, slugs, and earthworms are worthy quarry for some of these parasitic insects.

Many insects lay their eggs in the soil and leaf litter, and these immobile, nutrient-rich resources are prime targets for tiny wasps. There is no need for the mother

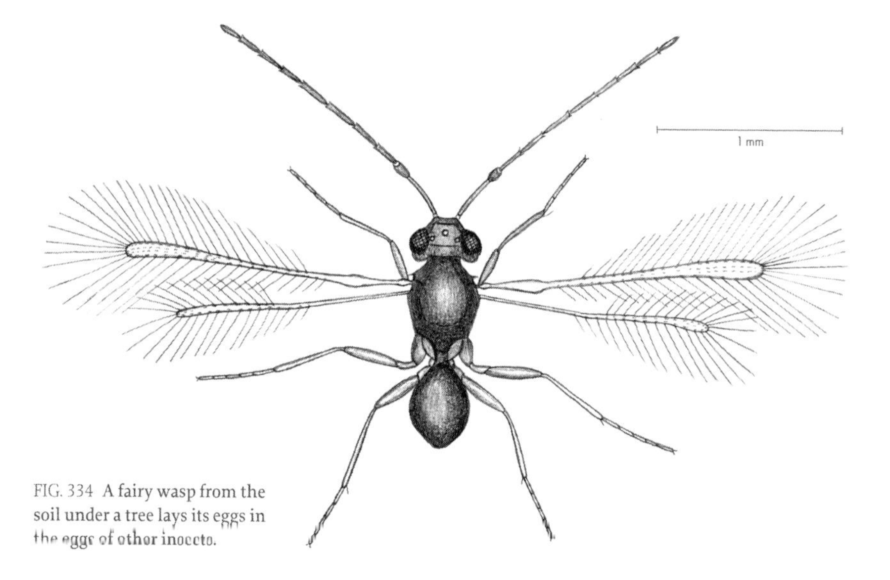

FIG. 334 A fairy wasp from the soil under a tree lays its eggs in the eggs of other insects.

wasp to chase down a host as a nursery for her offspring. Hunting passive eggs as do many platygastrid wasps (fig. 335) presents fewer hazards and is less taxing than hunting bigger, self-defensive game. Some species of these tiny wasps are also known to choose passive, sedentary larvae of gall midges as hosts (fig. 104; chapter 2).

— PLATYGASTRID WASPS, SUPERFAMILY PLASTYGASTROIDEA, FAMILY PLATYGASTRIDAE (*platy* = flat; *gastri* = belly) (4,000 species; 500 species NA; 0.5–2.5 mm).

— SNAILS AND SLUGS, PHYLUM MOLLUSCA, CLASS GASTROPODA (*gastro* = belly; *poda* = foot) (~ 80,000 species, of which about half are land dwelling; ~ 3,000 species NA; 1.0–100 mm) (Nekola 2014).

The ranks of decomposers are unquestionably dominated by the arthropods, but this one class of animals from an entirely different phylum—the mollusks—contributes mightily to the recycling of fallen leaves (BarkScape, 6; RootScape, 2). Most gastropods use their rasping tongues to scrape surfaces of decaying leaves for their nutrition. They pave the way for the handiwork of smaller arthropods, bacteria, and fungi that eventually reduce decaying leaves to humus.

A few gastropods, however, have taken up very different lifestyles as predators of their fellow gastropods. Snails and slugs are also prey for a few species of soil arthropods. Beetles known as snail–eating ground beetles in the family Carabidae and a few daddy longlegs have specialized as snail predators with long, forcepslike jaws that can gingerly enter the narrow aperture of a snail shell to cleanly extract their prey. Larvae of many firefly species grow up feeding on snails and slugs. The over 160,000 insect parasitoids almost invariably choose an arthropod for a host;

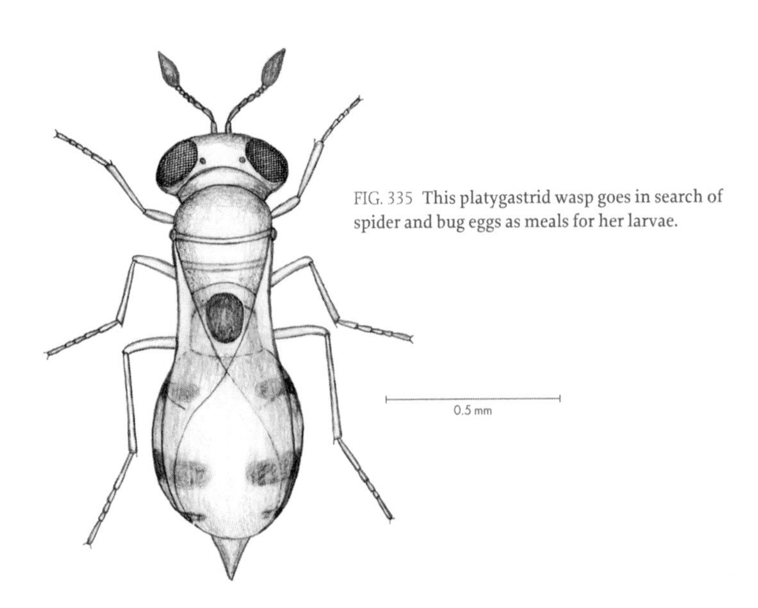

FIG. 335 This platygastrid wasp goes in search of spider and bug eggs as meals for her larvae.

however, the larvae of several parasitoid flies have adopted nonarthropod snails, slugs, and even earthworms as their exclusive hosts.

— SNAIL-KILLING FLIES, FAMILY SCIOMYZIDAE (*scio* = shadow; *myi* = fly) (630 species; 200 species NA; 3–15 mm larvae and adults).

The larvae of all the flies in this family prey on gastropods or tiny clams. A few species prefer to dine on snail eggs. Female flies often scatter their eggs in habitats frequented by snails or slugs. The newly hatched larvae are left to fend for themselves and track down one of their slow-moving hosts. The mother flies of some species leave less to chance and place their eggs directly on the shells or the eggs of their hosts.

The slender adult flies are known for their lovely wings with many dappled patterns (fig. 336). When Gerard Manley Hopkins began the poem "Pied Beauty" with the line "Glory be to God for dappled things," his praise could have been bestowed on the stunning wings of sciomyzid flies. They can be spotted slowly sauntering over vegetation and have a propensity for perching with their heads down.

Earthworms and their cousins the potworms are among the most familiar recyclers beneath a tree (RootScape, 23, 24). Both contribute efficiently to the decomposition of fallen leaves. The larger, more mobile earthworms are also great mixers of this nutrient-rich organic matter with the underlying mineral soil. They shred and chomp leaves and drag them into their burrows, swallowing bits of surrounding soil as they do so. After passing through each earthworm's gut, the mix of plant remains and soil is returned to the surface as droppings or casts in which nutrients are now concentrated and well mixed.

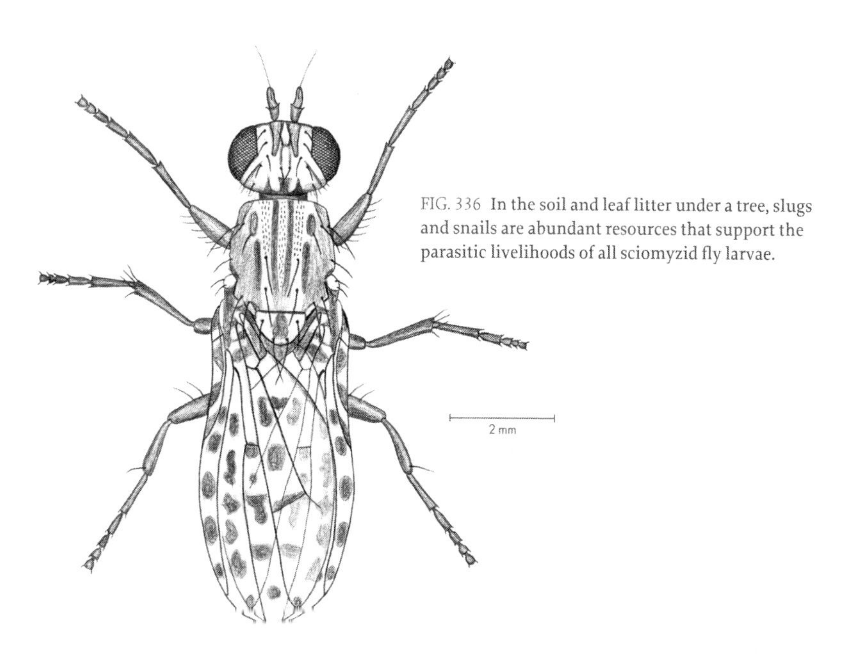

FIG. 336 In the soil and leaf litter under a tree, slugs and snails are abundant resources that support the parasitic livelihoods of all sciomyzid fly larvae.

— EARTHWORMS AND POTWORMS, PHYLUM ANNELIDA, CLASS OLIGOCHAETA (*oligo* = few; *chaeta* = bristle) (8,000 species earthworms; 200 species NA; 650 species potworms; 200 species NA).

Plenty of birds, toads, and mammals as well as arthropods such as centipedes and beetles dine on earthworms. Among all the species of insect parasitoids, the only ones that have chosen earthworms as their hosts, the cluster flies, are members of one fly family.

— BLOW FLIES, BLUEBOTTLE FLIES, GREENBOTTLE FLIES, FAMILY CALLIPHORIDAE (*calli* = beautiful; *phori* = carry) (1,500 species; 90 species NA; larvae 6–20 mm).

These colorful flies, often decked out in metallic greens and blues, show great versatility in the feeding habits of their larvae: the larvae can be predators or detritivores; some feed on decaying animal carcasses; some prey on grasshopper eggs; a few are parasites of vertebrates; and cluster fly larvae are parasitoids of earthworms (fig. 337).

Members of the family that lack the shiny luster of the more glamorous bluebottles and greenbottles are the cluster flies. This name given to the adult flies arose from their habit of passing the cold winter months clustering or congregating in hollow trees, under bark, or in buildings. Their previous summers, however, are spent growing up as larval parasitoids of earthworms. In anticipation of surviving the cold months in adult form, the larval flies stock up copious fat reserves from their earthworm meals. Homeowners who swat one of these overwintering flies in their attics or garages discover that the fly's fat reserves make it very greasy and leave behind a lingering odor reminiscent of buckwheat honey.

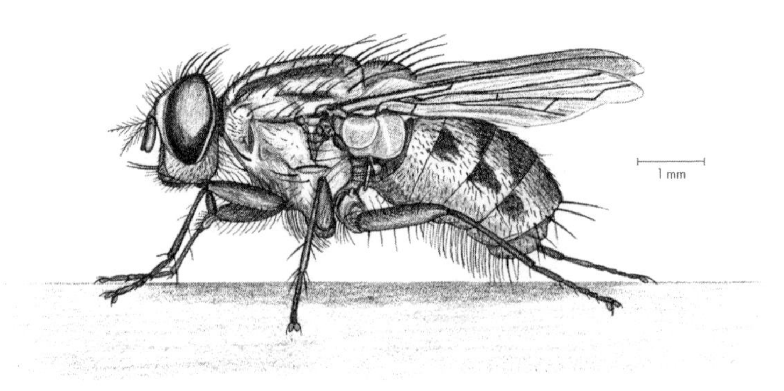

FIG. 337 Parasitic larvae of a few species of calliphorid flies choose earthworms as hosts.

ROOT FEEDERS: THEIR PARASITES AND PREDATORS

Leaves may come and go with the seasons, but roots provide shelter and nutrients throughout the year. In a tree's underground world, roots secrete sugars and hormones from their sap into the soil to attract beneficent microbes such as nodule-forming bacteria and mycorrhizae (Haney and Ausubel 2015; Erb et al. 2012). Whatever tree roots exude into their surrounding environment undoubtedly attracts larger organisms as well that can benefit from associating with tree roots. These organisms chew on roots, suck sap from roots, or offer meals to the numerous insects that are their parasites and predators.

Many herbivorous beetle larvae lead lives in the wood of trees or on their green leaves. But many others have taken up lives underground feeding on roots, often taking several years to complete their growth. Familiar beetles of the treetops begin their lives as root-chomping larvae under the tree. These beetles are members of four highly populated families: weevils; chrysomelid or leaf beetles; scarab beetles; and click beetles. With species numbering respectively 83,000, 50,000, 30,000, and 10,000, it is not surprising that the diverse beetles of these families have found homes and numerous occupations both belowground and aboveground. Here are four larval representatives of these families.

Like the legless larvae of flies, weevil larvae are among the very few beetle larvae that navigate without six legs (fig. 338). Although most soft-bodied chrysomelid larvae live among the leaves aboveground, the few that dwell in the soil fashion protective cases for themselves from their shed skins and droppings (fig. 339). The larva can withdraw into its hard case and plug the entrance with its hard, flat head capsule. The soft, vulnerable larva lies safe inside its fortified case.

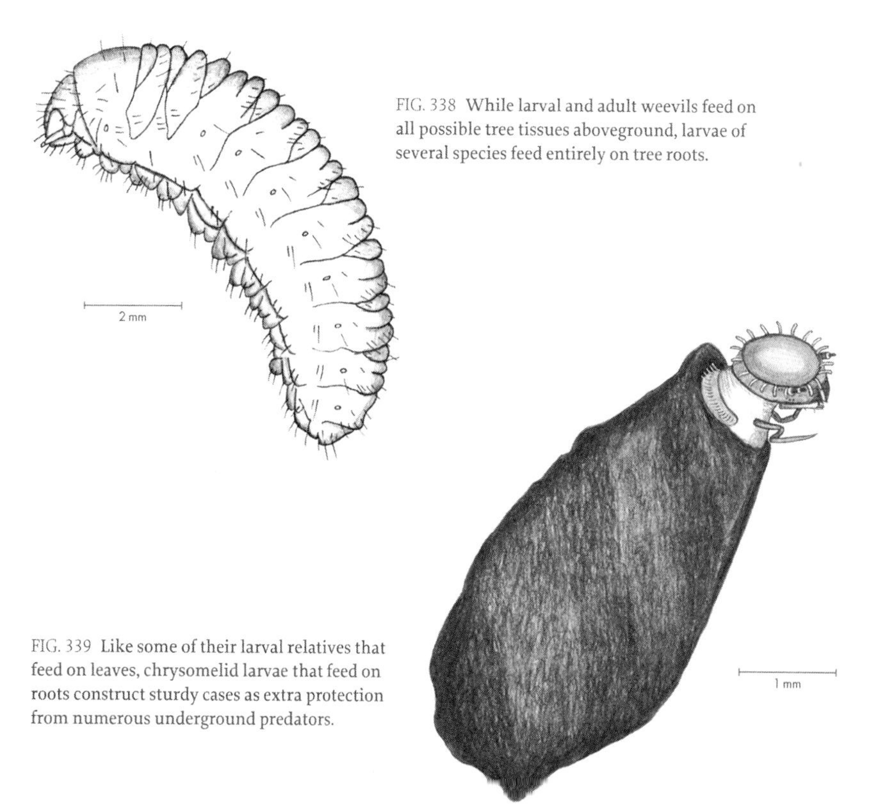

FIG. 338 While larval and adult weevils feed on all possible tree tissues aboveground, larvae of several species feed entirely on tree roots.

FIG. 339 Like some of their larval relatives that feed on leaves, chrysomelid larvae that feed on roots construct sturdy cases as extra protection from numerous underground predators.

The distinctive C-shaped, robust larvae of scarab beetles move around tree roots with a languid, lumbering gait (fig. 340; RootScape, 30). In contrast to these sluggish beetle grubs, the sleek, straight larvae of click beetles are hard, wiry wireworms (fig. 341; RootScape, 26). While the larvae of some species of click beetles gnaw on tree roots, the damage these root-feeding larvae inflict is mitigated by other species of click beetles whose larvae are their predators.

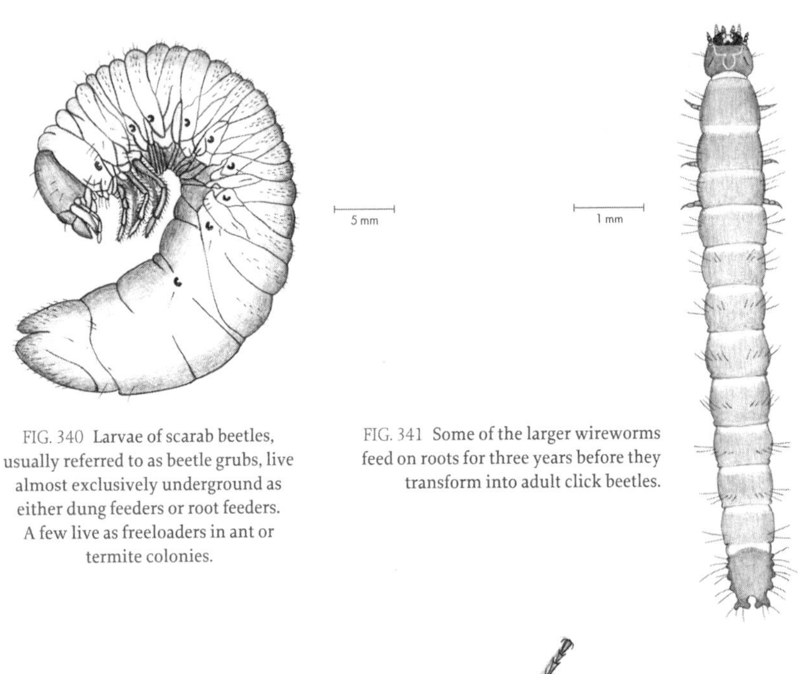

FIG. 340 Larvae of scarab beetles, usually referred to as beetle grubs, live almost exclusively underground as either dung feeders or root feeders. A few live as freeloaders in ant or termite colonies.

FIG. 341 Some of the larger wireworms feed on roots for three years before they transform into adult click beetles.

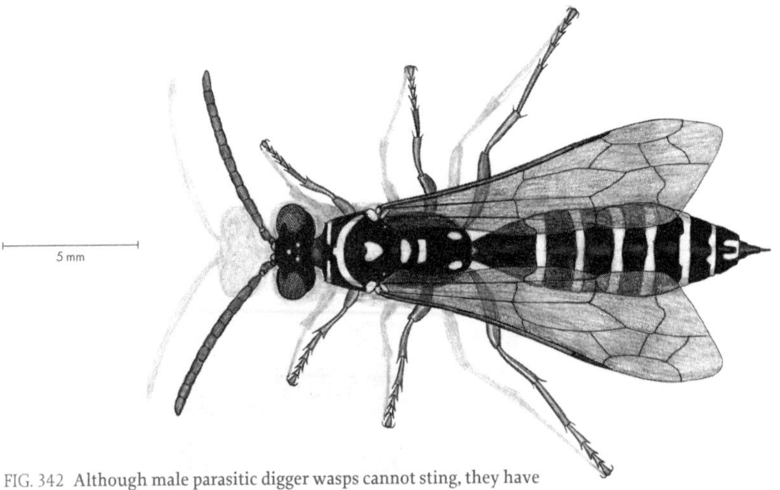

FIG. 342 Although male parasitic digger wasps cannot sting, they have a projection from the end of their abdomen that mimics a stinger, and they are decked out in the same bright warning colors of yellow and black that protect stinging female wasps from hostile encounters.

— PARASITIC DIGGER WASPS, FAMILY TIPHIIDAE *tipha* = kind of insect) (1,500 species; 200 species NA; 3–35 mm) and FAMILY SCOLIIDAE (*scoli* = curved, probably referring to their corrugated wing tips) (560 species; 25 species NA; 10–50 mm).

Subterranean beetle larvae are prime targets for parasitic digger wasps (fig. 342). With her sturdy legs, a mother wasp digs down to locate an underground larval host, paralyzes it with her stinger, and lays an egg on the side of the host. The wasp larva that hatches not only shares the beetle host's underground chamber but also devours its host from the outside in.

Insects that drain root sap face a different set of challenges from those encountered by beetles that chew roots. Root sap, like leaf sap, is deficient in many amino acids and vitamins. Sap-sucking insects have recruited gut microbes to supplement not only their nutrient-deficient diets but also their immune systems and digestive abilities. In the underground they must face predators and parasites that have a singular fondness for them.

— BURROWING BUGS, SUBORDER HETEROPTERA, FAMILY CYDNIDAE (*cydno* = splendid) (770 species; 45 species NA; 2–20 mm).

With spiny legs and a spade-shaped head, these bugs excavate tunnels around roots and feed on root sap (fig. 343). However, these creatures of dark tunnels often inexplicably emerge to congregate at bright lights on autumn evenings. Each digestive tract of a burrowing bug harbors a population of gut microbes that contribute to its health and well-being, and these microbes pass from mother to offspring. Like other subterranean mothers such as millipedes, centipedes, and earwigs that are surrounded by hostile insect-eating fungi, burrowing bugs

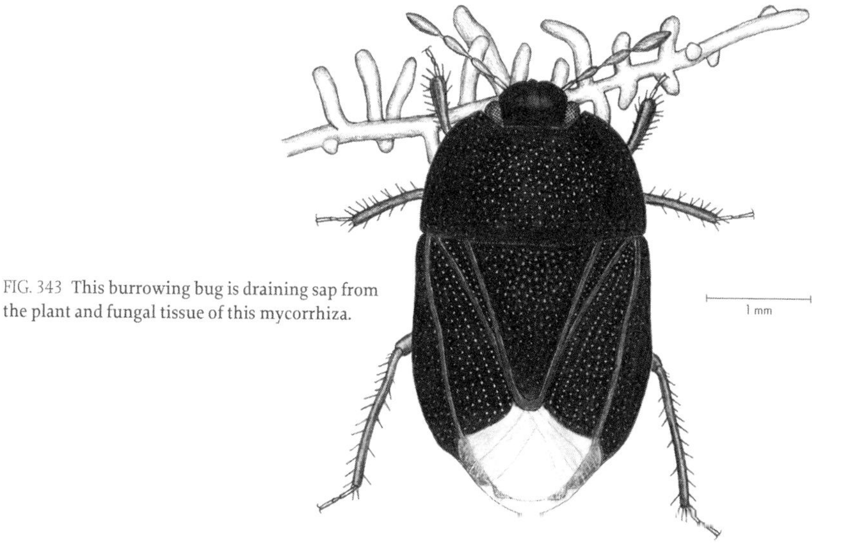

FIG. 343 This burrowing bug is draining sap from the plant and fungal tissue of this mycorrhiza.

watchfully hover over their eggs and newly hatched offspring to provide auspicious beginnings for them. In addition, the newly hatched burrowing bug nymphs fortify themselves with beneficial gut microbes that they pick up from their mother's droppings.

— CICADAS, SUBORDER AUCHENORRHYNCHA, FAMILY CICADIDAE (2,000 species described; 200 species NA; 20–60 mm).

Certain soil insects spend most of their exceptionally long lives surviving exclusively on the sap of roots. For many long years, the nymphs of North America's periodical cicadas inhabit the dark world of tree roots. Three species of periodical cicadas spend 13 years underground, and three other species spend 17 years among tree roots, feeding and growing on the sap that flows through even the tiniest roots. The spring eventually arrives when it is time for each nymph to dig its way to the surface. As the cicadas emerge from the soil, the ground beneath the trees on which they have been feeding is pitted with thousands of emergence holes, each about a half inch across. Each cicada sheds its nymphal skin and spreads the wings that have been forming during all those years underground, finally completing metamorphosis to the adult stage (fig. 344; RootScape, 29).

The colorful adult cicadas with stunning red eyes and orange wings climb nearby tree trunks and cover the leaves and branches in vast numbers. Male cicadas emerge from their nymphal shells shortly before female cicadas and quickly gather in groups, each consisting of as many as several thousand individuals. Each group of males is called a "chorus center"; once all males have assembled, they begin producing their whining and deafening *burr-r-r*. Male cicadas do not have vocal cords to produce their long, loud songs. Instead, they rapidly vibrate a pair of stretched membranes called drums that are found in an inner chamber of the abdomen. The sound produced in a chorus center is clamorous to human ears but is unquestionably appealing to the ears of female cicadas. Females flock to chorus centers to mate and then use sharp ovipositors to insert their eggs beneath the bark of twigs and branches. Each egg contains a packet of essential symbiotic microbes that will sustain the cicada nymph throughout all its years spent underground.

Within a matter of weeks, the millions of boisterous cicadas disappear from the trees as tiny nymphs of a new generation begin hatching from the eggs and falling to the ground. They now begin their long, hidden journeys through the soil for 17, or at least 13, more years. For anyone who has witnessed the sights and sounds of these hordes of cicadas that cover trees for this brief time, the experience is never to be forgotten.

Nymphs of other cicadas share root sap with the nymphs of periodical cicadas; these other nymphs spend at least 4 years but fewer than the 13 or 17 years that periodical cicadas spend underground. These shorter-lived cicadas are often larger than periodical cicadas, and they emerge from the soil in far fewer numbers and later in the summer. They too are handsome insects, but the eyes and wings of these cicadas are not the striking red and orange of periodical cicadas.

In their dark, sheltered worlds underground, one would expect cicada nymphs to lead idyllic lives with few concerns; however, one family of beetles is adept at

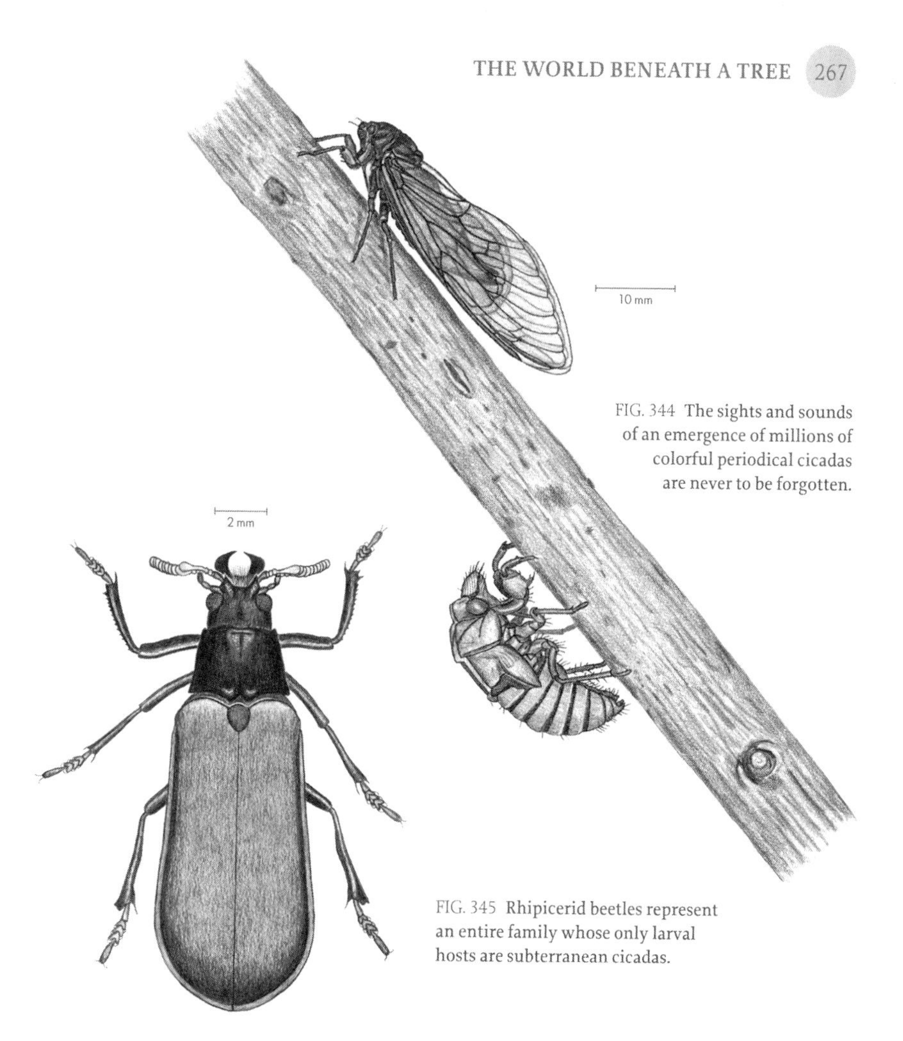

FIG. 344 The sights and sounds of an emergence of millions of colorful periodical cicadas are never to be forgotten.

FIG. 345 Rhipicerid beetles represent an entire family whose only larval hosts are subterranean cicadas.

locating and parasitizing subterranean cicadas. These beetles fill an uncommon profession in a tree's food web.

— CICADA PARASITE BEETLES, FAMILY RHIPICERIDAE (*rhipi* = fan; *cera* = horn) (70 species; 5 species NA; 11–25 mm).

Cicada parasite beetles of the family Rhipiceridae have larvae that lead hidden and mysterious lives underground as parasites of root-feeding cicada nymphs (fig. 345). The origin of the family name is obvious after seeing the ornate antennae of the male beetles. Male beetles of some tropical species have outlandish, branched antennae that fan out like peacock tails. These massive antennae are not only as long as their bodies but also wider. The adult beetles also sport large, sickle-shaped mandibles even though they supposedly do not feed. However, each autumn adults form mating aggregations on tree trunks, so these impressive jaws are probably used to impress male rivals and eligible females during courtship in the aggregation. The mated females then deposit their eggs in cracks and crevices of tree bark.

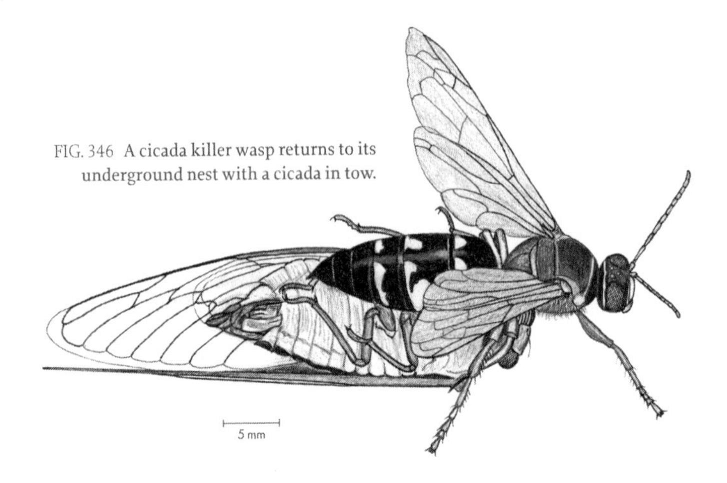

FIG. 346 A cicada killer wasp returns to its underground nest with a cicada in tow.

The newly hatched (first instar) rhipicerid larvae, like those of many insect parasites, must rely on their own initiative to find a host underground. The lively first-instar larvae of certain parasitic insects, like those of many tachinid flies, spider flies (chapter 1), bee flies (chapters 2 and 5), and epipyropid moths (chapter 3), are referred to as planidia and have reputations for wandering far and wide in search of their insect hosts. This beetle larva can move with speed and resolution down a tree trunk and into the soil to find its home on a cicada nymph. Once settled on a cicada host, the lively larva molts to a sluggish larva that lives out its languid subterranean days on the same cicada host.

— CICADA KILLER WASPS, FAMILY CRABRONIDAE (*crabron* = hornet), GENUS *SPHECIUS* (*spheci* = wasp) (21 species; 3 species NA; 15–50 mm).
Cicadas are the sole source of nutrition for the larvae of another insect. One group of digger wasps has capitalized on the abundance and ample size of cicadas by provisioning their underground nurseries with paralyzed cicadas for their meat-eating larvae. The relatively large cicadas provide abundant food for a larval wasp but present a challenge to a hunting wasp in tackling and maneuvering prey that is as large and bulky as the hunter herself (fig. 346).

After spending fall, winter, and spring being raised on cicada meat, adult cicada killer wasps appear aboveground each summer to excavate new burrows for a new generation. They visit flowers for nectar, which supplies them with the energy to capture, paralyze, carry, and bury cicadas for the next generation of cicada killers.

All creatures need energy and nutrients for their survival. Trees capture energy from the sun and share the sun's energy with all their companions. In turn, trees and all their companions depend on mineral nutrients recycled generation after generation by the companions of trees belowground; the decomposers that dwell under each tree make these nutrients available to everyone. Without the photosynthetic contributions of the tree aboveground and the recycling efforts of the decomposers belowground, the rich tapestry of life on a tree encompassing all its diverse companions would not be possible.

OBSERVING FIRSTHAND 7

Nature will bear the closest inspection. She invites us to lay our eye level with her smallest leaf and take an insect view of its plain.
—Henry David Thoreau

LOOKING OVER BARK, LEAVES, TWIGS, AND BRANCHES

Some of a tree's residents truly stand out by virtue of their stunning colors, calling attention to themselves as spiny, nasty tasting—even toxic—and certainly to be avoided. Spotting other creatures is more challenging: caterpillars and walkingsticks that masquerade as twigs, treehoppers that look like thorns, and insects of all shades of green. Not only do the colors of some leaf inhabitants such as katydids and planthoppers perfectly match their surrounding shades of green, but their surface patterns also match the intricate venation patterns of adjacent leaves.

In the dark and sheltered world beneath the bark of dead limbs, you will come across creatures that spend most or all of their lives on the underside of the bark. You will see the tracks that they and their forebears have left as records of their travels. They seem well suited for living in darkness and dampness. With a pocketknife, slowly and carefully pry loose a piece of bark to get some idea of what goes on in this concealed habitat. Why are these creatures shaped as they are? What do they feed on in this sunless realm? What will they become? After you have satisfied your curiosity, show respect for the creatures that live under the bark by gently replacing any bark that you may have removed.

The loss of hollow limbs and hollow trees not only deprives birds and mammals of winter refuges but also leaves certain insects without homes. Hollow limbs and tree trunks are used as winter homes by hornet queens, stink bugs, and ladybird beetles. Moths and butterflies also overwinter as adults in hollow trees. These include the well-known mourning cloaks, question marks, commas, and red admiral butterflies (fig. 268).

Pieces of tree trunks and branches with their bark still intact offer many points of entry to habitats under the bark. Gently peel away the bark from wood that has been left outdoors for a few months. Some creatures like centipedes or rove beetles will dash away, but most beetle and fly larvae will remain ensconced in their cozy chambers under the bark. Stacks of firewood offer ideal habitats for the larvae of many species of beetles and flies and for flat bugs, centipedes, millipedes, mites, springtails, and the fungi that keep them company.

Many animal patterns blend perfectly with tree bark. Since the texture as well as color of bark is distinctive for each tree species, you can often guess on which tree's bark certain animals are more often found. By running their fingers over the ridges and grooves of bark, some people can even identify trees with their eyes closed. To small insects and spiders, these ridges and grooves are part of a landscape in which hiding places abound. Even larger animals such as bats, mice, and flying squirrels find shelter under the shaggy slabs and strips of bark that project from the trunks of shagbark hickory. It is often a landscape where lichens and mosses sprout in scattered patches or cover large stretches of bark. As you look over this landscape, you may find a beetle, mite, or bark louse meandering over hills and valleys of bark while a moth, caterpillar, or fly larva may be nestled among the mosses and lichens.

To collect one of these inhabitants for closer inspection, you want to remove it as gently as possible and then return it gently to its home. A simple device called an aspirator makes this possible (fig. 347). By placing the intake tube next to a small creature and drawing in air from the other tube, you can apply suctioning action that will pick out tiny animals from recesses and retreats on and under the tree.

REARING HIDDEN COMPANIONS OF TREES

A less intrusive approach to exploring is to place pieces of wood with bark in a covered terrarium and then wait to see who emerges days, weeks, months—and for some long-lived larval insects like leopard moths, click beetles, or wood wasps—even years later. From limbs and branches in more advanced stages of decay, a different constellation of insects will come forth. In a large glass terrarium some

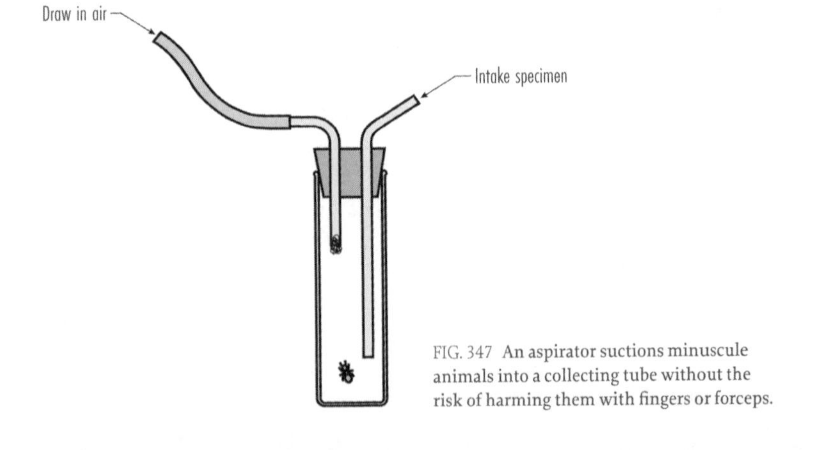

FIG. 347 An aspirator suctions minuscule animals into a collecting tube without the risk of harming them with fingers or forceps.

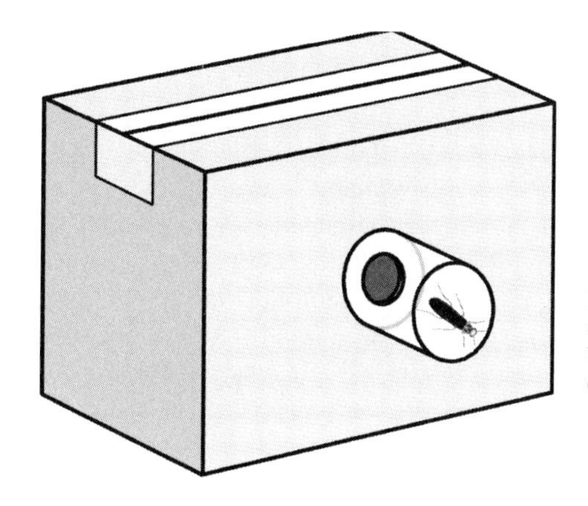

FIG. 348 Insects that emerge in the darkness of a box will most often beeline to the one source of light offered by an attached clear vial.

of the tinier insects might go unnoticed, so one clever method for bringing even the tiniest inhabitants into the limelight is to place pieces of branches in a dark, covered box with only one small opening to which you attach a clear vial or jar (fig. 348. Most beetles or flies newly emerged from their dark chambers in the wood avidly search for the brightest spot on their horizons. The clear vial or jar serves as a welcoming beacon to the insect that has spent so much of its life in darkness.

Every tree has its caterpillars—fuzzy, spiny, bumpy, and smooth—for wherever there are moths and butterflies, there must also be caterpillars. Inchworms may dangle from leaves by silken threads. Some caterpillars may pose as edges of leaves, bird droppings, or twigs, but the bright colors of other caterpillars stand out conspicuously against the green leaves.

A caterpillar can be just as content when placed in a large glass jar as it was on the tree where it was found. As long as a caterpillar always has a fresh supply of leaves from that tree, it will steadily chew away until it can grow no more. When that happens, the time has come for its metamorphosis. The caterpillar of a butterfly forms a chrysalis when it transforms from larva to pupa. Some moth caterpillars spin cocoons and pupate on the leaves where they fed for days or weeks; others wander far from their leafy homes and burrow into the ground to pupate without ever spinning a cocoon. In case one of these wandering caterpillars is living in your glass jar, it is probably a good idea to put two or three inches of soil in the bottom of the jar where it can retreat to molt in privacy.

The pupal stage may last only a few days or as long as several months. If a moth or butterfly does not appear after a month, it is not likely to appear until the coming spring or not at all. It could have been parasitized, and in that case, a parasitic fly or wasp could appear in the rearing jar instead. However, if the caterpillar was feeding on the tree when there were fewer than 12 hours of daylight, the caterpillar or its pupa will have entered a quiescent state known as diapause from which it will not come forth until it has experienced several weeks of winter temperatures.

During the winter months many cocoons are easy to spot on branches and twigs. Leave the cocoon, chrysalis, or naked pupa outside in some sheltered spot during the cold months of winter. When the first warm days of spring arrive, bring the jar inside again and wait a few more days or weeks while the insect completes its transformation. Be sure to place an upright stick or towel in the jar. The newly emerged moth or butterfly will need to climb up this stick or towel to spread and dry its wings properly.

Each September and October I gather hundreds of acorns for tree planting in sturdy cloth bags. As I finish planting them in November and come to the bottom of each bag, the last few brown acorns lie among chubby, wriggling white larvae. After reaching their full size on the nutrients stored in acorns, these weevil larvae and caterpillars have emerged from those acorns with exit holes. Under the oak tree where they originally fell, they would have crept into the soil to spend the winter sheltered beneath the forest litter. By placing these larvae and the debris they helped create in the bottom of a collecting bag in an enclosed container with a little soil and leaf litter, you can witness what occurs under oak trees every year between autumn and summer.

Galls can be collected from a tree or from the ground beneath a tree in any season of the year. If a gall is still attached to a green twig or leaf, cut the end of the twig or leaf and wrap it with a wet paper towel before placing it in a jar with a loose-fitting lid. You can place the twig or leaf in a vial of water with a cotton plug and then put the vial and its contents in a jar (fig. 349). As long as the green twig or leaf is kept wet, the inhabitants of the gall should continue developing inside. Galls collected in autumn or winter can be placed in a jar and left outdoors or in an unheated room. The creatures that emerge from the galls may include insects other than the gall maker itself. A tiny wasp that has parasitized the gall maker or another insect known as an inquiline, or squatter, may have moved into the spacious chamber offered by the gall to share its nutrients and shelter with the gall maker.

Most insects leave their spring and summer homes in leaf mines and leaf rolls to pupate on bark or in the soil, but some hang around in their cozy mines and rolls throughout the winter months. A notable feature of leaf mines is their translucency. When held up to the light, the silhouette of the miner can usually be seen beneath the clear epidermal layers on the top and bottom of the mine (fig. 350). To discover who resides in a leaf roll or leaf mine—or, in some cases, who has parasitized the miner or roller—you can place a well-hydrated leaf or small branch on which the mine or roll is found in an observation jar, as shown in fig. 349.

ATTRACTING HIDDEN COMPANIONS WITH LIGHTS AND BAITS

All sorts of lights on warm evenings—natural, incandescent, fluorescent, or LED (light-emitting diode)—will lure insects from their sanctums among leaves, bark, roots, and branches. Although fluorescent or incandescent black lights that emit ultraviolet light are probably the most enticing to insects, LED lanterns are the most portable and probably the safest and simplest to obtain and use. Moonless nights

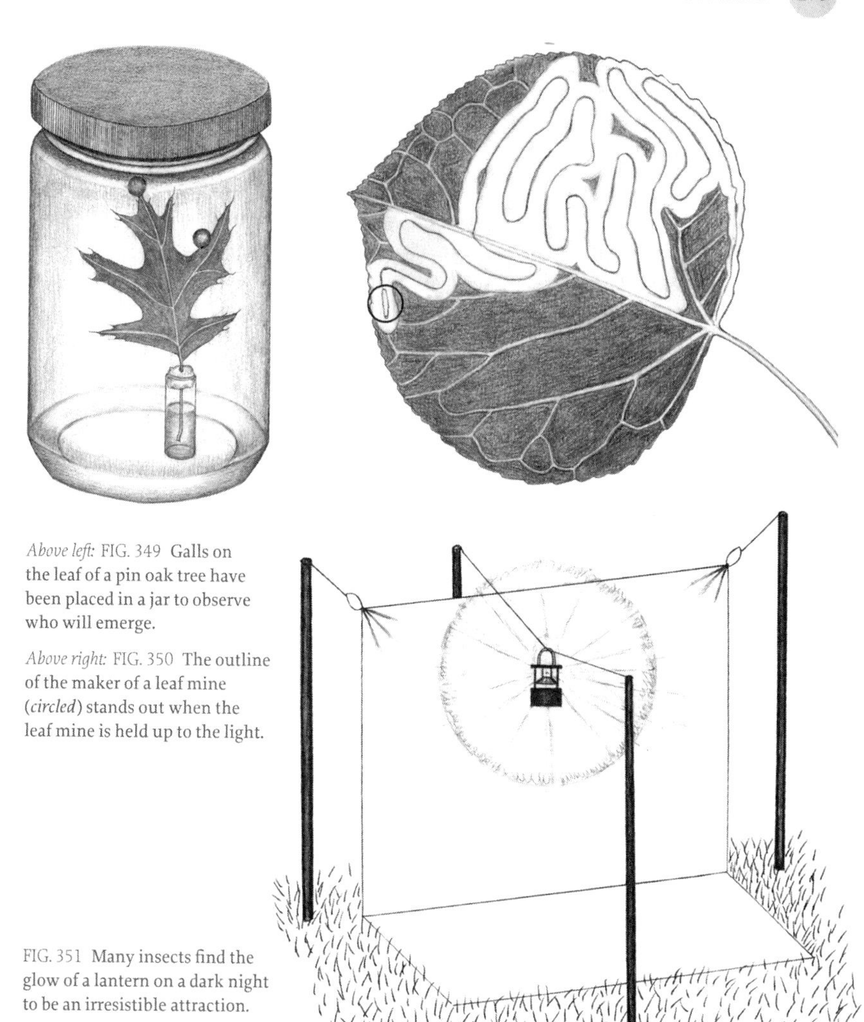

Above left: FIG. 349 Galls on the leaf of a pin oak tree have been placed in a jar to observe who will emerge.

Above right: FIG. 350 The outline of the maker of a leaf mine (*circled*) stands out when the leaf mine is held up to the light.

FIG. 351 Many insects find the glow of a lantern on a dark night to be an irresistible attraction.

are the best for attracting insects to lights, since the light from the light source does not have to compete with the appeal of moonlight. The light source will draw in the insects, but they will also need a landing strip and a stage on which to perform.

Stretch a white sheet vertically between two upright poles or trees. Spread another white sheet horizontally over the ground to ensure that insects that fall or drop to the ground below the light will not go unnoticed (fig. 351). The light source can be suspended between two poles on a line perpendicular to the surface of the sheet and about a foot away from it. Setting up this arrangement under a tree in your backyard or in a forest clearing should prompt a number of hidden creatures to make appearances on the white sheet.

The scent of rotting fruit or rotting fungi is irresistible to some residents who live under bark, in damp soil, or on tree branches. Insects that alight on mushrooms or overripe fruit placed at the bottom of a deep jar will hang around to savor the abundant flavor. Fermented, overripe fruit and mushrooms contain enough alcohol to inebriate birds and mammals that consume them. The nervous systems and neurotransmitters of smaller insects are just as susceptible to being inebriated by fermented fruit. A few drops of alcohol-infused bait are probably sufficient to sedate, relax, and calm their usual fears and suspicions of human observers. The besotted insects that come to the baits are more inclined to tolerate your snooping on their activities.

Sugaring for moths, beetles, and other insects is an old tradition among naturalists. In his classic volume *The Moth Book*, W. J. Holland (1903) describes in melodramatic prose the anticipation and thrills offered by a night of sugaring the trunks of trees along a winding woodland path. All sugaring recipes seem to contain the five basic ingredients of molasses, overripe bananas, brown sugar, rum, and stale beer; however, the basic recipe has been refined and modified over the years to cater to the tastes of particular insects. The main thing to remember is to concoct a mixture thick and fluid enough to spread on tree trunks with a paintbrush. To improve its consistency and enhance its intoxicating odor, a variety of rotting and fermenting fruits, brewer's yeast, and even stale sweet sodas can be added to the mix. Addition of turpentine seems to appeal to bark beetles and other wood-boring beetles. As Frank Lutz observed in his *Field Book of Insects* (1948), "There are about as many recipes for making the sugar mixture as there are for 'mother's biscuits.'"

Using red lights or white lights with red filters is the best way to observe insects and nocturnal animals at night. While insects can see blue and ultraviolet, they cannot see red light at the opposite end of the spectrum. Likewise, mammals such as flying squirrels and white-footed mice go about their business unperturbed when viewed with red lights.

URGING SOME OF THE SMALLEST COMPANIONS OF TREES TO LEAVE THEIR HIDEAWAYS

A simple funnel can provide an even better window on the hidden worlds of soil and tree bark as well as the mosses and lichens that cover bark. The Berlese funnel was named after the Italian scientist who devised it for his study of soil creatures. The funnel enabled him to cleanly separate the living, mobile creatures of soil from the nonmobile, inanimate components of soil. The funnel works equally well in revealing who lives not only in soil and leaf litter but also on tree bark, mosses, and lichens. When samples wrapped in cheesecloth are placed in the funnel, the cool, moist, dark world of the creatures is disturbed by light and heat. Cheesecloth retains soil, litter, and bark but allows living creatures to retreat from the bright light. They head toward the base of the funnel and drop into the collecting dish. You can also place a 40- or 60-watt light bulb directly over the cheesecloth bag to accelerate the escape of creatures from the sample.

Researchers use metal funnels supported on four sturdy metal legs or on a ring stand; however, funnels made from recyclable plastic jugs or large soft drink containers are equally effective in coaxing hidden creatures from soil or bark samples. They are certainly economical and simple to construct (fig. 352).

Homemade Berlese funnels can easily be modified into an equally useful apparatus—the Baermann funnel—for extracting the even smaller creatures that inhabit the moist surface films on leaf litter, mosses, and soil particles.

To convert a Berlese funnel to a Baermann funnel, leave the screw cap of the jug or container in place. Then add tap water from above to completely submerge the sample wrapped in cheesecloth (fig. 353, right). The water-loving creatures will slowly relinquish their hold on the inanimate particles of vegetation or soil and settle to the bottom of the funnel. An hour or two after the water is added, the creatures will have concentrated in the neck of the funnel. When most of the water is removed from the top of the funnel with a pipette or syringe, the water that remains in the neck of the funnel will contain an assemblage of microscopic organisms such as nematodes, tardigrades, rotifers, and protozoa (fig. 193). As the cap is unscrewed, the water and creatures will pour into the collecting dish.

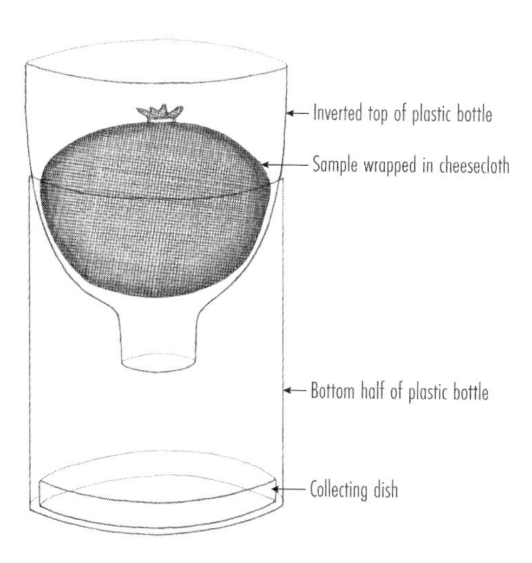

Left: FIG. 352 A simply constructed Berlese funnel will coax small creatures from their hideaways in leaf litter, bark, or moss on tree trunks.

Below: FIG. 353 A Baermann funnel reveals the microscopic creatures such as nematodes, rotifers, and tardigrades that inhabit the moist surfaces of bark, lichens, and roots.

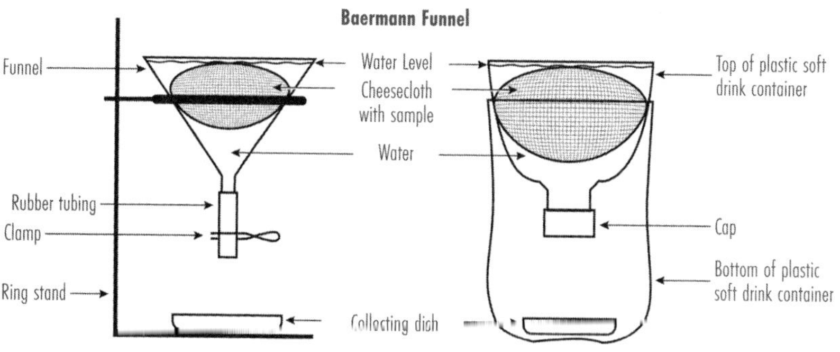

A fancier Baermann funnel can be assembled from a large glass funnel to which a short stretch of clamped rubber tubing has been added at its tapered end. Once the funnel is supported by a metal ring stand, water is added to the funnel to fill its clamped neck. The sample wrapped in cheesecloth is lowered into the wide portion of the funnel. More water is then added to completely submerge the sample. Over the next hour or two the creatures inhabiting the sample will leave the water surface films and settle to the neck of the funnel. When the clamp on the rubber tubing is opened, water containing the concentration of microbial life will pour out into a collecting dish (fig. 353, left).

Collecting dishes for Berlese funnels can come in many forms. My favorite is a jar or petri dish whose bottom is lined with a layer of plaster of Paris, anywhere from a quarter of an inch to an inch thick (fig. 354). When this plaster layer is saturated with water, the dish retains moisture for many hours even when uncovered. The damp, moist environment of the dish creates a hospitable habitat for moisture-loving creatures. After a collecting session, the dish can be covered with its clear lid. The creatures can be observed under magnification before being returned to their original home.

Artful deceivers and well-hidden occupants of tree branches certainly do not draw attention to themselves, and they sometimes need a little coaxing to reveal their whereabouts. A sudden jolt to a branch or cluster of leaves dislodges insect, spider, and mite residents of leaves and twigs. On the white background of a sheet or an enamel pan, these residents usually stand out clearly, especially as they begin to scurry across the surface. You will most likely dislodge a number of creatures you did not expect to see.

With one hand holding a stick or rod, firmly strike a tree branch. With the other hand, hold a white umbrella, a white enamel pan, or a spread white sheet to collect whatever creatures are dislodged from the branch (fig. 355). Catch some with an aspirator or an inverted jar before they scurry off. Others, such as caterpillars, move slowly and linger for some time as they are leisurely admired.

FIG. 354 Collecting dishes and observation jars with a bottom layer of plaster of Paris provide a temporary home for Berlese funnel refugees..

FIG. 355 The hidden residents of a tree branch can be encouraged to drop into a white umbrella by firmly striking the branch with a pole or cane.

WHAT GOES ON UNDERGROUND?

Hidden beneath a tree are the countless connections that the fine roots establish with underground fungi. The best way to observe this communication network among tree roots and fungi is to reach into the top 10 cm of leaf litter under an oak or a beech tree for a sample of the fine roots. When viewed under a magnifier or stereomicroscope, the tips of the fine roots do not taper to even finer tips but instead have stubby ends that are encompassed by a sheath of fungal filaments that have transformed a simple tree root into a mycorrhizal root (fig. 356; RootScape, 25). The fungal filaments radiate out into the surrounding leaf litter and soil, vastly increasing the root's surface area for absorption.

Many underground inhabitants must find a welcoming home among these worlds of fine roots and their fungal partners. Some of these creatures may be seen

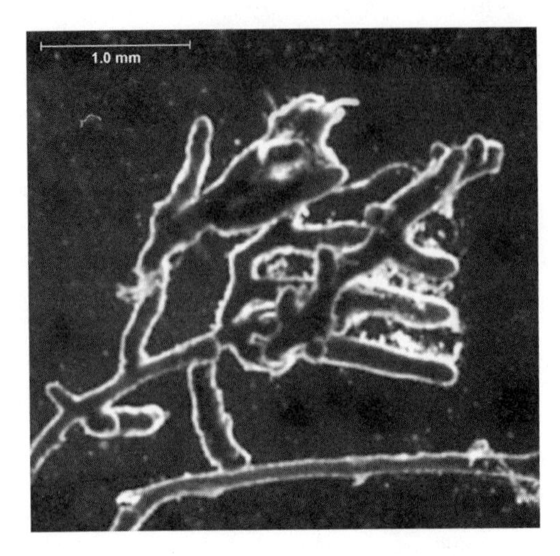

FIG. 356 The hyphal filaments of mycorrhizal fungi not only envelop the fine surface roots of an oak tree but also extend between root cells and throughout the surrounding soil.

wandering about under the microscope; even more will show up if the mycorrhizal roots are then placed in a Berlese funnel or Baermann funnel and encouraged to drop into the collecting dish below the funnel. All manner of arthropods depend on the shelter and food web provided by roots and their fungi.

TAKING A CLOSER LOOK WITH MICROSCOPES

Once they are exposed, most of the hidden companions of trees are visible to the unaided eye; however, their beauty and exceptional features can be fully appreciated only when magnification brings out details of their forms. No piece of equipment can provide more entertaining and informative views of these tiny organisms than a stereomicroscope. Many models of these microscopes are available and reasonably priced. They allow the observer not only to view specimens in three dimensions but also to quickly zoom in (50×) and zoom out (10×) on a particular structure once it is in focus (fig. 357). For all the enjoyment they bring to the passionate observer, they are well worth the investment. Most schools and nature centers have one or more of these microscopes.

Even more reasonably priced and more compact are digital microscopes that are used in conjunction with laptop computers. A compact laptop and digital microscope can be packed up for fieldwork and can provide better views of many structures than conventional handheld magnifiers.

The stereomicroscope and digital microscope reveal the complex topography on the lower surfaces of leaves. Here natural depressions and copses of leaf hairs (trichomes) form microscopic forests where tiny arthropods live their lives. In these leaf domatia, they lay their eggs, shed their exoskeletons at each molt, take refuge from predators, and ambush their prey (fig. 358). The leaf provides a welcoming habitat.

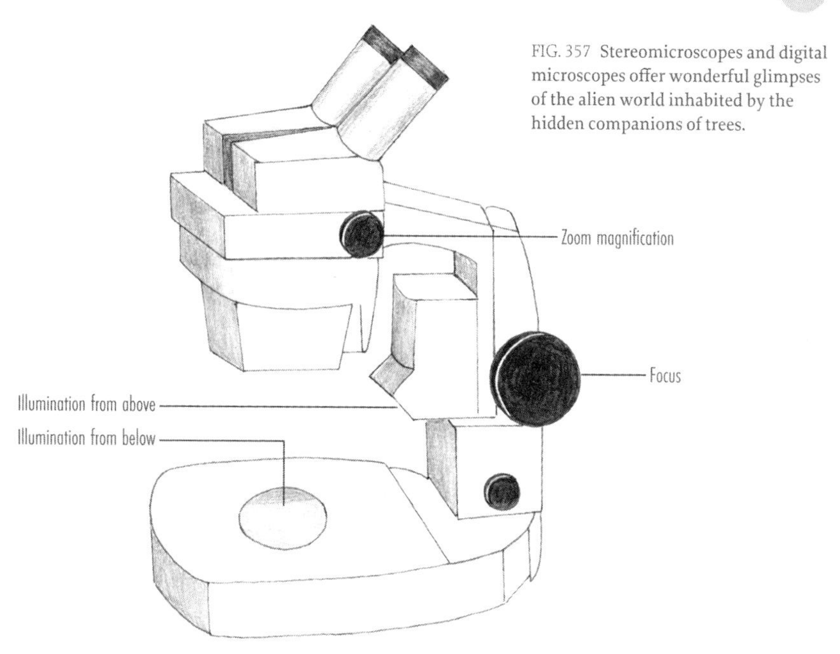

FIG. 357 Stereomicroscopes and digital microscopes offer wonderful glimpses of the alien world inhabited by the hidden companions of trees.

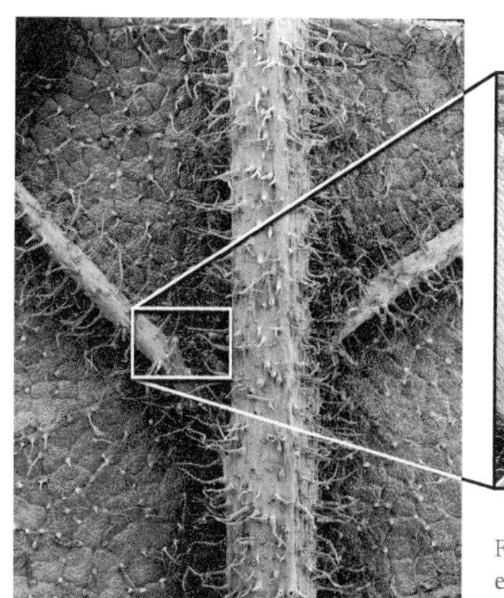

FIG. 358 The domatia on the lower surface of this slippery elm leaf provided a haven for a tiny arthropod's three eggs (*arrows*); the cap or operculum for one of the hatched eggs is marked with the arrowhead.

Whenever you can, return living creatures to their homes on or under a tree. Each creature is part of an intricate web of connections. Removing any member of the web distorts it in unforeseen ways, as John Muir so well expressed:

> *When we try to pick out anything all by itself,*
> *we find it hitched to everything else in the Universe.*

EPILOGUE

The trees in Joyce Kilmer Memorial Forest have endured and thrived in the company of innumerable animal and microbial allies. Over the years, their leaves have been chewed, their sap has been drained, and their branches have fallen, but the trees continue to stand proudly and majestically. All companions of trees contribute to the grandeur of the forest, even the leaf chewers, sap suckers, and wood borers. They not only provide sustenance to many other companions of trees but also eventually return nutrients they borrowed from the trees when at last they settle to the earth and their remains are recycled. The trees' alliances with microbes and animals from treetops to root tips remain strong, incalculable, and mutually supportive. The trees are telling us by their example what the poet Wordsworth espoused in few words: "Let Nature be your teacher."

Other observant naturalists have admonished us for our folly of attempting to counter the lessons that Nature presents. E. B. White observed that humans should "spend less time proving that [they] can outwit Nature and more time tasting her sweetness and respecting her seniority." Aldo Leopold lamented the challenges to Nature's seniority: "The last word in ignorance is the man who says of an animal or plant, 'What good is it?' If the biota, in the course of aeons, has built something we like but do not understand, then who but a fool would discard seemingly useless parts? To keep every cog and wheel is the first precaution of intelligent tinkering." We have discovered more cogs and wheels in the companions of trees than we ever imagined. Saving these "endless forms" to carry out their endless tasks will impart harmony and integrity to the "tangled banks" of life in our agricultural fields, urban landscapes, prairies, and forests. Thoreau summed up the timeless lessons in one prophetic sentence: "In wildness is the preservation of the world."

Wildness and wild places offer lessons to our ailing rural and urban worlds. The early advocates of organic agriculture realized the importance of following Nature's examples. Lady Eve Balfour wrote about the checks and balances of nature in *The Living Soil* (1943): "Under natural conditions such as obtain in virgin forest or prairie, a constant supply of raw material is provided by the native flora and fauna which live and die, so to speak, *in situ*. Their residues and remains are continually converted by the decomposers into humus from which fresh life can spring. A perfect balance between growth and decay is established, and the fertility of the soil is permanently maintained." In *Teaming with Microbes: A Gardener's Guide to the Soil Food Web* (2006), Jeff Lowenfels describes how rewarding gardening with Nature as a partner can be. "No one ever fertilized an old growth forest ... it operates just fine without interference from man-made fertilizers, herbicides, and pesticides." The contemporary practitioners of regenerative agriculture are discovering how they can create a variety of enticing habitats where all manner of benevolent companions—pollinators, predators, and parasites—are welcome. Let Nature work unencumbered by human contrivances such as pesticides and synthetic fertilizers.

We continue to find ways to improve our relationship with Nature by preserving and restoring welcoming habitats for the company that trees keep. In agricultural and urban landscapes, mycorrhizal inoculations, composting, elimination of tillage and pesticides, addition of cover crops, pollinator strips, hedgerows, and windbreaks have all been steps in preserving and restoring the wild features of our natural world. In wild landscapes, frequent low-intensity, controlled fires create openings in forest canopies. Our imitation of Nature's once natural conditions enhances not only the diversity of plants and microbes but also the appeal of the habitat to a greater variety of mammals, birds, amphibians, reptiles, insects, and other invertebrates. Ironically, the unfathomable complexity of Nature offers lessons in how working with Nature and following Nature's example can bring the simplicity and economy that we have so futilely sought to impose in our wars against Nature. Working in partnership with the company that trees keep promises to maintain the enduring alliances that enhance the lives of trees. In this way, we honor the hamadryad spirits of trees, so revered by our ancestors, and we delight in the discoveries that come our way.

REFERENCES

General

Arnett RH (1971). *The Beetles of North America.* American Entomological Institute, Ann Arbor, MI.

Balfour EB (1943). *The Living Soil.* Faber & Faber, London.

Beadle D, Leckie S (2012). *Peterson Field Guide to Moths.* Houghton Mifflin Harcourt, Boston.

Borror DJ, White RE (1970). *A Field Guide to the Insects of America North of Mexico.* Houghton Mifflin, Boston.

Eaton ER, Kaufman K (2007). *Kaufman Field Guide to Insects of North America.* Houghton Mifflin, Boston.

Evans AV (2014). *Beetles of Eastern North America.* Princeton University Press, Princeton, NJ.

Fuller HJ, Tippo O (1954). *College Botany.* Henry Holt, New York.

Gullan PJ, Cranston PS (2014). *The Insects: An Outline of Entomology.* Wiley-Blackwell, Hoboken, NJ.

Holland WJ (1903). *The Moth Book. A Guide to the Moths of North America.* Doubleday, Page, New York.

Howell WM, Jenkins RL (2004). *Spiders of the Eastern United States.* Pearson Education, Boston.

Jeffords MR, Post SL, Wiker JR (2014). *Butterflies of Illinois: A Field Guide.* Illinois Natural History Survey Manual 14. Champaign, IL.

Kendrick B (1992). *The Fifth Kingdom.* Focus Information Group, Newburyport, MA.

Krantz GW, Walter DE (2009). *A Manual of Acarology.* 3rd ed. Texas Tech University Press, Lubbock.

Kricher JC, Morrison G (1988). A Field Guide to Eastern Forests of North America. Houghton-Mifflin, Boston.

Lowenfels J (2006). *Teaming with Microbes: A Gardener's Guide to the Soil Food Web.* Timber Press, Portland, OR.

Lutz, FE (1948). *Field Book of Insects.* G. P. Putnam's Sons, New York.

Raven PH, Evert RF, Eichhorn SE (2012). *Biology of Plants.* 8th ed. W. H. Freeman, New York.

Stehr FW (1991). *Immature Insects.* Kendall/Hunt, Dubuque, IA.

Tallamy DW (2021). *The Nature of Oaks.* Timber Press, Portland, OR.

Wagner DL (2005). *Caterpillars of Eastern North America.* Princeton University Press, Princeton, NJ.

Walter DE, Proctor HC (1999). *Mites: Ecology, Evolution, and Behaviour.* CABI, New York.

Wohlleben P (2016). *The Hidden Life of Trees.* Greystone Books, Vancouver/Berkeley.

Chapter 1. The Many Forms and Functions of a Tree's Companions

Ananthakrishnan TN (1999). Behavioural dynamics in the biological control of insects: role of infochemicals. Curr Sci 77: 33–37.

Andreas PA, Kisiala A, Emery RJN, De Clerck-Floate R, Tooker JF, Price PW, Miller DG, Chen M-S, Connor EF (2020). Cytokinins are abundant and widespread among insect species. Plants 9(2): 208; https//doi.org/10.3390/plants9020208.

Arnold AE, Mejía LC, Kyllo D, Rojas EI, Maynard Z, Robbins N, Herre EA (2003). Fungal endophytes limit pathogen damage in a tropical tree. Proc Natl Acad Sci USA 100: 15649–15654.

Askew RA (1971). *Parasitic Insects.* Heinemann Educational Books, London.

Bari R, Jones JDG (2009). Role of plant hormones in plant defence responses. Plant Mol Biol 69: 473–488.

Bastias DA, Martinez-Ghersa MA, Ballaré CL, Gundel PE (2017). Epichloë fungal endophytes and plant defenses: not just alkaloids. Trends Plant Sci 22: 939–948.

Behie SW, Moreira CC, Sementchoukova I, Barelli L, Zelisko PM, Bidochka MJ (2017). Carbon translocation from a plant to an insect-pathogenic endophytic fungus. Nature Commun 8: 14245.

Berg G, Köberl M, Rybakova D, Müller H, Grosch R, Smalia K (2017). Plant microbial diversity is suggested as the key to future biocontrol and health trends. FEMS Microbiol Ecol 93; https://doi.org/10.1093/femsec/fix050.

Beston H (1928). *The Outermost House.* Henry Holt, New York.

Bonfante P, Anca I-A (2009). Plants, mycorrhizal fungi, and bacteria: a network of interactions. Annu Rev Microbiol 63: 363–383; https://doi.org//10.1146/annurev.micro.091208.073504.

Chanclud E, Morel JB (2016). Plant hormones: a fungal point of view. Mol Plant Pathol 17: 1289–1297.

Choi J, Choi D, Lee S, Ryu C-M, Hwang I (2011). Cytokinins and plant immunity: old foes or new friends? Trends Plant Sci 16: 388–394.

Conn CE, Bythell-Douglas R, Neumann D, Yoshida S, Whittington B, Westwood JH, Shirasu K, Bond CS, Dyer KA, Nelson DC (2015). Convergent evolution of strigolactone perception enabled host detection in parasitic plants. Science 349: 540–543.

Dermastia M (2019). Plant hormones in phytoplasma infections. Front Plant Sci; https://doi.org/10.3389/fpls.2019.00477.

Dudareva N, Pichersky E, Gershenzon J (2004). Biochemistry of plant volatiles. Plant Physiol 135: 1893–1902.

Erb M, Meldau S, Howe GA (2012). Role of phytohormones in insect-specific plant reactions. Trends Plant Sci 17: 250–271.

Etminani F, Harighi B (2018). Isolation and identification of endophytic bacteria with plant growth promoting activity and biocontrol potential from wild pistachio trees. Plant Pathol J 34: 208–217.

Feener, DH, Brown BV (1997). Diptera as parasitoids. Annu Rev Entomol 42: 73–97.

Feldman TS, O'Brien HE, Arnold AE (2008). Moths that vector a plant pathogen also transport endophytic fungi and mycoparasitic antagonists. Microbial Ecol 56: 742–750.

Forbes AA, Bagley RK, Beer MA, Hippee AC, Widmayer HA (2018). Quantifying the unquantifiable: why Hymenoptera—not Coleoptera—is the most speciose animal order. BMC Ecology; https://doi.org/10.1186/s12898-018-0176-x.

Fürstenberg-Hägg J, Zagrobelny M, Bak S (2013). Plant defense against insect herbivores. Int J Mol Sci 14: 10242–10297.

Gan H, Churchill ACL, Wickings K (2017). Invisible but consequential: root endophytic fungi have variable effects on belowground plant-insect interactions. Ecosphere 8(3); https://doi.org/10.1002/ecs2.1710.

Giron D, Huguet E, Stone GN, Body M (2016). Insect-induced effects on plants and possible effectors used by galling and leaf-mining insects to manipulate their host plants. J Insect Physiol 84: 70–89.

Griffin EA, Carson WP (2018). Tree endophytes: cryptic drivers of tropical forest diversity. In: Pirttilä A, Frank A (eds), *Endophytes of Forest Trees*. Forestry Sciences, vol. 86. Springer, Cham.

Hander T, Fernández-Fernández ÁD, Kumpf RP, Willems P, Schatowitz H, Rombaut D, Staes A, Nolf J, Pottie R, Yao P, et al. (2019). Damage on plants activates Ca^{2+}-dependent metacaspases for release of immunomodulatory peptides. Science 363: 1301.

Haney CH, Ausubel FM (2015). Plant microbiome blueprints: a plant defense hormone shapes the root microbiome. Science 349: 788–789.

Hardoim PR, van Overbeek LS, Berg G, Pirttilä AM, Compant S, Campisano A, Döring M, Sessitsch A (2015). The hidden world within plants: ecological and evolutionary considerations for defining functioning of microbial endophytes. Microbiol Mol Biol Rev 79: 293–320.

Hashem A, Abd_Allah EF, Alqarawi AA, Al-Huqail AA, Wirth S, Egamberdieva, D (2016). The interaction between arbuscular mycorrhizal fungi and endophytic bacteria enhances plant growth of *Acacia gerrardii* under salt stress. Front Microbiol 7: 1089; https://doi.org/10.3389/fmicb.2016.01089.

Heinrich B (2019). Summer leaf-fall. Nat Hist 127: 9–11.

Heinrich B (2021). Living with porcupines: peaceful but challenging. Nat Hist 129: 9–11.

Heinrich B, Collins SL (1983). Caterpillar leaf damage, and the game of hide-and-seek with birds. Ecology 64: 592–602.

Jung J, Kim J-S, Taffner J, Berg G, Ryu C-M (2020). Archaea, tiny helpers of land plants. Comput Struct Biotechnol J 18: 2494–2500.

Jung SC, Martinez-Medina A, Lopez-Raez JA, Pozo MJ (2012). Mycorrhiza-induced resistance and priming of plant defenses. J Chem Ecol 38: 651–664.

Lowenfels J (2017). *Teaming with Fungi: The Organic Gardener's Guide to Mycorrhizae*. Timber Press, Portland, OR.

Luna E, Bruce TJA, Roberts MR, Flors V, Ton J (2012). Next-generation systemic acquired resistance. Plant Physiol 158: 844–853.

Lyford WH (1980). Development of the root system of northern red oak (*Quercus rubra* L.). Harvard Forest Paper No. 21. Petersham, MA.

Marasco R, Rolli E, Ettoumi B, Vigani G, Mapelli F, Borin S, Abou-Hadid AF, El-Behairy UA, Sorlini C, Cherif A, Zocchi G, Daffonchio D (2012). A drought resistance-promoting microbiome is selected by root system under desert farming. PLoS ONE; https://doi.org/10.1371/journal.pone.0048479.

McDonald R. How urban trees can save lives. Nature Conservancy, October 2016.

Muday GK, Brown-Harding H (2018). Nervous system-like signaling in plant defense. Science 361: 1068–1069.

Musser RO, Hum-Musser SM, Eichenseer H, Peiffer M, Ervin G, Murphy JB, Felton GW (2002). Caterpillar saliva beats plant defences. Nature 416: 599–600.

Pearse F (2020). Weather makers. Science 368: 1302–1305.

Ramirez CC, Guerra FP, Zuniga RE, Cordero C (2009). Differential expression of candidate defense genes of poplars in response to aphid feeding. J Econ Entomol 102: 1070–1074.

Santoyo G, Moreno-Hagelsieb G, Orozco-Mosqueda MdC, Glick B (2016). Plant growth-promoting bacterial endophytes. Microbiol Res 183: 92–99.

Shrivastava G, Ownley BH, Augé RM, Toler H, Dee M, Vu A, Köllner TG, Chen F (2015). Colonization by arbuscular mycorrhizal and endophytic fungi enhanced terpene production in tomato plants and their defense against a herbivorous insect. Symbiosis 65: 65–74.

Simard S (2021). *Finding the Mother Tree: Discovering the Wisdom of the Forest.* Knopf, New York.

Stratton-Porter G (1910). *Music of the Wild.* Doubleday, New York.

Sutherland H (1902). *The Book of Bugs.* Street & Smith, New York.

Vega FE (2008). Insect pathology and fungal endophytes. J Invertebr Pathol 98: 277–279.

Walker AA, Weirauch C, Fry BG, King GF (2016). Venoms of heteropteran insects: a treasure trove of diverse pharmacological toolkits. Toxins 8: 43; https://doi.org/10.3390/toxins8020043.

Chapter 2. Out on a Limb: Living on Leaves, Buds, and Twigs

Agrawal AA, Karban R (1997). Domatia mediate plant-arthropod mutualism. Nature 387: 562–563.

Andreas PA, Kisiala A, Emery RJN, De Clerck-Floate R, Tooker JF, Price PW, Miller DG, Chen M-S, Connor EF (2020). Cytokinins are abundant and widespread among insect species. Plants 9(2): 208; https//doi.org/10.3390/plants9020208.

Bari R, Jones JDG (2009). Role of plant hormones in plant defence responses. Plant Mol Biol 69: 473–488.

Body M, Kaiser W, Dubreuil G, Casa J, Giron D (2013). Leaf miners co-opt microorganisms to enhance their nutritional environment. J Chem Ecol 39: 969–977.

Borkent A, Bissett J (1985). Gall midges (Diptera: Cecidomyiidae) are vectors for their fungal symbionts. Symbiosis 1: 185–194.

Brütting C, Crava CM, Schäfer M, Schuman MC, Meldau S, Adam N, Baldwin IT (2018). Cytokinin transfer by a free-living mirid to *Nicotiana attenuata* recapitulates a strategy of endophytic insects. *e*Life 7; https://doi.org/20.7554/eLife.36268.

Brütting C, Schäfer M, Vanková R, Gase K, Baldwin IT, Meldau S (2017). Changes in cytokinins are sufficient to alter developmental patterns of defense metabolites in *Nicotiana attenuata.* Plant J 89: 15–30.

Bultman TL, Faeth SH (1986). Selective oviposition by a leaf miner in response to temporal variation in abscission. Oecologia 69: 117–120.

Casacci LP, Bonelli S, Balletto E, Barbero F (2019). Multimodal signaling in myrmecophilous butterflies. Front Ecol Evol 29; https://doi.org/10.3389/fevo.2019.00454

Costa JT (1997). Caterpillars as social insects: largely unrecognized, the gregarious behavior of caterpillars is changing the way entomologists think about social insects. Am Sci 85: 150–159.

Dussourd DE (2017). Behavioral sabotage of plant defenses by insect foliovores. Annu Rev Entomol 62: 15–34.

Dussourd DE, Denno RF (1991). Deactivation of plant defense: correspondence between insect behavior and secretory canal architecture. Ecology 72: 1383–1396.

Eiseman C (2019). *Leafminers of North America.* Self-published, e-book.

Fabre JH (1916). *The Life of the Caterpillar.* Dodd Mead, New York.

Felt EP (1965). *Plant Galls and Gall Makers.* Facsimile of 1940 edition. Hafner, New York.

Frago E, Dicke M, Godfray HCJ (2012). Insect symbionts as hidden players in insect-plant interactions. Trends Ecol Evol 12: 705–711.

Frost SW (1959). *Insect Life and Insect Natural History.* Dover, New York.

Fürstenberg-Hägg J, Zagrobelny M, Bak S (2013). Plant defense against insect herbivores. Int J Mol Sci 14: 10242–10297.

Giron D, Huguet E, Stone GN, Body M (2016). Insect-induced effects on plants and possible effectors used by galling and leaf-mining insects to manipulate their host plants. J Insect Physiol 84: 70–89.

Hebert PDN, Ratnasingham S, Zakharov EV, Telfer AC, Levesque-Beaudin V, Milton MA, Pedersen S, Jannetta P, deWaard JR (2016). Counting animal species with DNA barcodes: Canadian insects. Phil Trans R Soc B 371: 20150333; http://dx.doi.org/10.1098/rstb.2015.0333.

Hosokawa T, Kikuchi Y, Shimada M, Fukatsu T (2007). Obligate symbiont involved in pest status of host insect. Proc Roy Soc B; https://doi.org/10.1098/rspb.2007.0620.

Joy JB (2012). Symbiosis catalyses niche expansion and diversification. Proc Roy Soc B 280: 20122820; http://dx.doi.org/10.1098/rspb.2012.2820.

Kaiser W, Huguet E, Casas J, Commin C, Giron D (2010). Plant green-island phenotype induced by leaf-miners is mediated by bacterial symbionts. Proc Roy Soc B; https://doi.org/10.1098/rspb.2010.0214.

Kuromori T, Seo M, Shinozaki K (2018). ABA (abscisic acid) transport and plant water stress responses. Trends Plant Sci 23: 513–522.

Lutz FE (1948). *Field Book of Insects.* G. P. Putnam's Sons, New York.

Needham JG, Frost SW, Tothill BH (1928). *Leaf-Mining Insects.* Williams and Wilkins, Baltimore.

Nyman T, Widmer A, Roininen H (2000). Evolution of gall morphology and host-plant relationships in willow-feeding sawflies (Hymenoptera: Tenthredinidae). Evolution 54: 526–533.

O'Dowd DJ, Willson MF (1997). Leaf domatia and the distribution and abundance of foliar mites in broadleaf deciduous forest in Wisconsin. Am Midl Nat 137: 337–348.

Price PW, Mattson WJ, Baranchikov YN (1993). *The Ecology and Evolution of Gall-Forming Insects.* USDA Forest Service, General Technical Report NC-174. St. Paul, MN.

Rohfritsch O (2008). Plants, gall midges, and fungi: a three-component system. Entomol Exp Appl 128: 208–216.

Santiago-Blay JA (2005). Leaf-mining chrysomelids. In: Jolivet P, Santiago-Blay J, Schmitt M (eds), *New Developments in the Biology of* Chrysomelidae, pp. 305–306. Brill Academic Publishing, Leiden.

Sugio A, Dubreuil G, Giron D, Simon J-C (2014). Plant-insect interactions under bacterial influence: ecological implications and underlying mechanisms. J Exp Bot 66: 467–478.

Takei M, Yoshida S, Kawai T, Hasegawa M, Suzuki Y (2015). Adaptive significance of gall formation for gall-inducing aphids on Japanese elm trees. J Insect Physiol 72: 43–51.

Tooker JF, DeMoraes CM (2011). Feeding by a gall-inducing caterpillar species alters levels of indole-3-acetic and abscisic acid in *Solidago altissima* (Asteraceae) stems. Arthropod-Plant Interact 5: 115–124.

Wakie TT, Neven LG, Yee WL, Lu Z (2019). The establishment risk of *Lycorma delicatula* (Hemiptera: Fulgoridae) in the United States and globally. J Econ Entomol; https://doi.org/10.1093/jee/toz259.

Walter DE, Behan-Pelletier VM (1999). Mites in forest canopies: filling the size distribution shortfall. Annu Rev Entomol 44: 1–19.

Walter DE, O'Dowd DJ (1992). Leaves with domatia have more mites. Ecology 76: 1514–1518.

Wilson LF (1972). Life history and outbreaks of an oak leafroller, *Archips semiferanus* (Lepidoptera: Tortricidae), in Michigan. Gt Lakes Entomol 55: 71–77.

Zhang H, Guiguet A, Dubreuil G, Kisiala A, Andreas P, Emery RJN, Huguet E, Body M, Giron D (2017). Dynamics and origin of cytokinins involved in plant manipulation by a leaf-mining insect. Insect Sci 24: 1065–1078.

Chapter 3. Tapping a Tree's Circulatory System

Askew RA (1971). *Parasitic Insects.* Heinemann Educational Books, London.

Bartlett CR, Adams ER, Anthony T (2011). Planthoppers of Delaware (Hemiptera, Fulgoroidea), excluding Delphacidae, with species incidence from adjacent states. Zookeys 83: 1–42.

Dermastia M (2019). Plant hormones in phytoplasma infections. Front Plant Sci; https://doi.org/10.3389/fpls.2019.00477.

Heinrich B (1992). Maple sugaring by red squirrels. J Mammalogy 73: 51.

Lohman DJ, Liao Q, Pierce NE (2006). Convergence of chemical mimicry in a guild of aphid predators. Ecol Entomol 31: 41–51.

Lukasik P, van Asch M, Guo HF, Ferrari J, Godfray HCJ (2013). Unrelated facultative endosymbionts protect aphids against a fungal pathogen. Ecol Lett 16: 214–218.

Mauck KE, De Moraes CM, Mescher MC (2010). Deceptive chemical signals induced by a plant virus attract insect vectors to inferior hosts. Proc Natl Acad Sci USA 107: 3600–3605.

Mayer CJ, Vilcinskas A, Gross J (2008). Phytopathogen lures its insect vector by altering host plant odor. J Chem Ecol 34: 1045–1049.

Staples J, Krall BS, Bartelt RJ, Whitman DW (2002). Chemical defense in the plant bug *Lopidea robiniae* (Uhler). J Chem Ecol 28: 601–616.

Starý P (1988). Aphidiidae. In: Minks AK, Harrewijn P (eds), *Aphids: Their Biology, Natural Enemies, and Control,* vol. 2B, *World Crop Pests,* pp. 171–184. Elsevier, Amsterdam.

Sugio A, Kingdom HN, MacLean AM, Grieve VM, Hogenhout SA (2011). Phytoplasma protein effector SAP11 enhances insect vector reproduction by manipulating plant development and defense hormone biosynthesis. Proc Natl Acad Sci USA 108: E1254–E1263.

Wakie TT, Neven LG, Yee WL, Lu Z (2019). The establishment risk of *Lycorma delicatula* (Hemiptera: Fulgoridae) in the United States and globally. J Econ Entomol; https://doi.org/10.1093/jee/toz259.

Chapter 4. The World between Bark and Heartwood

Adams AS, Currie CR, Cardoza YJ, Klepzig KD, Raffa KF (2009). Effects of bacteria and tree chemistry on the growth and reproduction of bark beetle fungal symbionts. Can J For Res 39: 1133–1147.

Barriault I, Barabé D, Cloutier L, Gibernau M (2010.) Pollination ecology and reproductive success in jack-in-the-pulpit (*Arisaema triphyllum*) in Québec (Canada). Plant Biol (Stuttg) 12: 161–171.

Bauer LS, Duan JJ, Gould JR, Van Driesche R (2015). Progress in the classical biological control of *Agrilus planipennis* Fairmaire (Coleoptera: Buprestidae) in North America. Can Entomol 147: 300–317.

Blackwell M, Vega FE (2018). Lives within lives: hidden fungal biodiversity and the importance of conservation. Fungal Ecol 35: 127–134; https://doi.org/10.1016/j.funeco.2018.05.011.

Eisner T, Schroeder FC, Snyder N, Grant JB, Aneshansley DJ, Utterback D, Meinwald J, Eisner M (2008). Defensive chemistry of lycid beetles and of mimetic cerambycid beetles that feed on them. Chemoecology 18: 109–119.

Forcella F (1982). Why twig-girdling beetles girdle twigs. Naturwissenschaften 69: 398–400.

Kaasalainen U, Fewer DP, Jokela J, Wahlsten M, Sivonen K, Rikkinen J (2012). Cyanobacteria produce a high variety of hepatotoxic peptides in lichen symbiosis Proc Natl Acad Sci USA 109: 5886–5891; http://dx.doi.org/10.1073/pnas.1200279109.

LeLannic J, Nénon J-P (1999). Functional morphology of the ovipositor in *Megarhyssa atrata* (Hymenoptera, Ichneumonidae) and its penetration into wood. Zoomorphology 119: 73–79.

Lu M, Wingfield Gillette NE, Mori SR, Sun JH (2010). Complex interactions among host pines and fungi vectored by an invasive bark beetle. New Phytol 187: 859–866.

Lutz FE (1948). *Field Book of Insects*. G. P. Putnam's Sons, New York.

Madden JL (1968). Behavioural responses of parasites to the symbiotic fungus associated with *Sirex noctilio* F. Nature 218: 189–190.

McLeod G, Gries R, von Reuß SH, Rahe JE, McIntosh R, König WA, Gries G (2005). The pathogen causing Dutch elm disease makes host trees attract insect vectors. Proc Roy Soc B 272: 2499–2503.

Mohammed WS, Ziganshina EE, Shagimardanova EI, Gogoleva NE, Ziganshin AM (2018). Comparison of intestinal bacterial and fungal communities across various xylophagous beetle larvae (Coleoptera: Cerambycidae). Sci Rep 8: 10073.

Morris EE, Kepler RM, Long SJ, Williams DW, Hajek AE (2013). Phylogenetic analysis of *Deladenus* nematodes parasitizing northeastern North American *Sirex* species. J Invert Path 113: 177–183.

Mosconi F, Campanaro A, Carpaneto GM, Chiari S, Hardersen S, Mancini E, Murizi E, Sabatelli S, Zauli A, Mason F, Audisio P (2017). Training of a dog for the monitoring of *Osmoderma eremita*. Nat Conserv; doi:10.3897/natureconservartion.20.12688.

Perotti MA, Young DK, Braig HR (2016). The ghost sex-life of the paedogenetic beetle *Micromalthus debilis*. Sci Rep 6; https://doi.org/10.1038/srep27364.

Polidori C, Garcia AJ, Nieves-Aldrey JL (2013). Breaking up the wall: metal-enrichment in ovipositors, but not in mandibles, co-varies with substrate hardness in gall wasps and their associates. PLoS ONE 8(7): e70529.

Ranius T (2002). *Osmoderma eremita* as an indicator of species richness beetles in tree hollows. Biodivers Conserv 11: 931–941.

Scott-Chialvo, CH, Chialvo P, Holland JD, Anderson TJ, Breinholt JW, Kawahara AY, Zhou X, Liu S, Zaspel JM (2018). A phylogenomic analysis of lichen-feeding tiger moths uncovers evolutionary origin of host chemical sequestration. Mol Phylogenet Evol 121: 23–34.

Scott JJ, Oh D-C, Yuceer MC, Klepzig KD, Clardy J, Currie C (2008). Bacterial protection of beetle-fungus mutualism. Science 322: 63.

Yamamoto S (2019). The extant telephone-pole beetle genus *Micromalthus* discovered in mid-Cretaceous amber from northern Myanmar (Coleoptera:Archostemata:Micromalthidae). Hist Biol; https://doi.org/10.1080/08912963.2019.1670174.

Yan EV, Beutel RG, Lawrence JF, Yavorskaya MI, Hörnschemeyer T, Pohl HW, Vassilenko DV, Bashkuev AS, Ponomarenko AG (2019). *Archaeomalthus*—(Coleoptera, Archostemata) a "ghost adult" of Micromalthidae from upper Permian deposits of Siberia? Hist Biol; https://doi.org/10.1080/08912963.2018.1561672.

Chapter 5. In the Company of Flowers and Fruits

Bello C, Barreto E (2021). The footprint of evolution in seed dispersal interactions. Science 372: 682–683.

Fricke EC, Ordonez A, Rogers HS, Svenning J-C (2022). The effects of defaunation on plants' capacity to track climate change. Science 375: 210–214.

Frost SW (1959). *Insect Life and Insect Natural History*. Dover, New York.

Kricher JC, Morrison G (1988). *A Field Guide to Eastern Forests of North America*. Houghton-Mifflin, Boston.

Ostaff DP, Mosseler A, Johns RC, Javorek S, Klymko J, Ascher JS (2015). Willows (*Salix* spp.) as pollen and nectar sources for sustaining fruit and berry pollinating plants. Can J Plant Sci 95: 505–516.

Ostfeld RS, Jones CG, Wolff JO (1996). Of mice and mast: ecological connections in eastern deciduous forests. Bioscience 46: 323–330.

Rodriguez A, Alquezar B, Pena L (2013). Fruit aromas in mature fleshy fruits as signals of readiness for predation and seed dispersal. New Phytol 197: 36–48.

Shields O (1989). World numbers of butterflies. J Lepid Soc 43: 178–183.

Yang Y, Wang Z, Yan C, Zhang Y, Zhang D, Yi X (2018). Selective predation on acorn weevils by seed-caching Siberian chipmunk *Tamias sibiricus* in a tripartite interaction. Oecologia; https://doi.org/10.1007/s00442-018-4161-z.

Chapter 6. The World beneath a Tree

Duarte M, Robbins RK (2010). Description and phylogenetic analysis of the Calycopidina (Lycaenidae, Theclinae, Eumaeini): a subtribe of detritivores. Rev Bras Entomol 54(1): 45–65; https://doi.org/10.1590/S0085-56262010000100006.

Erb M, Meldau S, Howe GA (2012). Role of phytohormones in insect-specific plant reactions. Trends Plant Sci 17: 250–271.

Haney CH, Ausubel FM (2015). Plant microbiome blueprints: a plant defense hormone shapes the root microbiome. Science 349: 788–789.

Haritos VS, Home I, Damcevski K, Glover K, Gibb N, Okada S, Hamberg M (2012). The convergent evolution of defensive polyacetylenic fatty acid biosynthesis genes in soldier beetles. Nat Commun 3: 1150; https:// doi.10.1038/ncomms2147.

Langor DW, deWaard JR, Snyder BA (2019). Myriapoda of Canada. Zookeys; https://doi.org/10.3897/zookeys.819.29447.

Maraun M, Schatz H, Scheu S (2007). Awesome or ordinary? Global diversity patterns of oribatid mites. Ecography 30: 209–216.

Marchenko II, Bogomolova IN (2015). Spatial-typologic organization of populations of soil gamasid mites (Acari, Mesostigmata) in northern Altai Mountains. Contemp Probl Ecol 8: 202–210.

Marek PE, Buzatto BA, Shear WA, Means JC, Black DG, Harvey MS, Rodriguez J (2021). The first true millipede—1306 legs long. Sci Rep 11: 23126; https://doi.org/10.1038/s41598-021-02447-0.

Nardi JB (2007). *Life in the Soil: A Guide for Naturalists and Gardeners.* University of Chicago Press, Chicago.

Nekola JC (2014). Overview of the North American terrestrial gastropod fauna. Am Malacol Bull 32: 225–235; https://doi.org/10.4003/006.032.0203.

Saporito RA, Donnelly MA, Norton RA, Garraffo HM, Spande TF, Daly JW (2007). Oribatid mites as a major dietary source for alkaloids in poison frogs. Proc Natl Acad Sci USA 104: 8885–8890.

Shear WA (2015). The chemical defenses of millipedes (Diplopoda): biochemistry, physiology and ecology. Biochem Syst Ecol 61: 78–117.

Turnbull MS, Stebaeva S (2019). Collembola of Canada. Zookeys; https://doi.org/10.3897/zookeys.819.23653.

INDEX